ORNAMENTAL LAKES

Ornamental Lakes traces the history of lakes in England, from their appearance in the early eighteenth century, through their development in the 1750s, and finally to their decline in the nineteenth century. Apart from the natural lakes in the Lake District, the bodies of water we see in England today are man-made, primarily intended to ornament the landscapes of the upper classes.

Through detailed research, author Wendy Bishop argues that, contrary to accepted thinking, the development of lakes led to the dissolution of formal landscapes rather than following changes in landscape design. Providing a comprehensive overview of lakes in England, including data on who made these lakes, how, and when, it additionally covers fishponds, water gardens, cascades and reservoirs.

Richly illustrated and accompanied by case studies across the region, this book offers new insights in landscape history for students, researchers and those interested in how landscapes evolve.

Wendy Bishop. After a career in teaching, including Malawian secondary schools, and teaching children with behavioural difficulties, she received a stern health warning and retrained, doing degrees in garden design and garden history. This was followed by a PhD on ornamental lakes at the University of East Anglia. She is now an independent researcher. Alongside lakes, her research interests include the history of reservoirs and their evolving uses and modes of construction.

ORNAMENTAL LAKES

Their Origins and Evolution in English Landscapes

Wendy Bishop

Routledge
Taylor & Francis Group

LONDON AND NEW YORK

First published 2021
by Routledge
2 Park Square, Milton Park, Abingdon, Oxon OX14 4RN

and by Routledge
52 Vanderbilt Avenue, New York, NY 10017

Routledge is an imprint of the Taylor & Francis Group, an informa business

British Library Cataloguing-in-Publication Data
A catalogue record for this book is available from the British Library

Library of Congress Cataloging-in-Publication Data
Names: Bishop, Wendy, 1955– author.
Title: Ornamental lakes : their origins and evolution in English landscapes / Wendy Bishop.
Description: Abingdon, Oxon ; New York, NY : Routledge, 2021. | Includes bibliographical references and index.
Identifiers: LCCN 2020054108 (print) | LCCN 2020054109 (ebook) | ISBN 9780367894191 (hardback) | ISBN 9780367894184 (paperback) | ISBN 9781003019053 (ebook)
Subjects: LCSH: Water gardens—England—History. | Lakes—England—History. | Reservoirs—England—History. | Landscape architecture—England—History.
Classification: LCC SB423 .B597 2021 (print) | LCC SB423 (ebook) | DDC 635.9/674—dc23
LC record available at https://lccn.loc.gov/2020054108
LC ebook record available at https://lccn.loc.gov/2020054109

ISBN: 978-0-367-89419-1 (hbk)
ISBN: 978-0-367-89418-4 (pbk)
ISBN: 978-1-003-01905-3 (ebk)

Typeset in Bembo
by Apex CoVantage, LLC

Access the Support Material: www.routledge.com/9780367894184

I dedicate this work to my husband, Dr. David Bishop, who saved my life, and Prof. Tom Williamson. Their support was crucial to the success of this venture. Likewise, Catherine Bishop's help with photography and images was invaluable.

An Englishman thinks nothing of a garden without water.
— Count Kielmansegge, *Diary of a Journey to England
in the years 1761–2*, trans. Countess Kielmansegg

CONTENTS

List of figures and tables   viii
Acknowledgements   xv
Abbreviations and notes   xvi
Glossary   xvii

1   Introduction: setting the scene   1

2   The ancestors of lakes   17

3   The emergence of lakes   62

4   The making of lakes   100

5   The lake makers   134

6   Why lakes emerged   179

7   Lakes in the nineteenth century   209

8   Conclusion   253

Appendices   261
Bibliography   266
Index   275

FIGURES AND TABLES

Figures are by the author unless otherwise stated in source notes.

Figures

1.1   A detail of Charles Bridgeman's plan of Stowe, 1739, engraved
      by Jacques Rigaud, showing the Octagonal Lake, which was
      c. 1 h and made in the 1720s. The lakes to the left and right of
      the octagon, as depicted here, are semi-geometric lakes   4

1.2   A diagram of a 1732 map of Wolterton, Norfolk, by J. Corbridge.
      A *c.* 4 h lake was made by 1732   4

1.3   Kimberley Park, Norfolk: a mid-eighteenth century plan. The
      8.5 h lake was made by 1739   5

1.4   A 1785 plan of Stourhead, Wiltshire. The main lake of 7.5 h was
      made in 1754   6

1.5   Horton Park, Northamptonshire, OS 6" map, 1899. The river-lake
      (1.8 h), made *c.* 1760, is not much wider than the natural river at
      the beginning and end   7

1.6   The Serpentine, Hyde Park by J. Rocque, 1762   8

1.7   Cross-section of a typical eighteenth-century dam with a barrel sluice.
      Not to scale   9

1.8   The dam at Clumber Park, Nottinghamshire, from the downstream
      side, showing the modern spillway which has been inserted into the
      c. 1774 dam   10

1.9   A. The bird's-eye-view, or three-quarter elevation: the viewer is
      three-quarters of the way up an imaginary vertical axis. B. The
      one-quarter elevation: the viewer is a quarter of the way up the
      vertical axis   13

1.10	Melton Constable, Norfolk, by Knyff and Kip, *Britannia Illustrata* 1708	14
1.11	Melton Constable, *c.* 1725, by Edmund Prideaux	14
2.1	Diagram of fourteenth-century fishponds adjacent to the Bishop of Lincoln's palace at Lyddington, Rutland, with probable areas of water indicated	19
2.2	An anonymous painting of Margam House, Glamorgan, *c.* 1700	20
2.3	Plan of Bodiam Castle, Sussex, built by Sir Edward Dalingridge in the 1380s, showing the lake-moat and possible fishponds	23
2.4	Bodiam Castle, Sussex	24
2.5	An interpretation of the landscape of Kenilworth Castle, Warwickshire, *c.* 1575	25
2.6	Caerphilly Castle, Glamorgan, defences	26
2.7	Sir William Knight's Ambulatory, *c.* 1521, built to overlook his fishponds at Horton Court, Gloucestershire	30
2.8	An engraving of Villa D'Este, Tivoli, 1573, by Étienne Dupérac	31
2.9	A digitised copy of the 1611 plan of Theobalds by J. Thorpe. The New River was the work of Robert Cecil	33
2.10	Layout of the gardens of Theobalds based on the Thorpe survey of 1608. A. Fishpond. B. Horsepond. C. Conduit Court. D. Great Garden. The approach from the London road can be seen on the right of the gardens	34
2.11	A detail of Ralph Treswell's 1580 map of Holdenby, Northamptonshire	36
2.12	A detail of Ralph Treswell's 1587 map of Holdenby, Northamptonshire	36
2.13	A 1634 map of Gorhambury estate, Hertfordshire, by Benjamin Hare	37
2.14	A detail of the 1634 map of Gorhambury Estate, Hertfordshire	38
2.15	Raglan Castle, Monmouthshire: The Deplorable Mapp of 1652 by Laurence Smythe. North is towards the top left	40
2.16	An artist's recreation of the castle and grounds of Raglan, Monmouthshire, in the early seventeenth century	42
2.17	Raglan Castle, Monmouthshire: detail from the Deplorable Mapp of 1652 by Laurence Smythe	43
2.18	A 1732 copy of *A Mapp of part of the Manors of Earls Court, Kensington and Abbots 1694/5* by E. Bostock Fuller. North is to the right	44
2.19	A plan of fishponds in Gervaise Markham's *Cheape and Good Husbandry,* 1623	44
2.20	A 1630s design for the gardens at Wilton House, Wiltshire, by Isaac de Caus	47
2.21	Colen Campbell's 1725 plan of Longleat, Wiltshire. It is essentially the same as Knyff and Kip's 1708 plate in *Britannia Illustrata*	48
2.22	Longleat, Wiltshire, in *Britannia Illustrata* 1708	50
2.23	Different types of water features found in *Britannia Illustrata* 1708	51
2.24	A painting of the house and gardens of Bowood Park, Wiltshire, *c.* 1725	53
2.25	A diagram of the pre-1763 estate survey of Bowood	53

2.26   Temple Newsam, Yorkshire, in *Britannia Illustrata* 1708   55
2.27   Pieter Tillemans's *c.* 1729 painting of Burley-on-the-Hill, Rutland   56
2.28   Tortworth, Gloucestershire, in *The Ancient & Present State of
     Glostershire* by Sir Robert Atkyns, 1712   57
3.1   A plan of Welford Park, Berkshire, *c.* 1700   63
3.2   Colen Campbell's plan of Boughton Park, Northamptonshire,
     in *Vitruvius Britannicus* 1725   64
3.3   Staunton Harold, Leicestershire, in *Britannia Illustrata* 1708   66
3.4   Blenheim Palace: diagram based on a *c.* 1728 sketch by a French
     visitor, Fougeroux, showing Armstrong's lake of *c.* 3.5 h. A. is
     Vanbrugh's bridge. B. is a cascade   67
3.5   A design for an estate by Stephen Switzer in *Ichnographia Rustica*
     1742 ed., Vol. II, plate opposite title page   70
3.6   The earliest irregular lakes. A. Thoresby, Nottinghamshire, *c.* 1719
     (20 h) for the 1st Duke of Kingston. B. Londesborough, Yorkshire,
     1728–30 (4.5 h) for Lord Burlington. C. Castle Howard, Yorkshire,
     1724, (2.5 h) for the 3rd Earl of Carlisle. D. Holkham, Norfolk,
     by 1729 (8 h) for Thomas Coke   71
3.7   A 1690 estate map of Thoresby Park, Nottinghamshire   73
3.8   A plan of Thoresby Park in *Vitruvius Britannicus* 1725   74
3.9   A 1738 estate map of Thoresby Park, Nottinghamshire   75
3.10   An estate map of Holkham Park, Norfolk, *c.* 1745   77
3.11   A detail of a 1727 estate map of Castle Howard, Yorkshire,
     showing South Lake   78
3.12   South Lake, Castle Howard, Yorkshire, OS 6" map, 1950   79
3.13   A 1697 painting of Moor Park, Surrey   80
3.14   Vanbrugh's planned lake at Blenheim, *c.* 1705 (early), with an
     elaborate cascade. The lake was *c.* 3 h. Emphasis added by author   81
3.15   Vanbrugh's plan for a lake at Blenheim, *c.* 1705 (late)   82
3.16   Vanbrugh's plan for Blenheim in *Vitruvius Britannicus*, dating from
     1715 or earlier   83
3.17   A 1739 estate survey of Londesborough Park, East Yorkshire   85
3.18   A plan of Claremont, Surrey, in *Vitruvius Britannicus* 1725   87
3.19   Rocque's 1745 map of Wanstead, Essex   88
3.20   A 1747 estate map of Longleat, Wiltshire, by John Ladd   89
3.21   A diagram of the serpentine canal on an estate survey of Badminton,
     1750, by Robert Whittlesey   90
3.22   The water at Beachborough, Kent, OS 6" map, 1872. Note the
     round structure on the western side, which is probably the temple
     shown in Edward Haytley's painting   91
3.23   Ditchley, Oxfordshire, OS 6" map, 1900   92
3.24   A chart of the numbers of all types of lakes, 1700–99   93
3.25   The lake at Stourhead, Wiltshire, OS 6" map, 1962   94
3.26   A 1722 estate map of Stourhead, Wiltshire   95
3.27   A 1792 painting of the bottom of the lake at Stourhead, Wiltshire   96

4.1   A plan of a fish 'pond' (*vivarium*), based on Roger North's instructions
       in *A Discourse on Fish and Fishponds*   102
4.2   A cross-section for a dam at Grimsthorpe, Lincolnshire, by
       John Grundy, 1747   103
4.3   At Stourhead, Wiltshire, the course of the stream can be seen
       winding along the lake bed, 2016   105
4.4   A cross-section of the dam at Cusworth, Yorkshire, 1764,
       by Richard Woods. The water is flowing from left to right   107
4.5   Part of the 1779 Crow survey of Petworth Park, Surrey, West
       Sussex Record Office   109
4.6   The Grecian Valley, Stowe, Buckinghamshire, 1805, by J. C. Nattes   111
4.7   Drainage patterns. The river is at the bottom of the valley, and
       all the sub-surface water in the drainage area is moving in the
       direction of the arrows. The dotted line is the planned lake
       surface and the solid line is the original river   112
4.8   Bretby, Derbyshire, depicted in *Britannia Illustrata* 1708   114
4.9   The hybrid lake at Wollaton Park, Nottinghamshire, OS 6"
       map, 1899   115
4.10   A 1758 estate map of Raynham Park, Norfolk. The house is at
       the top and north is to the left of the plan. The water shown is
       the same as that on a *c*. 1730 map of the estate   116
4.11   The tithe map for West Raynham, 1838   117
4.12   A diagram of pieces of water at Trentham Park, Staffordshire, from
       a plan dating from the 1700s. North is to the right   118
4.13   Chatsworth, Derbyshire, engraving by W. Watts, 1779   120
4.14   Kedleston Park, Derbyshire, OS 6" map, pre-1930 '55   122
4.15   The bridge-weir at Kedleston, designed by Robert Adam in 1764
       but not built until *c*. 1771   123
4.16   A diagram of a basic weir construction   124
4.17   John Smeaton's 1776 design for a weir on the River Coquet,
       Northumberland. Note the stone crest 'B' and the stone facing 'a'.
       The water is flowing from left to right   124
4.18   The weir, probably modern, retaining the lake at Kimberley Park,
       Norfolk   124
4.19   Detail of the 1699 map of Bramshill Park, Hampshire, by Isaac
       Justis, showing the lake and three ponds   126
4.20   The north-west dam at Bramshill House, Hampshire   127
4.21   Highclere Castle and Park, Hampshire, OS 6" map, 1909   130
5.1   Part of Brown's second plan for Kimberley, dated 1778. North is
       at the bottom   135
5.2   A detail of Brown's second plan for Kimberley, dated 1778. North
       is at the bottom   136
5.3   A detail of Brown's sketch plan for Packington, Warwickshire,
       of *c*. 1750, showing the planned lake. North is at the top   137
5.4   Brown's design for a cascade at Packington, 1751   138

5.5 Packington Park, Warwickshire, OS 6" map, 1886 140

5.6 Brown's 1751 plan of Packington, Warwickshire. 1. The House 2. Courts 3. Kitchen Garden 4. Seats Firs Chasse? 5. Wooden Bridge 6. A Mill 7. A seat 8. Gatehouse? 10. Oler Plantation 11. Bridge 13. The Great Road 14. Cascade 16. The Lake 17. My Lady's Lodge 19. A Seat? 20. The Church 141

5.7 Longleat, Wiltshire: High Wood is *c.* 50m behind the viewer; the hall door is just up the steps 143

5.8 One of Brown's fishponds at Longleat. The angle of the slope indicated is typical of Brown's landscaping 144

5.9 Bowood House, Wiltshire, from the east side of the lake 145

5.10 An adjacent view of Bowood from the east, showing the dam 145

5.11 Belhus Park, Essex, OS 6" map, 1862. Richard Woods modified the ends of the lake in 1770 149

5.12 The 'lake' at Prior Park, Somerset, with an extract from the OS 6" map, 1902 151

5.13 A 'lake' in the Elysian Fields, Stowe, Buckinghamshire 152

5.14 The Five Arch Bridge, Wotton Underwood, Buckinghamshire 152

5.15 Supply inlet for the lakes, Wotton Underwood, Buckinghamshire, OS 6" map, 1898 153

5.16 An open spillway in the Warrells dam at Wotton Underwood, Buckinghamshire 154

5.17 Bridge-dam at Bowood, Wiltshire, upper side 156

5.18 Bridge-dam at Bowood, Wiltshire, lower side 156

5.19 Brown's 1763 plan for Bowood Park, Wiltshire 157

5.20 A 1789 plantation management plan for Bowood Estate, Wiltshire 158

5.21 View from Clark's Hill back to the house and lake at Bowood The star marks the site of the original house, demolished in 1955 159

5.22 The lake at Trentham Park, much as Brown planned it, OS 6" map, pre-1932–55 160

5.23 The 1727 'Coppy Map' of Trentham, Staffordshire. The house is just off the top right corner 161

5.24 Boydell's 1752 view of Blenheim, Oxfordshire, from the north-east 163

5.25 Boydell's 1752 view of Blenheim, Oxfordshire, from the south-west 163

5.26 Richard Woods' plan for Wardour Castle, Wiltshire, 1764 166

5.27 Wivenhoe Park, Essex, by Constable, 1816 167

5.28 Nathaniel Richmond's 1763 plan (attributed) for Danson Park, Kent 168

5.29 Danson Park, Kent, OS 6" map, 1908. The lake is basically the same shape as shown on the 1799 OS drawing 169

5.30 Shardeloes Park, Buckinghamshire, OS 1:10,000 map, 1930–59 169

5.31 Estate plan of Kedleston Park, Derbyshire, by George Ingman, 1764 170

5.32 Locko Park, Derbyshire, 2020 172

5.33 A plan of Erlestoke Park, Wiltshire, by William Emes, 1786 173

5.34 An 1825 estate map of Erlestoke Park, Wiltshire 174

5.35	A detail of an 1825 map of Erlestoke Park, indicating the cascades	175
6.1	Belton House, Lincolnshire, in *Vitruvius Britannicus*, 1725	180
6.2	Belton Park, Lincolnshire, OS 1:10,000 map, pre-1930–55	180
6.3	Claude Lorrain *Landscape with the Rest on the Flight into Egypt* 1654	185
6.4	The Temple of Flora and the Palladian Bridge, Stourhead, Wiltshire	186
6.5	A reconstruction of Pliny's Tuscum Villa by Castell, 1728	188
6.6	J. Rocque's plan of Claremont, Surrey, 1738	189
6.7	Pieter Tillemans's painting of Thoresby Park, Nottinghamshire, *c.* 1725	193
6.8	A detail showing boats in Pieter Tillemans's painting of Thoresby Park, Nottinghamshire, *c.* 1725	194
6.9	View towards the temple: *The Brockman family at Beachborough* Kent, by Edward Haytley, 1745	197
6.10	Men fishing: *The Brockman family at Beachborough* Kent, by Edward Haytley, 1745	197
6.11	*View from the head of the lake at Stowe* by Rigaud & Baron, 1733	199
6.12	*The three masted ship on the Eleven Acre Lake at Stowe* by Chatelain and Bickham, 1753	201
6.13	Chatsworth House, Derbyshire, with James Paine's bridge in the foreground	203
6.14	The Pantheon reflected in the lake at Stourhead, Wiltshire	204
7.1	A chart of the numbers of all types of lakes, 1700–1899	210
7.2	Lake shapes in the eighteenth and nineteenth centuries	212
7.3	Dunorlan Park, Kent, OS 6" map, 1873	213
7.4	Rood Ashton Park, Wiltshire, OS 6" map, 1885	214
7.5	Fonthill, Wiltshire: Bitham Lake, OS 6" map, pre-1930–61	215
7.6	Repton's plan for Thoresby Hall, Nottinghamshire, in his 1791 Red Book for Thoresby. Repton's cascade is on the left, adjacent to the lake	219
7.7	Repton's 'slide' of the river and approach at Thoresby, Nottinghamshire, before improvement	220
7.8	Repton's 'slide' of the river and approach drive at Thoresby, Nottinghamshire, after improvement	220
7.9	The cascade at Thoresby Hall, designed by Repton, engraved by J. Peltro, 1801, after Repton	221
7.10	Humphry Repton's plan for a lake at Panshanger, Hertfordshire, 1799.	222
7.11	Dauntsey lake, Wiltshire, possibly made *c.* 1865, OS 1:10,000 map, pre-1930–59	226
7.12	A diagram of an ideal lake shape, by J. C. Loudon, based on his designs in *An Encyclopaedia of Gardening* 1825	227
7.13	W. S. Gilpin's illustration of the architectural separation of gardens and park at Heanton, Devon, in *Practical Hints Upon Landscape Gardening* 1835	229
7.14	W. S. Gilpin's late 1820s sketch of proposed changes to the south of Wolterton Hall, Norfolk	230

7.15 Birkenhead Park, Liverpool, OS 6" map, 1897. The lakes are 1.7 h and 1.2 h   231

7.16 A design for a pond by Edward Kemp, in *How to Lay Out a Small Garden* 1850   233

7.17 A cascade by James Pulham II, Dunorlan Park, Kent   237

7.18 The Pulham cascade at Sheffield Park, Sussex, 1988   238

7.19 Harrow Manor, Middlesex, by John Glover, *c.* 1820   239

7.20 The lake at Kimberley Hall, Norfolk, seen from the terrace, 2014. Possibly designed by Nesfield, the terrace was originally gravelled   240

7.21 Knostrop Hall, Yorkshire, by J. A. Grimshaw, 1870   240

7.22 Brown's landscape at Benham Park, Berkshire, 2015   241

7.23 W. A. Nesfield's parterre at Crewe Hall, Cheshire, *c.* 1870   242

7.24 A garden plan for Sudbury Hall, Cheshire, by W. A. Nesfield, 1852. North is towards the lower right   243

7.25 View from the south parterre at Castle Howard over South Lake to the Mausoleum, 2017   244

7.26 Charles Barry's terrace overlooking the lake at Harewood, Yorkshire, 2019   245

7.27 Harewood, Yorkshire, OS 6" map, 1906–9   246

7.28 Holkham Hall, Norfolk, showing the terraces by W. A. Nesfield and W. Burn   248

8.1 View across the lake to Holkham Hall, Norfolk, 2013. The house can just be seen at the end of the lake   256

Tables

1 An example from the Landscape Database   15

2 A chronological list of geometric lakes   65

3 The chronological beginnings of irregular lakes   69

4 Brown's 'top ten' lakes   162

5 Lakes by Francis Richardson   164

6 Lakes by Nathaniel Richmond   164

7 Lakes by Richard Woods   164

8 Lakes by William Emes   165

9 Repton's works with water   217

10 Edward Kemp's works with water   232

11 Works by James Pulham II   236

Appendices

Appendix I A chronological table of semi-geometric lakes   263

Appendix II A chronological table of river-lakes   264

ACKNOWLEDGEMENTS

A number of people played pivotal roles in the conception and delivery of this book. In particular, I would like to mention Prof. Tim Mowl, Prof. Robin Simon and Sandy Haynes.

Many other people also gave generous and unstinting help: Andy Amor, Jeri Bapasola and Blenheim Palace, Charles Boot, Bill Brogden, Dr. Katie Campbell, Elke and Roger Cortis, Fiona Cowell, Lord Egremont, Sir Patrick and Lady Elias, Jonathan Finch, the staff of the Historic England library, Claude Hitching, Charles, 9th Marquis of Lansdowne, the Marc Fitch Foundation, Hugh and Ranji Matheson, Paul Methuen-Smith, Alan Murray Rust, Trevor Nicholson/Harewood House Trust, Gregor Pierrepont, James Puxley, Anne Rowe, Stowe School/SHPT, Prof. Michael Symes, John, 7th Earl of Verulam, Robert, 10th Baron Walpole, Stephen Whitehead.

ABBREVIATIONS AND NOTES

OS	Ordnance Survey
WANHS	Wiltshire Archaeology and Natural History Society
WSHC	Wiltshire and Swindon History Centre
OED	*Oxford English Dictionary*
VCH	Victoria County History
RCHME	Royal Commission on the Historical Monuments of England
ODNB	*Oxford Dictionary of National Biography*
h	hectares

North is at the top of maps unless otherwise stated.

The Landscape Database can be found at www.routledge.com/9780367894184.

GLOSSARY

Many terms in garden history are date specific. For example, a 'canal' in the seventeenth century means a linear ornamental pond, but from the later eighteenth century and onwards it means an industrial waterway.

Bank	Dam
Barrel sluice	a small tunnel passing through the dam, with a sluice gate
Bridge-dam	a dam made in the form of a bridge
Bridge-weir	a weir with a bridge on top
Canal	ornamental linear water feature (unless used in an industrial context)
Clay wall	wall of clay built inside a dam, starting below ground level, to make the dam waterproof
Conduit	variously, a pipe for carrying a water supply, or a fountain
Crest	top edge of a dam or weir
Cut off trench	a trench dug into the bedrock below a dam, filled with clay to stop the water leaking under the dam
Greenhouse	a building like an orangery for keeping tender evergreen plants
Head	Dam
Lake	an ornamental piece of water of 1 h or more
Lake-moat	a large, spreading moat around a castle or other elite building
Leat	a side channel for diverting water, either to prevent flooding, or to supply another area of water, or to drive a mill wheel
Overtop	water flowing over the top of a dam, with the likelihood of breaching it
Parterre	a geometric area of formal planting and paths, usually combining lawns and beds, but occasionally just lawns 'cut' into shapes
Pitched stone	stones laid to make a paved area

Plug	apparatus in a dam to control water and also allow a lake to be drained
Pond	a piece of water less than 1 h
Pond back	trap water behind a dam or weir so that it makes a pond or lake
Puddled clay	clay which has been mixed with some water and 'kneaded', so it cannot absorb more water, which means it has become impervious to water
Servatoria	small fishponds for storing fish ready to eat, often situated near the house
Sluice	a device for shutting off a flow of water
Spillway	a mechanism for controlling water levels in a lake; in the eighteenth century usually a sluice or valve of some kind
Terrier	a map or plan of an estate
Vivaria	fishponds for breeding fish, usually situated in the park; singular, *vivarium*
Water carriage	a 'pipe' often made from stone or elm wood, for conducting water from a source to supply ornamental water features
Wilderness	a relatively formal planted area of woodland, with hedge-lined walks through it, and flowering shrubs and trees

1

INTRODUCTION

Setting the scene

'What is a lake?' It sounds like a very simple question, with a straightforward answer, but a moment's reflection throws up more questions. 'What is the difference between a lake and a pond?' 'How big are they?' Another less obvious question is 'When did lakes begin?' Underlying these questions is the distinction between man-made lakes and naturally occurring lakes. England has a large number of lakes, and many people assume they have always been there, if they consider the matter at all. In fact, England's only natural lakes are in the Lake District. All the other bodies of water are man-made – for industrial or ornamental purposes.

Our concern is with ornamental lakes, which are familiar in well-known landscapes like Blenheim, Oxfordshire, Stourhead, Wiltshire or Studley Royal in Yorkshire. It is less well-known that there were no ornamental lakes in England in 1700, but by 1750 there were many. Furthermore, the concept of an ornamental *lake* simply did not exist in 1700. To contemporaries, 'lake' meant a large, naturally occurring piece of water, and only took on its current meaning in the late eighteenth century. So what happened? This book answers that question, and tells the story of ornamental lakes.

Despite a large body of information about eighteenth-century landscapes, very little is known about their ornamental lakes. They were one of the main features of landscapes for most of the century yet information about when they were made, and by whom, can be very scarce. Often, it is not known when a lake[1] was made, and frequently the designer is also a mystery. This is reflected in the Historic England listings, where a lake may be mentioned, but often no other information is given – because it is not known. Although there were bodies of ornamental water in the seventeenth and early eighteenth centuries, such as ponds and canals, these were geometric in style, as were the landscapes of those times. The informal lakes of the eighteenth century did not fit easily into the geometric landscapes of the 1700s and by the mid-eighteenth century, the prevailing landscape style had also become informal. What role did lakes have in that change, and when precisely did these lakes first appear? What were the factors which led to ornamental water becoming irregular? In order that the answers can be based on solid evidence, a statistical approach has been adopted.[2] There can also be a tendency to divide landscape evolution into distinct styles or phases, and lakes do not fit neatly into those categories. Instead, focussing on the lakes themselves can produce new concepts about them, and the landscapes they inhabited, as well as new information about how and where they were made.

Etymology

The words used by contemporaries in the eighteenth century to talk about water can be confusing. 'Pond' was used to mean a piece of water of any kind, big or small, in the park or in the gardens. This was the case well into the eighteenth century, and people also used the terms 'pool', 'mere' or 'water' when referring to water in ornamental landscapes. Some 'ponds' for example, were sizable geometric shaped pieces of water.

The etymology of the word 'lake' is very revealing about attitudes to water and, until around the 1740s or '50s, it meant a natural body of water. Generally, in estates, the only man-made pieces of water of any size were fishponds for breeding fish (*vivaria*), and these were in the park. The concept of a *man-made* 'lake' simply did not exist before *c*. 1720, and the etymology confirms this. When lakes did begin to be made, various terms were used such as 'The Great Water', 'the Piece of Water' (1724, Vanbrugh about Castle Howard),[3] 'The Bason' (1738, on Roque's plan of Claremont), 'The Intended Water' (1756, Brown's lake at Wimpole),[4] 'The Broad Water'. Vanbrugh was the first person known to use the term 'lake' as we use the word today, in 1709, in a letter about the water at Blenheim: "The Water (where it will appear to best Advantage, whether Lake or River) is full in View".[5] An exception is a comment by Sir Godfrey Copley, in a letter of 1703:

> I am glad the Canalls & Ponds go on so Well, but I am told great Lakes are now the mode. Vanbrook set out one for the D: of Newcastle to front his new house of 40 acres.[6]

However, as Vanbrugh was the person planning this lake, it suggests the word originated with him. It appears that he was instrumental in promulgating the concept of a lake, as we shall see. Plans of Blenheim drawn by him in *c*. 1705 show that he was trying to persuade Sarah, Duchess of Marlborough, to let him make a lake of 3–4 h in the valley to the west of the house (Figures 3.14 and 3.15). She subsequently used the word in letters to various people in 1723–5 when she was planning a piece of water to the east of Vanbrugh's bridge:

> tho Sir John formerly sett his heart upon turning that [the west side] into a lake, as I will do it on the other side; & I will have swans & all sorts of things in it.[7]

The irony is that the Duchess, who hated Vanbrugh, popularised a term he introduced. It was then used by Stukeley, who sketched Blenheim, and by Switzer, who worked there, in his 1727 *Universal System of Water and Water-works*, and it became attached to large bodies of ornamental water. However, even as late as the 1770s, 'lake' was not a term routinely used, though both Thomas Whateley and Arthur Young occasionally employed it. Repton commonly used it in the 1790s, in his Red Books, and by *c.* 1800 large areas of ornamental water were generally referred to as lakes.

Definitions

As the discussion of etymology shows, it is important to have clear definitions of what is meant by a lake, and these are set out in what follows.

Partly because of the difficulty in distinguishing between a lake and a pond, an ornamental lake is defined as a body of water of 1 hectare or more; it is also intended to be primarily ornamental and to provide aesthetic pleasure, although it could be used to stock fish, as well as being used recreationally. It may be argued that this is an arbitrary size to choose. It has the advantage, however, that a piece of water of this size cannot be mistaken for a pond, using today's concept of that word. Another factor is that it is also large enough to appear on most maps.[8] A hectare (2.4 acres) is about the same size as Trafalgar Square in London. The 1 hectare definition does not include islands, so an island counts as part of the area of the lake. This is partly a practical decision, to avoid detailed and lengthy calculations, but it is also because often lakes started life without islands, then acquired them, and then had them altered or removed, so a comparison of the size of a lake as it evolved would be very complex, and the amount of space the lake occupied in the landscape did not necessarily change. Perhaps the most important reason, however, for adopting the 1 hectare criterion is that it means that lakes can be rationally compared with each other, in terms of numbers and also size.

Ornamental lakes are characterised by having a visual relationship with the house, and by having ornamental walks or drives relating them to the house, especially if they lie at some distance from it. In order to provide a framework for discussing the changes in the chronology and form of lakes, a number of sub-categories are used.

A **geometric lake** is one where all the sides are straight, or an arc or a circle, and the shape is symmetrical in plan view (Figure 1.1). Planting around it is formal in style, and the lake fits into the overall geometry of the landscape design.

A **semi-geometric** lake is a sub-set of this, with straight sides, or geometric arcs, but the shape is *asymmetrical* in plan view (Figure 1.2). It does broadly fit into the geometry of the overall design.

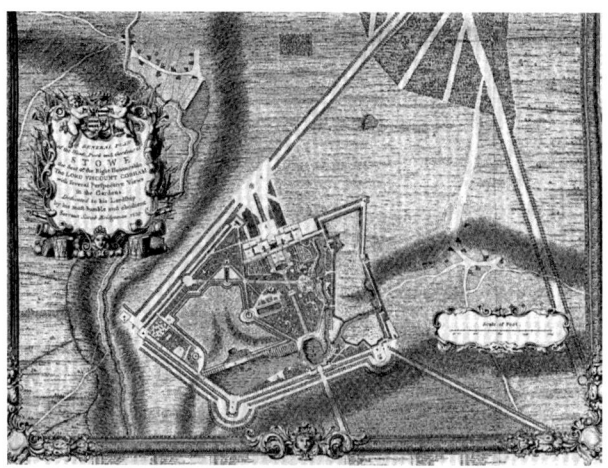

FIGURE 1.1 A detail of Charles Bridgeman's plan of Stowe, 1739, engraved by Jacques Rigaud, showing the Octagonal Lake, which was *c.* 1 h and made in the 1720s. The lakes to the left and right of the octagon, as depicted here, are semi-geometric lakes. *Stowe School/SHTP.*

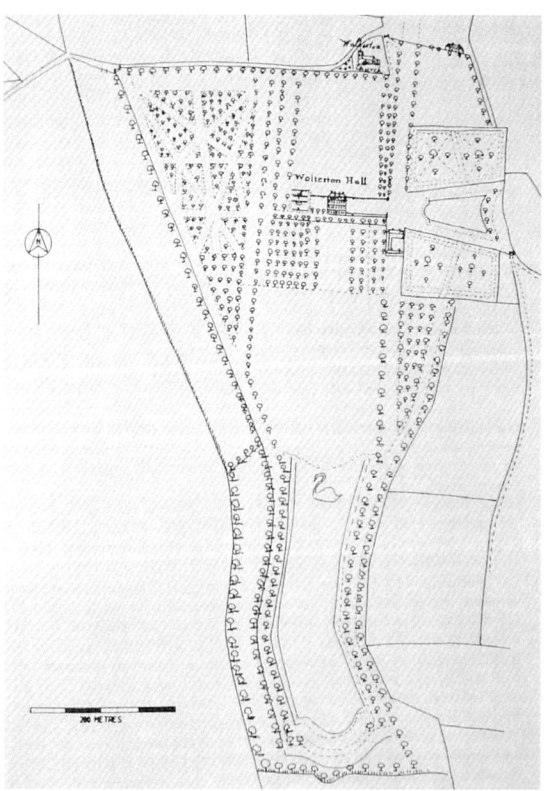

FIGURE 1.2 A diagram of a 1732 map of Wolterton, Norfolk, by J. Corbridge. A *c.* 4 h lake was made by 1732. *Courtesy of Tom Williamson.*

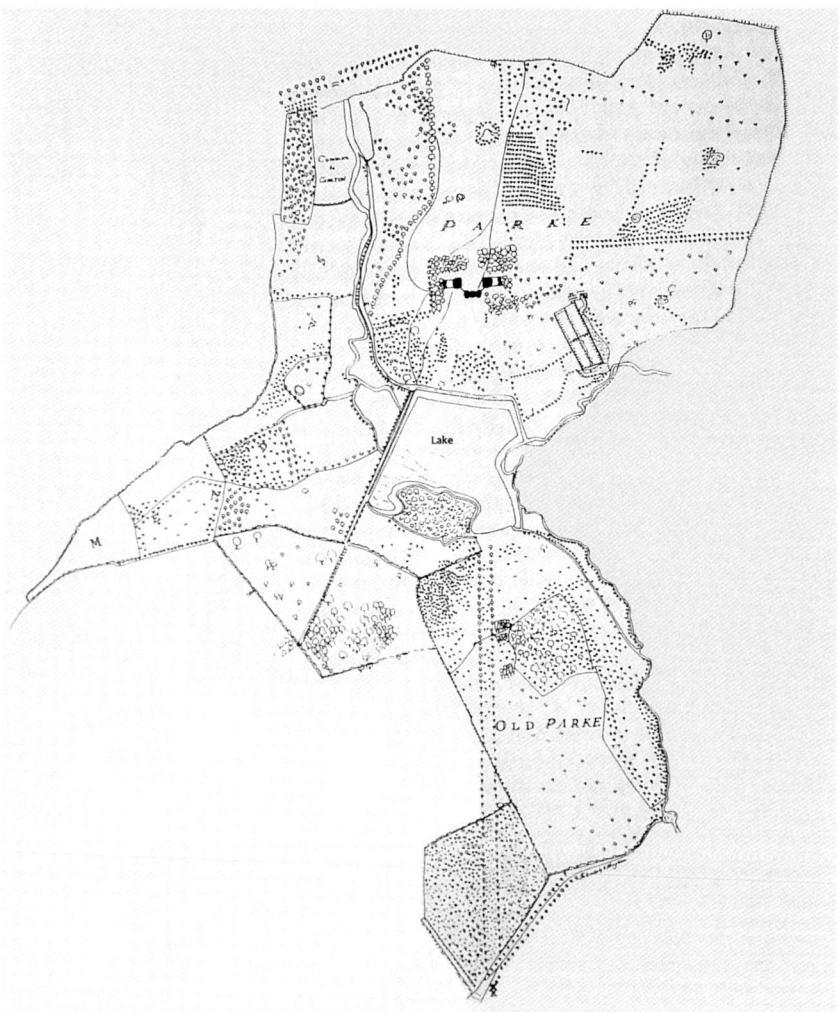

FIGURE 1.3 Kimberley Park, Norfolk: a mid–eighteenth century plan. The 8.5 h lake was made by 1739. *Courtesy of Tom Williamson.*

A **hybrid lake** has two or more straight sides, but one or more sides are wavy, and the plan view reflects this asymmetry (Figure 1.3). The associated planting varies in formality, and the lake may partly fit into the geometry of the overall design – often one straight side (or more) is on an axis of the design.

An **irregular lake** has an irregular or wavy outline in plan view, and informal planting is associated with it (Figure 1.4). It is not part of a symmetrical scheme of design. A sub-set of irregular lakes is the **river-lake**, which is consistently narrow along the entire length, and the shape very much echoes the original course of the

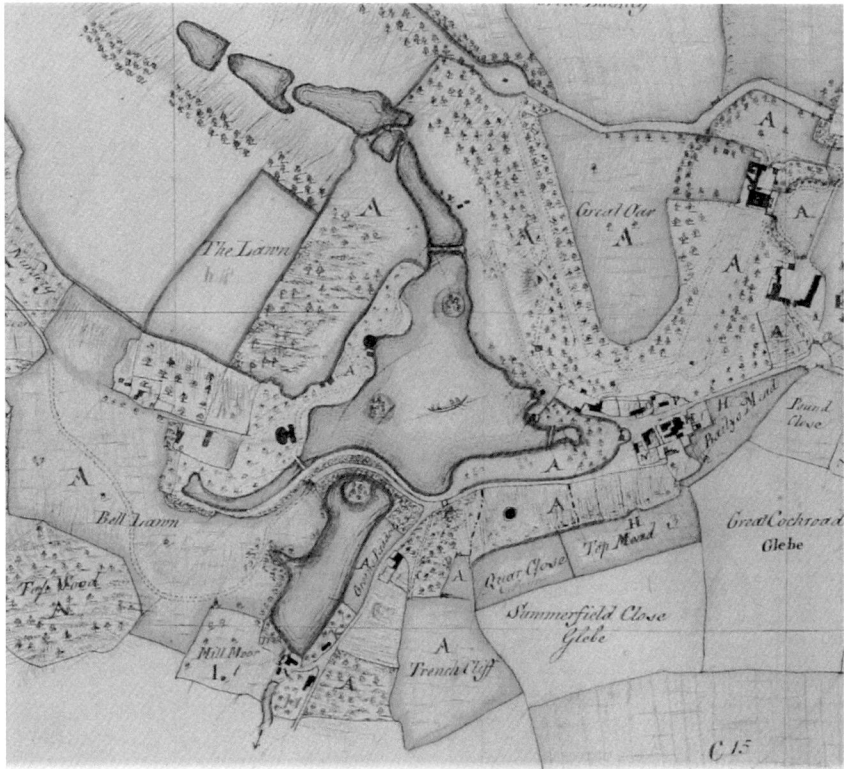

FIGURE 1.4 A 1785 plan of Stourhead, Wiltshire. The main lake of 7.5 h was made in 1754. *Reproduced with permission of Wiltshire and Swindon Archives. Ref. 135/4H.*

river it was made from (Figure 1.5). Also, it is flowing, and visibly so, and islands are not common. Associated planting is informal, and the lake is not part of a symmetrical scheme of design.

The term 'serpentine lake' is a popular one, but this is a description of a particularly sinuous form of an irregular lake, not a definition in its own right. Arthur Young appears to have been the first person to use the term in 1771, in *A Six Months Tour Through the North of Britain*, after visiting the Lake District. He used it to describe the lake at Ditchley, Oxfordshire:

> The gardens are disposed with taste; the sloping banks scattered with wood, and hanging to the serpentine lake, with the rotunda, finely placed on a rising ground among the trees, is a very beautiful landscape.[9]

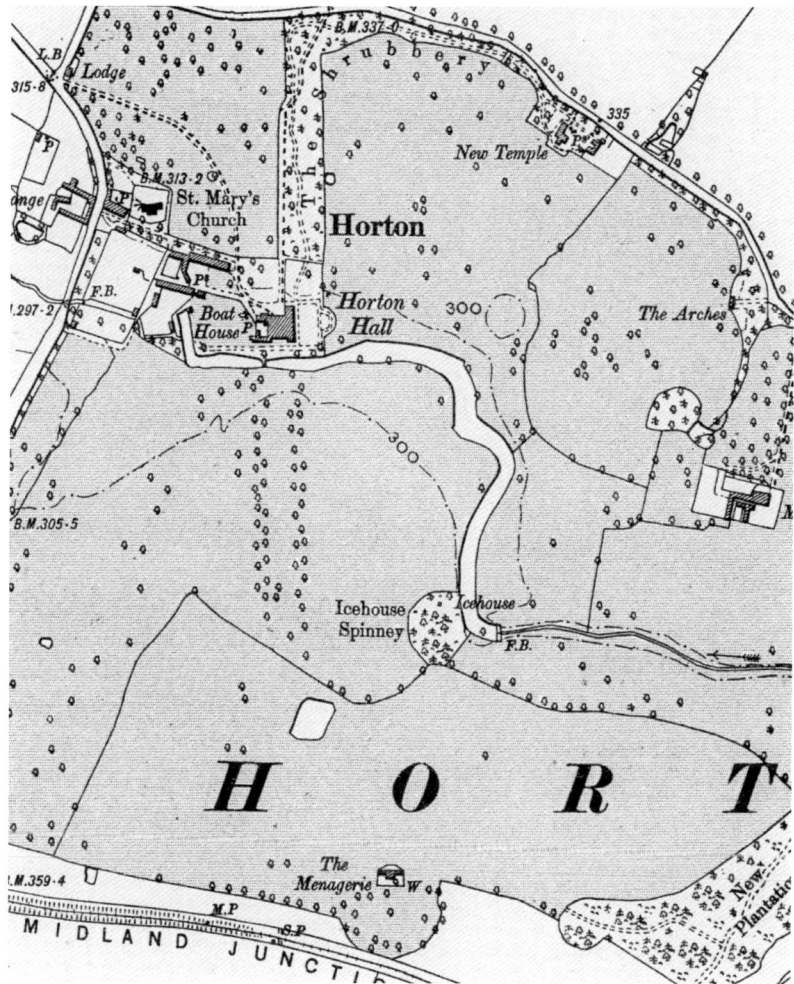

FIGURE 1.5 Horton Park, Northamptonshire, OS 6" map, 1899. The river-lake (1.8 h), made *c.* 1760, is not much wider than the natural river at the beginning and end. *Reproduced with permission of the National Library of Scotland.*

However, the term does not appear to occur again until 1948 when Christopher Hussey used it.[10] The origin of this descriptor is The Serpentine itself (Figure 1.6), and was actually short for 'The Serpentine River', made by Charles Bridgeman for Queen Caroline in 1731.[11] This will be discussed in more detail later, but the salient fact here is that, for contemporaries, it was considered to be a form of canal or river, and the category of 'serpentine canal' applies not

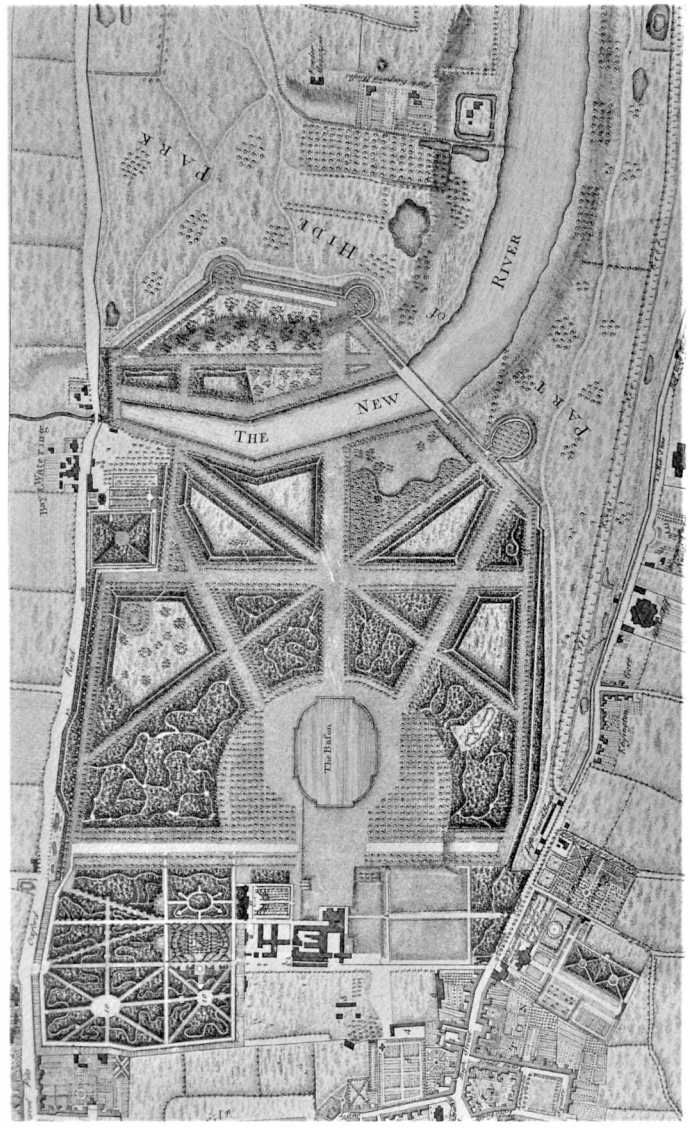

FIGURE 1.6 The Serpentine, Hyde Park by J. Rocque, 1762. *Reproduced with permission of Cambridge University Library. Ref. Maps.bb.18.G.140.*

just to The Serpentine (in Kensington Gardens), but also to water at Longleat, Wiltshire, in the 1730s, and at Foot's Cray Place, Kent. A handful of sinuous lakes, such as Luton Hoo and Syon House, or Morton Hall (Norfolk), may be described as 'serpentine' but they are not significantly different from other irregular lakes to merit a separate category.

Lakes evolved by several different routes, but before examining what those routes were, an overview of how lakes were constructed will help in understanding the factors governing how and where they were made.

Lake construction

One of the essential components in the construction of lakes is a source of water such as a stream, river or spring, which constantly replenishes it. This is what really distinguishes a lake from a pond, as well as questions of size. It is this constant refilling which means that lakes can usually be much bigger than ponds. It also means that lakes contain water that is flowing, although this is not necessarily apparent at first glance.

A lake is made by damming a water source such as a river. The water accumulates behind the dam – it is 'ponded back'. Because water levels can vary considerably throughout the year, dropping during droughts, or rising dramatically in times of flood, they require measures to control them: sluices. They can be closed to conserve water, and replenish the lake, or opened to allow flood water to escape. They are a vital element of lake construction, often as part of the dam, although the word refers to any mechanism for controlling the water. They were used in medieval fishponds, for example. One of the greatest dangers with lakes is flood water. If water reaches the top of a dam and pours over, known as overtopping, the current will start to erode the dam itself, and it is likely to be breached, or collapse altogether.[12] Sluices can be of various kinds. The most common type in the early modern period was the barrel sluice. Figure 1.7 shows a simple sluice construction, but many variations were possible in this basic design – sluices could

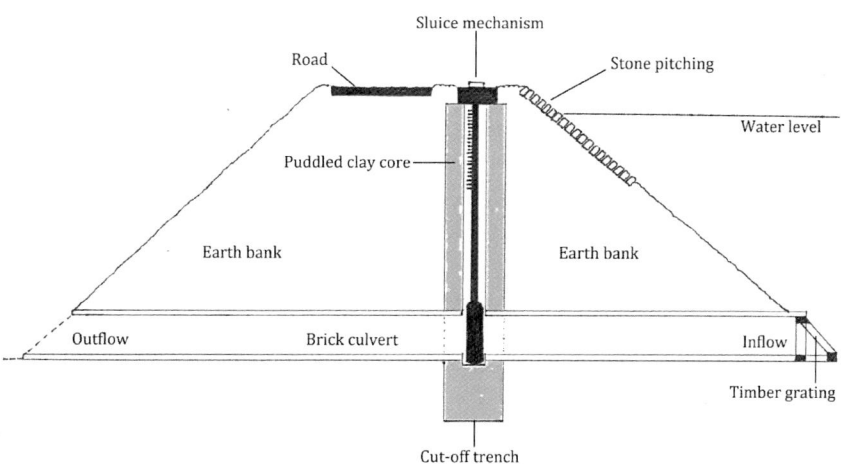

FIGURE 1.7 Cross-section of a typical eighteenth-century dam with a barrel sluice. Not to scale.

be placed in the middle or at the ends of dams. The barrel of the sluice (brick tunnel) passed through the dam at right angles to it, and was approximately large enough – around a metre high – to allow a man to crawl through it for maintenance purposes. Incorporating sluices in a dam was a delicate matter, for there was a strong potential for water to leak around or under the sluice, with potentially drastic consequences. The lake level might start to drop, or the dam itself might collapse.

Another mechanism for controlling the water was a by-pass channel. This involved making a channel leading from the side of the lake upstream of the dam, round the end of the lake to re-join the river below the dam. This kind of channel (a leat) was commonly used with mill ponds, where diverting large amounts of water on a daily basis was necessary. A combination of sluices and by-pass channels could be used, or more than one sluice might be incorporated in a dam. The particular form of design was governed by many factors, such as topography and rates of inward flow. By the nineteenth century, open spillways appeared, as materials more robust than earth were used. This kind of spillway had a small lower section in the crest of the dam which was designed to allow water to pour over it at times of increased flow, ensuring that the lake did not overtop the main part of the dam (Figure 1.8).

These basic construction criteria apply to all lakes, both geometric and irregular. However, different types of lake were constructed slightly differently, which had an effect on where they could be made, as well as the costs of making them, and these factors are discussed in Chapter 4.

FIGURE 1.8 The dam at Clumber Park, Nottinghamshire, from the downstream side, showing the modern spillway which has been inserted into the *c.* 1774 dam.

Sources

The most valuable primary sources for studying lakes are maps. These could be estate maps or terriers, or the independent county maps made in the late eighteenth century by men like Andrews and Dury, or Eyre and Jefferys. Following those, the Ordnance Survey (OS) drawings of the early eighteenth century, and the OS maps of the later eighteenth century are of great use. Tithe maps, usually mid-nineteenth century, are also valuable, as well as documentary evidence such as estate accounts and letters.

Very little has been written about the use, design or construction of ornamental lakes, but the most significant primary sources are Roger North's *A Discourse of Fish and Fishponds*, 1714, and Stephen Switzer's writings: *Ichnographia Rustica, or the Nobleman, Gentlemen, and Gardener's Recreation* 1718, *An Introduction to a General system of Hydrostaticks & Hydraulics* 1729 and *A Universal System of Water & Water-works* 1734. Attitudes to ornamental water, as well as factual information, can be gleaned from the diaries of travellers, both British and foreign, such as Baron Waldstein, who travelled round England visiting great houses and palaces in 1600–1, Celia Fiennes, the antiquarian William Stukeley, Daniel Defoe, Bishop Pococke, and Arthur Young, among others. Similarly, the letters of Sir John Vanbrugh and Lady Mary Wortley Montagu (who knew Vanbrugh) shed light on the actual making of lakes, and relationships between the people who made them.[13]

Apart from a few articles in the twentieth century, little has been written on the subject of lake-making,[14] although there are various books on modern reservoirs, and two seminal books on the history of dams, which give considerable information on their construction: G. M. Binnie *Early Dam Builders in Britain* 1987, and N. Smith *A History of Dams* 1971. Although both are primarily concerned with reservoirs and water supply, their technical approach is valuable, as well as their historical perspective.

Modern writings on ornamental landscape styles also provide insight regarding lakes, particularly those of Tom Williamson and Tim Mowl. However, the dawning of *design* in landscapes is a somewhat contentious subject, as Christopher Taylor sets it in the medieval period.[15] It may certainly be traced back as far as men such as William and Robert Cecil, and Sir Francis Bacon. As concepts of landscape design developed, they were promoted in the writings of people such as Thomas Whateley, Humphry Repton, the polemics of Sir Uvedale Price and Richard Payne Knight, and more considered works by Edward Kemp and J. C. Loudon, to name some of the most important. However, it was the men who actually made lakes who are of the greatest interest in this context, and it would be impossible to write anything on the subject without mentioning Lancelot 'Capability' Brown. His general contribution to landscape-making in the eighteenth century is well known, initially through Dorothy Stroud's ground-breaking book. However, other studies such as Fiona Cowell's book on Richard Woods and David Brown's work on Nathaniel Richmond help to put Brown's contribution into context, demonstrating that he was not the only person to construct lakes in the eighteenth century.

Methodology

Only lakes in England are being considered here because once natural lakes are included, like those in Wales and Scotland, it becomes necessary to determine whether a lake is natural or not, or whether it started off as a natural lake but was subsequently modified.

What made an ornamental lake different to a very large fishpond in the eighteenth century? Many looked very similar to each other. The answer lies in their context, and this has to be borne in mind constantly when considering large pieces of water. If it is in the park, with no connection with the house, it is a large fishpond. If it is within the designed landscape, and has a physical connection with the house such as a walk or drive, or a visual connection, then it is an ornamental lake, probably also stocked with fish. The nature of the planting adjacent to it also tells us, in many instances, whether it is ornamental or not. Examining these factors allows us to reach conclusions about what the lake-makers intended their water to be – ornamental or functional.

As mentioned earlier, maps of various sorts are intrinsic to this study, but this map-based assessment of lakes has limitations. Whilst estate maps tended to be fairly accurate in their surveying, they were sometimes made to show *planned* improvements, rather than what was actually in place. The early county maps of the later eighteenth century, and the OS drawings of the early nineteenth century could be somewhat impressionistic where lakes were concerned, and accuracy in the county maps often appears to be linked to the importance of the subscriber. The tithe maps of *c.* 1840 were generally very accurate, though quite variable in the amount of information they depicted, and the First Edition OS maps were extremely accurate in relation to water, though with a tendency to label most large bodies of water as fishponds.

The dating of lakes is frequently difficult. In the absence of data such as an early estate map, or a textual reference, there is often no definite information relating to the size or appearance of a lake before the First Edition OS maps, as the county maps and the OS drawings are particularly unreliable as regards smaller lakes, as well as the precise shape of larger lakes. The practice for dating lakes which was been adopted was to use both a date range, and an 'actual' date, which has been created by taking the mid-point between the last known date when a lake did not exist and the first known date when it did exist. For example, the 3 h lake at Canon's Ashby does not appear on the Eyre and Jefferys map of Northamptonshire surveyed in 1775, but it does appear on the 1812 OS drawing, so a date of 1794 was used for the 'actual' date, whilst the date range was 1775–1812. This meant that a database of all lakes could be compiled, known as the Landscape Database,[16] which can be found at:

www.routledge.com/9780367894184

The database was then used to analyse trends in the evolution of lakes.

Given that maps and paintings were often being used to help date landscape features, another factor which had to be considered was the 'lead' time involved in their production. Eyre and Jefferys conveniently state when their map of Northampton-shire was surveyed and engraved – 1775 and 1779 – but this is by no means common. Similarly, how long did it take for Knyff and Kip to produce their topographical pic-tures of landscapes in the early 1700s? All the available evidence has been considered when using these means of dating lakes. That might consist of information that the artist or surveyor was in the area at the time, possibly at another property.

There was a heavy reliance for information about lakes on visual as well as tex-tual sources, in the form of pictures, plans and maps, and also books of engravings such as Knyff and Kip's *Britannia Illustrata* 1708.[17] In the late seventeenth and early eighteenth centuries, the 'bird's eye view', or topographical view, was very popular (Figure 1.9). This was a view as if from a hot air balloon – or about three-quarters of the way up an imaginary vertical axis (a three-quarter elevation). As the eigh-teenth century progressed, the viewpoint became generally lower, at around one quarter, and nearer eye level by the end of the century (Figures 1.10 and 1.11), as can be seen in John Harris's *The Artist and the Country House*. The viewpoint of an image has a considerable impact on how much information about the landscape is conveyed, and what can be determined about lake sizes and shapes, as well as their proximity to the house. A map gives a full plan view of the landscape but little idea of physical topography or landforms. (Maps and surveys rarely included hachur-ing, and in 1700 the concept of contour lines was still over a hundred years in the future.) The three-quarter bird's-eye view probably owed its popularity to the way in which it gave a wide-ranging view of the landscape as well as some idea of the topography. This could be represented thus:

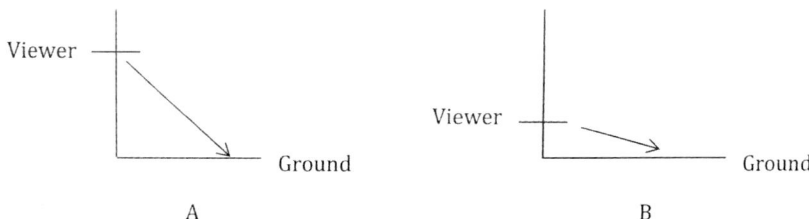

A   B

FIGURE 1.9 A. The bird's-eye-view, or three-quarter elevation: the viewer is three-quarters of the way up an imaginary vertical axis. B. The one-quarter elevation: the viewer is a quarter of the way up the vertical axis.

In a plan view, the viewer is at the top of the vertical axis. The engravings in *Britannia Illustrata* were all bird's-eye views, and this was very much the favoured format for views of landscapes in the later seventeenth century and early 1700s (Figure 1.10). By the 1760s, bird's-eye views seemed old fashioned. The one-quarter elevation view, as used by Edmund Prideaux in his sketches, for example (Figure 1.11), eventually became more common. This viewpoint, whilst being more realistic and intimate to modern eyes, incorporated much less information about the landscape as the view could literally not encompass it. Why did views change?

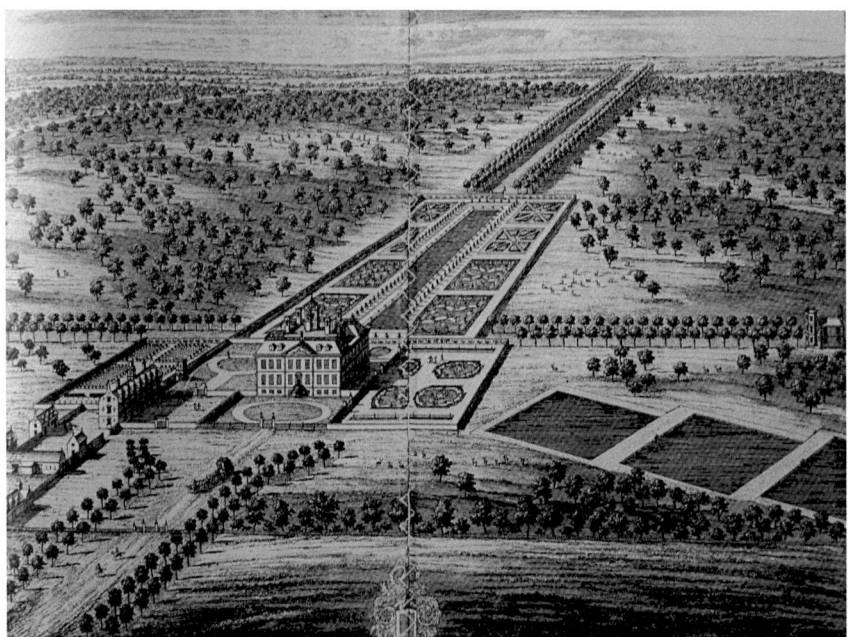

FIGURE 1.10 Melton Constable, Norfolk, by Knyff and Kip, *Britannia Illustrata* 1708. *Courtesy of Historic England.*

FIGURE 1.11 Melton Constable, *c.* 1725, by Edmund Prideaux. *Courtesy of Historic England.*

The answer may lie in the pivotal relationship between house and landscape: one function of house design was to enable the surrounding gardens and landscape to be seen and, in some measure, houses changed according to how those landscapes were best viewed. Geometric landscapes were best viewed from above, so that the extent and geometric pattern of the design could be fully appreciated. Depictions of houses and landscapes mirrored these requirements. Hence the bird's-eye view was the best way of depicting geometric landscapes. Once this rationale disappeared, views in landscape paintings also changed, to reflect the changes in those landscapes. The desire to go into the landscape to view it became the dominant factor.

The main book of engravings of early eighteenth-century landscapes is Leonard Knyff and Johannes Kip's *Britannia Illustrata*. Then there are various county books, such as Sir Robert Atkyns's *The Ancient & Present State of Glostershire* published in 1712 (Atkyns); Sir Henry Chauncy's *The Historical Antiquities of Hertfordshire* published in 1700 (Chauncy); and Colen Campbell's *Vitruvius Britannicus* Vol. 3, published in 1725. In spite of how much these illustrations are used by historians today – particularly *Britannia Illustrata* and *Vitruvius Britannicus* – surprisingly little attention has been paid to the details they show of the ornamental water, whereas they have much information to reveal about the landscapes of the early eighteenth century and the areas of ornamental water depicted.

Data collected from all these sources have been collated in the Landscape Database (Table 1). As well as basic information about when a lake was made, what type it is, and who made it, there is also information such as size and references. It is also possible to interrogate the database, for example asking how many river-lakes were made, or which lakes William Emes designed.[18] Collating the data like this enables the development of lakes to be evaluated chronologically, quantitatively and stylistically, which is a useful new approach to the study of a specific form or element of landscape design.

TABLE 1 An example from the Landscape Database[19]

Place	County	Dating	Date	Feature	Reference	Size (h)
(Welbeck Abbey)	Notts.		1703	Lake irregular	Copley letter	16
Abberley Hall	Worcs.	1867–80s	1872	Cascade informal	Hitching p 296	–
Acton Court	Gloucs.	1500s	1520	Fishpond+island	Mowl G. p 20	–
Albury Park 1	Surrey	1630s	1637	Pond geometric	Harris p 30	–
Albury Park 2	Surrey	1677–84	1680	Canals	Small article	–
Albury Park 3	Surrey	By 1761	1761	Lake irregular	Rocque 1761	1
Aldenham Park 1	Herts.	1840–95	1885	Pond informal	1840 tithe, OS 1895	2
Aldenham Park 2	Herts.	1883–98	1895	Lake irregular	OS 1883 & '98	2
Althorp Park 1	N'hants.	By 1678	1670	Moat	Harris p 67	–
Althorp Park 2	N'hants.		1707	Canal ornamental	B I Paintings	–
Althorp Park 3	N'hants.	By 1730s	1737	Lake geometric	H E	c. 4

Note: Brackets indicate a planned feature which was not constructed.

To set the context in which ornamental lakes began to appear, the landscapes preceding the eighteenth century will be explored in the next chapter. They contained large bodies of man-made water, albeit in the park not the gardens, such as moats, mill ponds, and fishponds for breeding fish. These components made up the waterscapes of the medieval period, and the way in which they influenced subsequent water features will be discussed. This will illuminate the origins of ornamental lakes and chart the course of their evolution.

Notes

1 The default meaning of 'lake' in this book is 'ornamental lake'.
2 For a detailed exposition, see W. Bishop, *The Origins and Evolution of Ornamental Lakes in English Designed Landscapes*, PhD thesis, UEA, 2017.
3 Letter, 11th February, 1724, from Vanbrugh to the 3rd Earl of Carlisle, in H. Dobrée & G. Webb, eds., *The Complete Works of Sir John Vanbrugh*, Vol. 4 (London: The Nonesuch Press, 1928) p 156.
4 G. Jackson-Stops, *An English Arcadia 1600–1990* (London: National Trust Enterprises, 1991) p 85.
5 Letter of 18th July, 1709, from Vanbrugh, possibly to Lord Ryalton, the Duke of Marlborough's son-in-law, in Dobrée & Webb, op. cit., pp 34–35.
6 BL. MS. Stowe 748, ff. 9–10 quoted in Kerry Downes, *Sir John Vanbrugh, A Biography* (London: St. Martin's Press, 1987) p 461. Sir Geoffrey's comment is decidedly tongue in cheek; the two men were certainly acquaintances.
7 Duchess of Marlborough in a letter to the Duke of Somerset, 1723, BL, Add. 61457 quoted in C. Dalton, *Sir John Vanbrugh and the Vitruvian Landscape* (Abingdon: Routledge, 2012) p 115.
8 Some flexibility was used with regard to this size and any body of water which measured 0.9 h was also included in the definition. This was because achieving a completely accurate measurement of a piece of water depends on the scale and accuracy of the maps being used.
9 A. Young, *A Six Months Tour Through the North of Britain* (London, 1771) p 327.
10 *Oxford English Dictionary* "serpentine adj. 3a" (Oxford University Press, online 2015), henceforward quoted as OED.
11 P. Willis, *Charles Bridgeman and the English Landscape Garden* (London: A. Zwemmer, 1997) p 96.
12 The crisis with the Toddbrook Reservoir, Derbyshire, in the autumn of 2019 is a graphic reminder of the dangers of overtopping.
13 Dobrée & Webb, op. cit., and R. Halsband, ed., *The Complete Letters of Lady Mary Wortley Montagu*, Vols. I–III (Oxford: Oxford University Press, 1965).
14 C. Currie's article 'Fishponds as Garden Features, *c.* 1550–1750', *Garden History*, Vol. 18, No. 1 (Spring, 1990), pp 22–46, and Judith Roberts' article '"Well Temper'd Clay": Constructing Water Features in the Landscape Park', *Garden History*, Vol. 29, No. 1 (2001) pp 12–28.
15 C. Taylor, 'Medieval Ornamental Landscapes', *Landscapes*, Vol. 1, No. 1 (April, 2000) pp 38–55.
16 See Bishop, op. cit., for a detailed exposition.
17 L. Knyff & J. Kip, *Britannia Illustrata, Nouveau Theatre de Grande Bretagne* (London, 1708), eds. Harris, J. & Jackson-Stops, G. (Bungay: Paradigm Press, 1984). All available texts were also used.
18 Lakes in some areas (mainly Cheshire) were excluded as they were found to be affected by salt extraction: subsidence could cause lakes to appear.
19 A list of abbreviations used in the Landscape database is included with the database.

2

THE ANCESTORS OF LAKES

The lakes which are a familiar part of eighteenth-century English landscapes did not just suddenly appear, but evolved from the water features which preceded them, and it is helpful to consider exactly how those features influenced lakes. That evolution did not rigorously follow broader developments in landscape style, but there was a general relationship between the two. There are also questions about whether the aesthetics relating to water features in general, as well as the technologies which were in use, affected how lakes developed. Although ornamental lakes were a new feature in the eighteenth century, large bodies of water in the landscape were not: fishponds, moats and millponds populated the medieval landscape, conferring status on their owners. In the medieval period, the emphasis was on function, such as food production or military capability, but by the eighteenth century, the emphasis leaned towards ornament and beauty in landscape parks. This changing relationship between function and ornament is relevant because, as we shall see, it affected both the form and the position of water features.

The precursors of lakes will be considered in three parts: the medieval era, the Tudor and Stuart period, and landscapes in *c.* 1700.

Medieval water features

The closest 'ancestor' of the informal or irregular lake of the eighteenth century is the large fishpond found in medieval parks as part of systems for producing fresh-water fish to eat. Only the elites of society – royalty, the nobility and monastic establishments – were wealthy enough to set up these systems. Oliver Creighton talks of 'elite residences' and this is a useful term in the medieval context, encompassing the habitations of royalty, aristocracy, and great magnates, as well as the top echelons of the ecclesiastical world.[1] These elite residences could have designed gardens in the medieval period – herbers, privy gardens, 'queens' gardens – and the concept of design in relation to them is widely accepted. It is less certain that medieval landscapes were also being consciously designed, as that concept is primarily about ornament, not function.[2] Christopher Taylor mentions the fishponds flanking the approaches to the Bishop of Ely's palace at Somersham as an example of design in a medieval landscape, pointing out that they could have been located elsewhere.[3] However, although we can say that they were being displayed to impress visitors, it is less certain that they were 'designed' as ornaments.

The concept of design is important because it helps to determine whether later bodies of water were actually lakes, or merely fishponds. 'Design' embraces two concepts. One concept is the organising of elements according to a predetermined overview of an area. In this context, perhaps it would be more useful to talk about 'manipulated' landscapes in the medieval period, rather than designed landscapes. Elements such as water, approaches, castles, *pleasaunces*, were being manipulated to maximise their effect in the landscape, but without an over-arching view of the landscape as a whole. The second concept is that of visual appeal – aesthetics. Here, the *intention* behind the construction of an element is of primary importance. Fishponds provide a good illustration of how this works. Large fishponds (*vivaria*) were usually situated in the park in the medieval period. They were functional. They may have seemed beautiful to those people who did see them, but they were not usually intended to be seen by many people, or by high status visitors. However, if the same fishpond subsequently had careful planting associated with it, and an access route from the residence, the intention has changed, and being ornamental has become part of its essence. It is the embellishing, or decorating, of the fishpond with planting, plus linking it to the residence, that signifies the change of intention so that the water is no longer purely functional, but has become ornamental, at least in part. It may be the same pond, in the same place, with the same number of fish, but the context has changed. It is also being incorporated into an overall view of the landscape. Many water features, especially geometric fishponds, existed at different points on the functional – ornamental spectrum, at different points in time, as intentions changed.

The fishponds, moats and mills which were the main water features in medieval landscapes continued to be the main watery elements in landscapes up to the time that lakes appeared, but what was their role in the emergence of those lakes?

Fishponds

The fishponds of the medieval period were often several hectares in size, and often closely resembled the ornamental lakes of the eighteenth century in both size and shape. The construction techniques used to make them were also similar, so a discussion of fishponds is highly relevant to the evolution of lakes. Most importantly, they acted as 'transfer agents': the status which they had conferred in the medieval period was transferred to lakes, as we shall see.

The earliest evidence of fishponds comes from the Roman era, though most of the examples which have been excavated tend to be geometric ornamental ponds which were also used for keeping fish.[4] The system of producing freshwater fish using large breeding ponds (*vivaria*) and small store ponds (*servatoria*) was commonly used in royal and monastic establishments throughout the medieval period. Despite the availability of sea fish, freshwater fish was preferred and elaborate ponds were constructed to breed and store fish. Producing freshwater fish was an expensive process and consequently, it was a high-status food, with both fish and ponds acting as status symbols: royalty and aristocracy bestowed freshwater fish as gifts.[5] The smaller fishponds were frequently near the house – as in the case of Acton Court, Gloucestershire – but also in very many other places such as Kenilworth, Warwickshire, and Rhuddlan, Denbighshire.[6] These smaller holding ponds (*servatoria*) acted as fridge equivalents. They were often secured in some way to prevent theft, and this was probably a factor in siting them in gardens adjacent to the house.[7] As well as being functional, it is possible that they also had an aesthetic role: there were seats around a new fishpond in the Queen's Garden made for Eleanor of Castile at Rhuddlan, in the late thirteenth century,[8] and it was in a turfed courtyard enclosed by staves.[9] This strongly suggests that the pond, as well as storing fish, was appreciated in its own right, as an ornament worth contemplating. Position in relation to the residence must be the guide here: small ponds in service areas were unlikely to have been ornamental whilst those in full view of high status apartments may have had ornamental value as well as conveying status, as at Somersham.[10]

The smaller ponds for storing fish were often constructed by digging a usually rectangular cavity and lining it with clay.[11] Alternatively, they could be made by embanking the soil. A series of ponds was normally required for successful fish production as fish of different ages were kept separately, to ensure they were readily accessible when mature, and to stop larger fish preying on small fry. Roach, bream and carp were some of the commonest types that were kept, along with tench and perch, whilst pike were kept in special ponds because they prey on other fish. Figure 2.1 illustrates a typical medieval ecclesiastical fishpond arrangement.

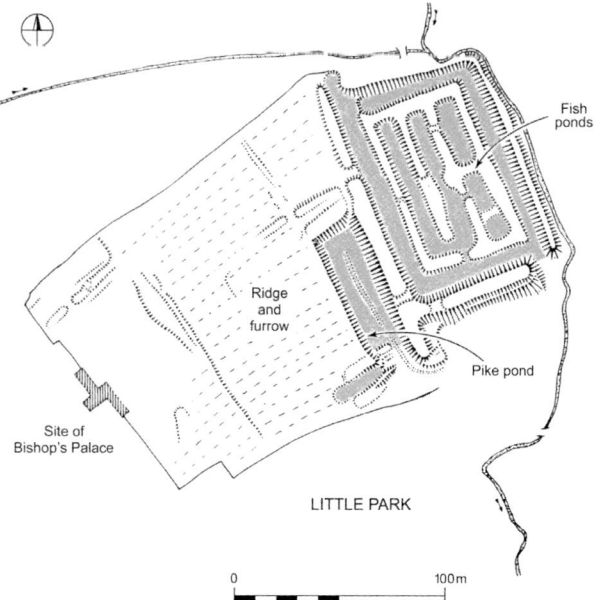

FIGURE 2.1 Diagram of fourteenth-century fishponds adjacent to the Bishop of Lincoln's palace at Lyddington, Rutland, with probable areas of water indicated. *O. Creighton.*

FIGURE 2.2 An anonymous painting of Margam House, Glamorgan, *c.* 1700. *By permission of the National Library of Wales, Cardiff.*

The ponds proudly displayed at Margam House, Glamorgan, *c.* 1700 (Figure 2.2), protected by ornamental gates, indicate how by the end of the seventeenth century fishponds had moved further along the spectrum from purely functional items towards water features which also had an aesthetic value.

Breeding ponds (*vivaria*) were large and often some distance away, in the park. Many were 'contour ponds', which were made by damming a stream, such as the 300 m long pond of the abbey at Cirencester, Gloucestershire,[12] and Waverley Abbey's pond at Tilford, Surrey, of 5.7 h.[13] Our knowledge of them is important because of what they can tell us about the form and construction to the ornamental lakes which appeared in the eighteenth century. At Sulby Abbey, Northamptonshire, a massive dam retained a fishpond of possibly 6.7 h as part of a monastic fish production system, so much the same size as an ornamental lake.[14] The making of these dams was a skilled job, involving moving large amounts of earth, and ramming it very firmly to produce a long dam which was triangular in cross-section, being wider at the bottom than the top (Figure 1.7).[15] These were all embankment dams: they worked simply because the weight of the earth piled up was sufficient to retain the water, and their construction is described in more detail in Chapter 4.[16] Often these fishponds were relatively shallow, to facilitate catching the fish, but some could be very large indeed, and deeper. The Bishop of Winchester's pond at Alresford is a case in point. Made in the flat valley of the Arle in *c*. 1190, it was extremely large – 80 h.[17] The dam which retained it was 6 m high, and subsequently carried the London to Southampton road, which crossed marshy land there.[18] Large dams across a valley, built in this way, were common to both fishponds and eighteenth-century lakes. A significant difference between the majority of *vivaria* and ornamental lakes, however, was their relationship with the house. *Vivaria*, being in the park, were not usually linked with the house in any way or visible from the house, in many instances. Conversely, lakes were often visible from the house, and were linked to it physically, often by an approach drive or other carriageway.[19] Despite these differences, an underlying similarity remains: both were large bodies of water which were irregular in shape, made by damming a water source, and were stocked with fish.

Sluices for medieval fishponds, as in the eighteenth century, were vital components, to enable the control of flood water as well as the draining of pools for catching fish and, periodically, to recondition the pool itself. Recommended by people like Roger North (1714) and John Taverner (1600), this was the practice of draining the pond for a season and ploughing it, about every five years.[20] Usually, a crop was grown on the site while it was empty, as the land was very fertile. Documents from the reign of Henry III (1216–72) reveal evidence of specialist 'pondmen': Brother John of Waverley (1247–51) was sent to oversee repairs and works in Darnhall, and other places in Cheshire, as well as further afield at Woodstock.[21] Other experts of the time were Henry de Lacy and Robert le Parker, who were royal fishermen.[22] Understanding the difficulties involved helps to explain why freshwater fish were a high status commodity, and the sums paid to these men to undertake repairs substantiate this: £40 for the repair of a fishpond at Feckenham, Worcestershire (1203–4), for example, and £20 for blocking up and repairing 'the great stew' of *c*. 4 h for the Bishop of Winchester at Bishop's Waltham, Hampshire (1251–60).[23] These examples throw some light on the kind of expertise which was needed to make dams and lakes in succeeding centuries, including the eighteenth.

Fishponds continued to be important components of food production well into the nineteenth century, as the fifth edition of the Rev. C. Marshall's *A Plain and Easy Introduction to the Knowledge and Practice of Gardening, with hints on Fishponds* 1813, implies. These larger ponds were in parks as this was where the space and water supplies usually were, but the park also acted as a 'security zone', providing a buffer area which protected not just the fish but everything else in the park – deer, game, rabbits, pasture, timber: if you were an unauthorised person inside the park, you were up to no good.[24]

Fish production systems such as these required considerable investment in making the ponds initially, and in their maintenance, not to mention the stocking and care of the fish, all of which were skilled jobs. Estate owners also needed to have sufficient land for such a system. These factors help to explain why freshwater fish was highly regarded throughout the medieval and early modern periods, and why fishponds were symbols of status in elite landscapes. Fish were kept in most areas of water, both in parks and gardens, as we shall see, and so that status was transferred to water in any form, both in medieval and modern times.

Moats and lake-moats

The other main water features in medieval landscapes were moats, the most familiar being moats around castles. Like fishponds, moats also became symbols of status, through their association with elite buildings, and moved water up the ladder of importance, as it were. It was this 'progression' of water features which contributed significantly to many estate owners in the eighteenth century wanting large bodies of water near their houses.

Oliver Rackham states that moats 'began' in *c*. 1150 and were most popular in the thirteenth century, but were out of fashion by 1350.[25] He is talking about primarily defensive castle moats, though he acknowledges Taylor's argument that even these moats were predominantly status symbols, rather than solely military features.[26] Moats were also made around monastic establishments and high status residences in the medieval period, as well as around sixteenth- and seventeenth-century houses.[27] In addition to their functions of defence and deterrence, moats had a number of other advantages. They were intrinsically connected with elite buildings, and this meant they conferred status, as well as delineating ownership and marking the centre of power. They were also useful for drainage, for keeping fish in, as sewers, and to keep vermin out. Roger North makes it clear that moats were stocked with fish: "I am an advocate for Moats. . . . They shall nourish a World of Fish".[28] By the sixteenth century, some had evolved into moated gardens: Lord Burghley's new estate of Theobalds (Figures 2.9 and 2.10) had a moated garden on the site of Cullings Manor, which he added to his estate, whilst the house and gardens had 'canals' which were moat substitutes, perhaps.

A variation of the normal moat shape, occurred around a number of high-status places and consisted of a large, spreading area of water surrounding the building. Bodiam Castle, built in the 1380s by Sir Edward Dalingridge, is a good example (Figure 2.3). This is a *lake-moat* – basically a lake around a castle which also acts as a moat. Leeds Castle, Kent, Caerphilly Castle and Tabley Old Hall, Cheshire are also good examples. Three places with partial lake-moats were Kenilworth and Raglan Castles, and Dunham Massey, Cheshire. It is possible that these lake-moats were not purely military, but were multi-functional.[29] As well as being beautiful to look at, they reflected the castle well. This was undoubtedly important in the medieval period and, as we shall see, also relates to eighteenth-century lakes.

0m   100m   200m   300m

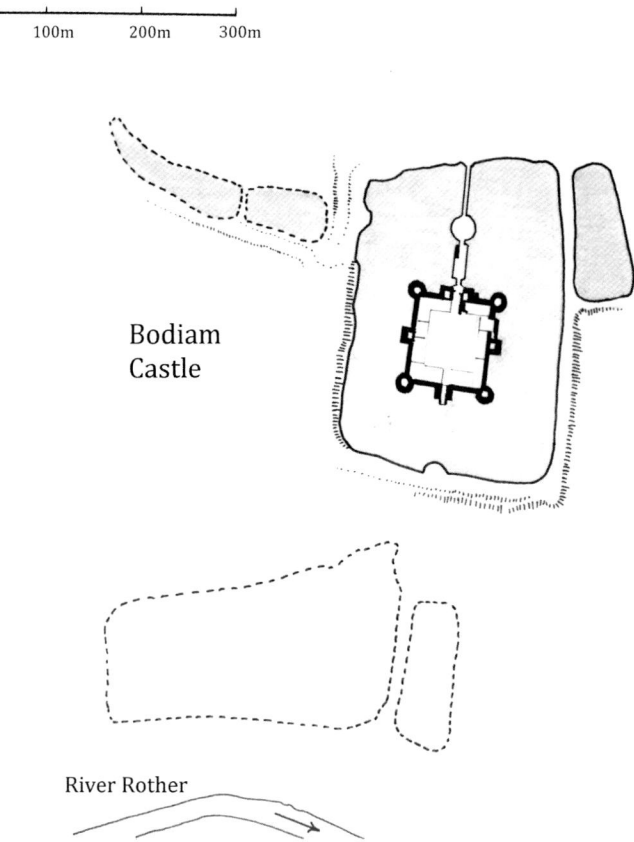

Bodiam
Castle

River Rother

FIGURE 2.3 Plan of Bodiam Castle, Sussex, built by Sir Edward Dalingridge in the 1380s, showing the lake-moat and possible fishponds.

Lake-moats were relatively uncommon and were usually only found around the most elite residences. Notable examples are:

Kenilworth, Warwickshire, *c.* 1125, possibly a thirteenth-century mere,
Framlingham, Suffolk, possibly early thirteenth century,
Leeds Castle, Kent, *c.* 1283,
Caerphilly, Glamorgan, late thirteenth century,
Ravensworth, Yorks, eleventh to fourteenth century,
Scotney Castle, Kent, 1370s,
Nunney, Somerset, 1373 (castle with a mere),
Bodiam, Kent, 1387,
Hertsmonceux, Sussex, 1440s,
Raglan, Monmouthshire, possibly fifteenth century.[30]

Lake-moats effectively prevented the undermining of the walls, a factor which is not sufficiently emphasised, but they also increased the sense of power and status of the castles they surrounded. They acted as mirrors, reflecting the castle or palace and, despite the reflected image being upside down, the overall effect was of the castle being twice as big as it really was (Figure 2.4), doubling the illusion of power. Lake-moats also provided a dramatic approach across the water to the castle itself – once you started along the causeway to enter Bodiam for example, you were very much at the mercy of the owner! This use of water, in the form of lake-moats, or maybe large fishponds, to make the approaches to elite buildings more dramatic seems to occur not just at noble or royal places, but also at monastic sites. At the Bishop of Lincoln's palace of Stow, Lincolnshire (1180s) for example, a causeway flanked by ponds may have been a planned entrance route to the castle, in a similar way to Somersham.[31] It seems that the fishponds, which were of high-status in their own right, were being displayed in this landscape, as they could have been positioned elsewhere, and were being used to enhance their owners' status, as well as to provide fish. We have seen from North that ordinary moats were stocked with fish, and monastic moats also often served as fishponds, so it seems probable that lake-moats were as well.[32] These two aspects remained important into the early modern period,

FIGURE 2.4 Bodiam Castle, Sussex. *Courtesy of Stephen Whitehead.*

with implications for the desirability and positioning of ornamental lakes in the eighteenth century.

In the late medieval period, we can see how the role of water developed by looking at places such as Kenilworth, Caerphilly, and Leeds Castle, Kent. Geoffrey de Clinton built the original castle at Kenilworth in *c*. 1125, and possibly created the mere. It was certainly enlarged several times in the twelfth century, and formed part of the castle's defences, along with a substantial fish-pond to the south-west, and the river to the east (Figure 2.5).[33] In 1279, the castle and its environs were used for a major tournament, and the function of the mere then was to enhance the status of the owner, as well as to indicate military strength perhaps.[34] By 1417, the mere had evolved again, into a lei-sure arena, as Henry V constructed a double moated *plesaunce* about half a mile from the castle, which was reached by boat – a 'voyage' of about half a mile across the mere. Secluded and exclusive leisure seems to have been the aim,

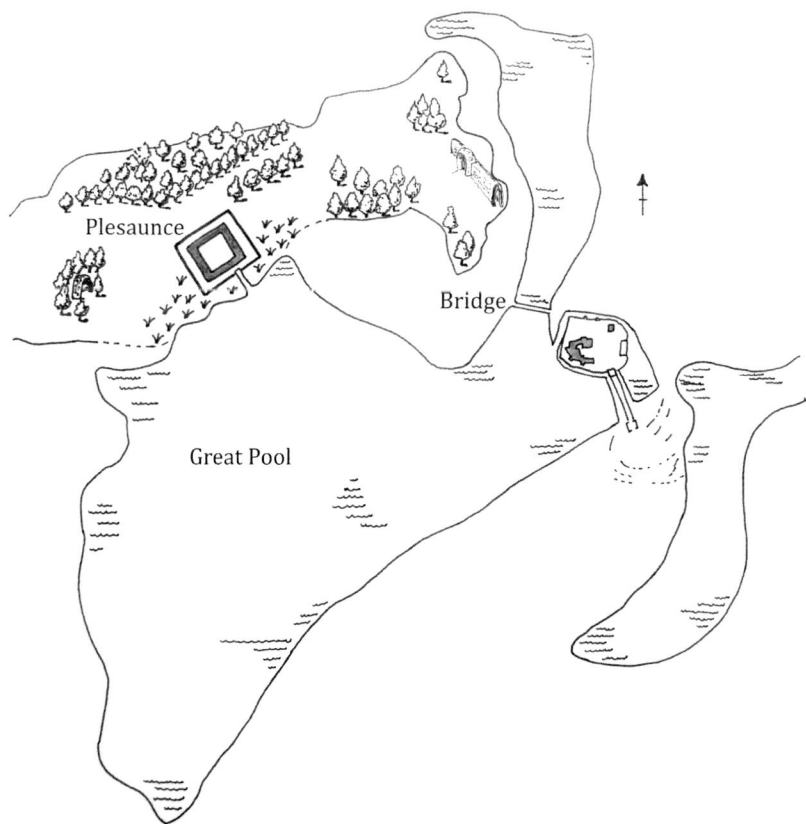

FIGURE 2.5 An interpretation of the landscape of Kenilworth Castle, Warwickshire, *c*. 1575.

as the four-acre enclosure was walled.[35] The plesaunce was located to entail a 'journey' by boat, so it seems likely that the experience of the water itself was the focus, suggesting that in the late medieval period, water was beginning to have an aesthetic role.

The *gloriette* looking over the lake-moat at Leeds Castle, built by Edward I in 1279–88 for his queen, Eleanor, also supports the theory that lake-moats were made to be looked at and admired, rather than to defend. The lack of a proper curtain wall, and the presence of vineyards and extensive fishponds, bear this out.[36] In contrast, Caerphilly Castle (Figure 2.6) was built in 1272–1307 by Gilbert de Clare, in direct response to the threat of attack from Llewelyn ap Gruffudd. Situated in a valley, it is flanked by two enormous lakes. Whilst its primary function was military, the castle was also an emphatic statement of status and this message was driven home by the elaborate water defences. Whilst defence was probably the primary concern, the lake-moat also emphasised and reflected the castle.

The role of the partial lake-moat at Raglan Castle (Figure 2.15) is less easy to establish. The castle, with its Great Tower and 'pitched stone' court (paved), was built by the Herberts, later Earls of Pembroke, in the early fifteenth century. 'The Great Poole' on the north-west side is of unknown date and it does not appear to have had a military role, though it would have kept the enemy at a distance on

FIGURE 2.6 Caerphilly Castle, Glamorgan, defences. *Courtesy of Andy Amor.*

that side. It would also have reflected the castle well, and was stocked with fish.[37] The main entrance was flanked by two fishponds (1.2 h) and included the Red Gate overlooking the bridge over those ponds. The gate was constructed later by Henry Somerset, the 5th earl (d. 1646).[38] A multi-functional role for lake-moats – defence, fish, reflection, status – is the most likely explanation, as with fishponds, but on a much larger scale, as befitted a castle.

These examples emphasise that, where possible, a large area of water immediately adjacent to elite buildings was highly desirable. Following the lead of royalty and the aristocracy, wealthy owners also aspired to this status symbol – a prominent piece of water in their landscapes. 'Ornament' for its own sake did not arrive until later, and the frontiers of man's *aesthetic* use of water remain in the early modern period.

In the sixteenth century, new houses were still being built on existing moated sites, such as Beckley, Oxfordshire, Blickling, Norfolk, and Quarrendon, Buckinghamshire. Existing moats were adapted for other functions, such as gardens, fishing platforms, and also orchards, as at Cullings Manor at Theobalds (Figure 2.9), and at Cope's Castle, Middlesex (Figure 2.18). It appears that where we have evidence of landscapes being created, or redeveloped, moats were still valued, and were incorporated into them, in contrast to the eighteenth century, when moats might be filled in, or drained.[39] An important part of the attraction of moats in the early modern period was that they had a validating role: they conferred authenticity by implying that the owner's family was long established. As in the medieval period, this conferred status, but at one remove, as it were. Like parks, they also demarcated areas: the gardens, fruit trees or fish contained within were marked as special, and out of bounds. Moats such as these were not regarded just as relics of a bygone age in the landscapes of magnates, but could evolve into moated gardens which were valued as secluded areas for high status individuals: by the mid-1620s, James I had created,

> an island planted with cherry, plum and other fruit trees while strawberries, primroses and violets were set round about the border of the Pallisadoe at this pond,[40]

in the gardens of Theobalds, which he acquired from Robert Cecil in 1607.

Mills and millponds

Like fishponds and moats, mills and millponds were functional water features in medieval landscapes, but they also increased the regard for water in the early modern period, and so merit consideration. Only the lords of the manor were allowed to own mills, so they were potent symbols of lordly authority, and in the early medieval period there seems to have been a widespread attitude that no manor was complete unless it had a mill.[41] Whilst there is a good body of knowledge about mills themselves, the same cannot be said about millponds.

Millponds used the same basic principles of construction as fishponds to control the flow of water such as dams, sluices and leats, and knowledge of these devices and techniques formed part of the building blocks used to make lakes in the eighteenth century. The most common source of power for a watermill is a river or stream, though tide mills were not uncommon in the medieval period.[42] The siting of watermills is very dependent on topography, as Leslie Syson makes clear: "The simplest method of obtaining power, if the site was right, was to use the natural fall of the river or stream."[43] Where the water source was insufficient, such as in a relatively flat area, or where demand for milling was high, it would be necessary to construct a millpond. This would retain a larger amount of water, which could be released as required. In other words, the millpond stored power for use by the miller, like a giant battery. He controlled this through the use of sluices and leats; being able to control excess water was vital to avoid damage occurring to the mill wheel, and to nullify fluctuating water-levels.[44]

There is an argument for saying that millponds could be partly ornamental, rather than purely functional, if they were being deliberately placed in view of an elite building. This seems to have been the case at Nappa Hall in Yorkshire, although the mill itself was tucked away out of site down the hill.[45] Stokesay Castle, Shropshire, originally built in the late thirteenth-century by Lawrence of Ludlow, also illustrates this. A millpond (0.1 h) near the castle was supplied by a leat taking water from the River Onny to a mill south east of the Castle. There was also a large pond south of the castle, fed by another stream, which probably supplied the castle moat, and it is likely that both were used for fish.[46] The position of the millpond and fishpond is the key factor here. They are both situated within full view of the castle, which was on a moated platform raised above the surrounding ground.[47] The mill required a long leat to supply it, which suggests the positioning was deliberate. It could have been situated nearer the main river, and the fishpond could have been located further away in the park. Instead, the main approach from Ludlow passed across the dam of the fishpond, illustrating the desire to display the pond as a status symbol. Possibly, the aesthetic impact was also valued, as at Nappa Hall.

Whilst millponds were usually small in general, they were clearly valued not just as economic units but also, by the early modern period, just as much for the status they conveyed on the owner.[48] The mills themselves were also a recurring motif in the waterscape, retaining their status well into the eighteenth century: Cuttle Mill at Rousham, Oxfordshire, was remodelled by William Kent in *c.* 1738 as an eye catcher. Similarly, at Chatsworth, Derbyshire, when the mill and fishponds adjacent to the house were demolished in the eighteenth century, an ornamental mill was then constructed in the gardens.[49] Likewise, at Bowood, Wiltshire, in the 1760s, Brown planned a mill at the north end of the lake, though it was not constructed.

Ornamental water in the Tudor period

Gardens as we might recognise them started to evolve in the Tudor period, but they had virtually no ornamental water in the first half of the sixteenth century: possibly one central fountain, but they were not common. As Roy Strong points out, it was only after the Battle of Bosworth (1485) and the beginnings of the 'Tudor pax' that the arts, including garden making, began to flourish again in England.[50] The purposes for building elite residences had changed, and so had their dynamics: they demonstrated power, wealth and status, which implied military potential, rather than actually providing military strength. 'Castle-palaces' began to develop and though these residences might well have parks, with large fishponds, it was the gardens which evolved, rather than the parks. A good example is the 'castle' built by the Duke of Buckingham at Thornbury, Gloucestershire, where he created a privy garden, following the example of Henry VII in his palace of Richmond (1501).[51] He also had a complex of fishponds in the park. This format was associated with the 'castle-palace' of the first half of the sixteenth century, and was characterised by fountains, usually a single one in the centre. One of the most notable at this time was Henry VIII's tiered fountain at Whitehall, depicted by Anton van den Wyngaerde (*c.* 1560).[52] Another was made for Sir William Fitzwilliam at Cowdray House, Sussex, in *c.* 1536.[53] These fountains were the first specifically ornamental water feature to appear in gardens or landscapes in England.[54]

In the later sixteenth century, there were some moves to use water in landscapes in a more conscious, designing way, but in the first half of the century, only two people made any impact with water: Cardinal Wolsey and Sir William Knight. There is convincing evidence that Wolsey constructed galleries which deliberately overlooked water at his palaces, notably at The More, Hertfordshire, where he utilised the existing moat and built a gallery linking two parts of it to his house in *c.* 1521.[55] This is apparently the first example of a process – utilising medieval water features – which would continue well into the eighteenth century. These galleries were almost a hallmark of Wolsey's gardens – Esher, Surrey, York Place (later named Whitehall) and Bridewell, both in London – were his. Similarly, either he or Henry VIII placed a gallery looking onto a 'pond garden' at Hampton Court. Also, at about this time (1521), Sir William Knight, a priest who had represented Henry abroad, built an Ambulatory at right angles to his house, Horton Court in Gloucestershire (Figure 2.7). It faced east across the valley, and overlooked the largest of three ponds.[56] Knight had spent some years in Italy and the style of the architecture and the positioning of the Ambulatory hints strongly at a harking back to terraces and loggias on Italian hillsides.[57] The suggestion is that Knight was consciously imitating the Italian experience: having a house on a hillside with an impressive view.[58] Lacking the rugged Italian scenery, fishponds were regarded as sufficiently interesting to be the focus in an English landscape, and perhaps we can say that, in this context, the fishpond is acting as a 'proto-lake'. This is also some of the earliest evidence that Italy had an influence on English landscapes, but in general, Henrician gardens did not embrace these views, remaining much more intimate, and displaying their single fountains.

FIGURE 2.7 Sir William Knight's Ambulatory, *c.* 1521, built to overlook his fishponds at Horton Court, Gloucestershire. *Courtesy of Tim Mowl.*

The Italian influence which led Knight to look out over the landscape developed significantly in the late sixteenth century as landscapes began to be embellished and consciously brought into view. However, the ideas of the Renaissance were not adopted wholesale in Britain – climate and morphology made that impossible – but they were influential. The *perceived* design of Italian gardens opened up the view of the landscape, and this led to a much wider landscape being appraised for its potential as ornament. In Britain, this was a landscape which included fishponds as well as rivers, moats and mills. These were not ornamental water features, of course, but the desire to view them was the first step towards large pieces of water being introduced into designed landscapes.

Italian influence

This increased awareness of the wider landscape led to a number of gardens being constructed in the later sixteenth century which displayed a consciousness that the landscape surrounding a great residence could be manipulated not only to enhance the setting of the building but also to provide areas for pleasure, relaxation and entertainment, both at an intimate level and also on a grander scale. Water parterres became fashionable in the later part of the century and the 1600s. However, no lakes were made. Rather, there was an acknowledgement that water in a landscape was interesting and visually appealing, and that it had valuable aesthetic qualities. Water had been appreciated in the medieval period, as we have seen with lake-moats, but in *c.* 1600 there was a conscious recognition that water *ought* to be included in a landscape for its aesthetic qualities. In landscapes, the impact of the Italian Renaissance often worked through indirect channels. Guidebooks for travellers were full of imaginary views of Roman gardens, and people attached enormous importance to what they thought were classical garden remains.[59] Gardens constructed in the first half of the sixteenth century in France and Italy were beginning to have an impact in England. The gardens at Blois (Louis XII), Gaillon (Cardinal d'Amboise) and Fontaine-bleau (Francis I) influenced royal gardens at Richmond, Hampton Court and Whitehall.[60] The influence of Italy worked through these gardens – the Italian designer, Pacello de Mergogliano, laid out Blois for Louis XII in 1500–10 – and also through publications. Works by men such as Jacopo Barozzi da Vignola (1507–73), Andrea Palladio (1508–80) and Jacques Androuet Du Cerceau, (1510–84) extended the Italian influence through northern Europe. One effect of Palladio's concepts was that the house and landscape began to be regarded as one unit – the villa – and that the landscape was to be valued for its visually appealing qualities, not just its productivity. The wider view became important, and this included the rivers, fishponds, mills and moats of the vernacular landscape peculiar to Britain.[61]

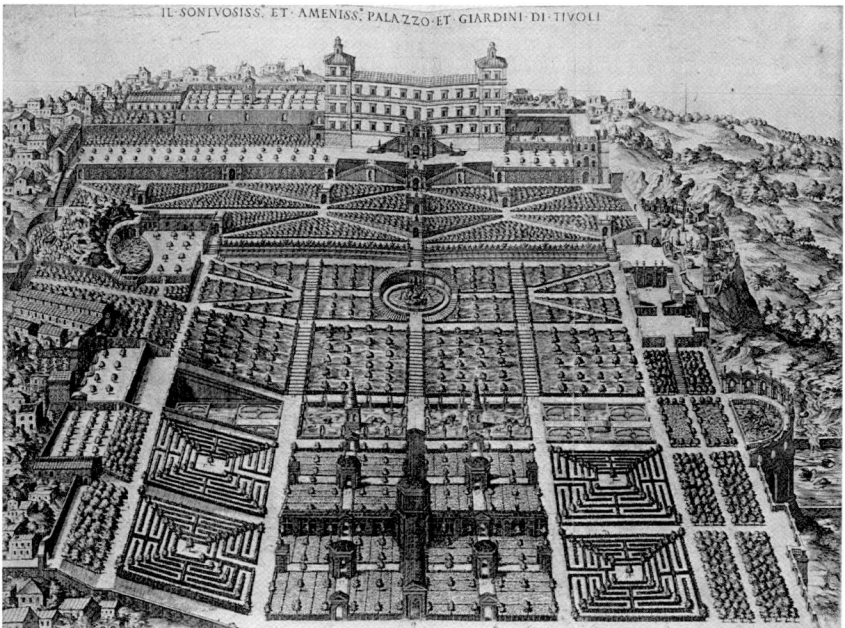

FIGURE 2.8 An engraving of Villa D'Este, Tivoli, 1573, by Étienne Dupérac. *The Metropolitan Museum of Art, New York.*

As well as Palladio's writings, an engraving by Étienne Dupérac of the Villa d'Este (attributed to Vignola), published in 1573 was influential on the group of gardens which was laid out during Elizabeth's reign, from the 1570s onwards (Figure 2.8).[62] They moved from being manipulated landscapes with formal courtyard gardens, and the associated background elements of moats, fishponds and deer parks, to landscapes which linked the wider view – of those same water features – to the house. It is important to note here that lakes did not feature in Italian gardens, although some notable Italian gardens bordered lakes, and that lakes are completely absent in English gardens of the period, but we can see the beginnings of the eighteenth-century formula of park plus water viewed from the house. The influence of the Italian Renaissance was channelled through the talented and powerful men of the era such as the Earl of Leicester, William Cecil, the Earls of Somerset, and Thomas Arundel, often via France. The foremost of these gardens were Kenilworth, Warwickshire, Theobald's, Gorhambury and Hatfield in Hertfordshire, Burghley, Lincolnshire, Holdenby and Lyveden, Northamptonshire, Raglan, Monmouthshire, and of those, Theobalds led the way.

Perhaps the earliest was Leicester's at Kenilworth (Figure 2.5), and Robert Laneham's description of it, written in 1575, is well-known. However, his description does stand out dramatically because it is the only one of its kind that we have. Consequently, other landscapes seem less important, less vivid, and thus less significant, which may not have been the case. Robert Dudley, Earl of Leicester (*c.* 1532–88), Elizabeth I's favourite, created a garden to entertain her in 1575. As well as details of an elaborate fountain, Laneham also describes how Dudley incorporated the mere into his display for Elizabeth.[63] Not only did he have pageants on the water, but also fireworks, viewed from a 600 foot long timber bridge like a "beautiful bracelet" across the northern arm.[64] The bridge also provided a view back to the castle, with its reflections, and we shall see that this ability to look back over water to the residence was to become an important feature in landscapes of the second half of the eighteenth century. The bridge also provided a view over the park, which was,

> Beautified with many delectable, fresh, and shaded bowers, arbours, seats and walks, that with great art, cost, and diligence, were very pleasantly appointed.[65]

Leicester's use of his deer park in this way is one step in the process by which deer parks evolved from hunting grounds, food larders and timber stores into ornamental landscape parks.[66] From the description, it is clear that he was redesigning not just the gardens of the castle but also the landscape surrounding it in a very sophisticated way. His use of water for dramatic effect was echoed in 1591 when the Earl of Hertford created a substantial artificial 'lake' at Elvetham, for a day, in order to stage an allegorical water battle for the queen. The use of water for theatricals in this way – pageants and naumachia – marks the beginnings of water being used for leisure activities, as well as its development as an ornamental component of the landscape, a thread which is taken up again in Chapter 6.

Water featured prominently at the new house which William Cecil built for himself at Theobalds (*c.* 1567–98), near Cheshunt (Figures 2.9 and 2.10). He consciously used that water in a variety of ways: as an ornament, for leisure and to impress. Theobalds was on a palatial scale, with elaborate and extensive gardens,

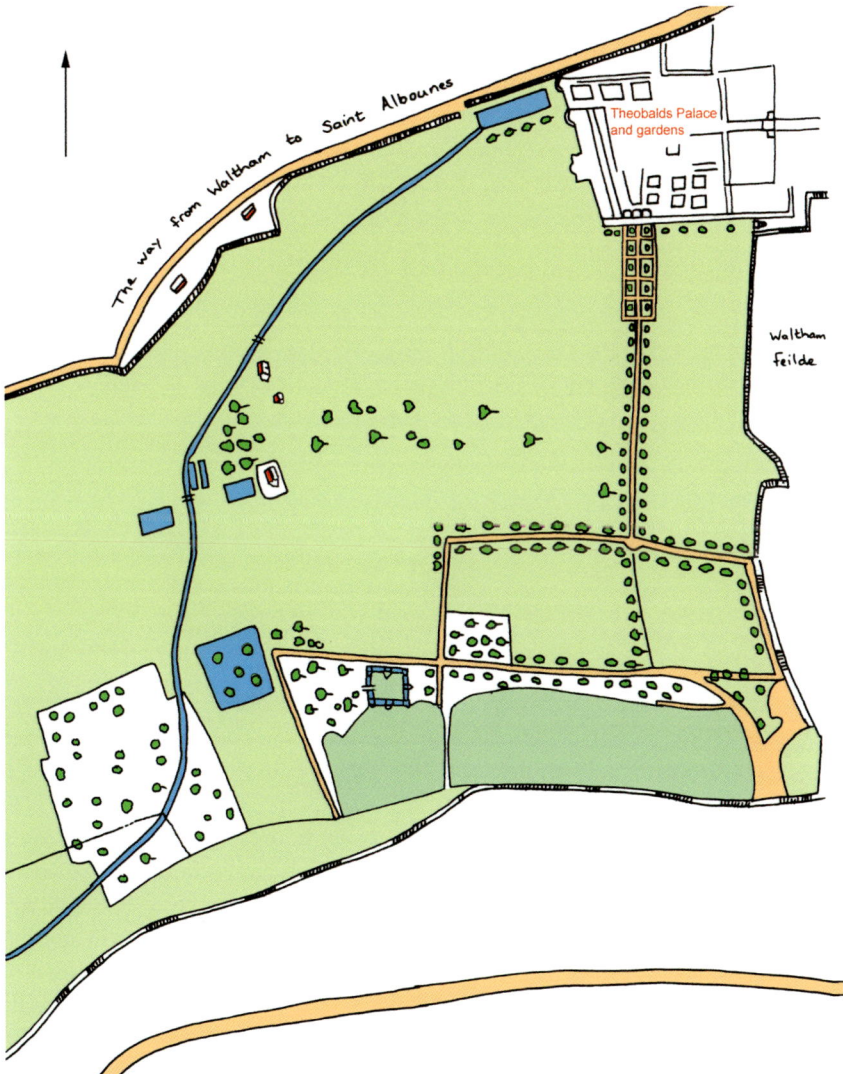

FIGURE 2.9 A digitised copy of the 1611 plan of Theobalds by J. Thorpe. The New River was the work of Robert Cecil. *Courtesy of Anne Rowe.*

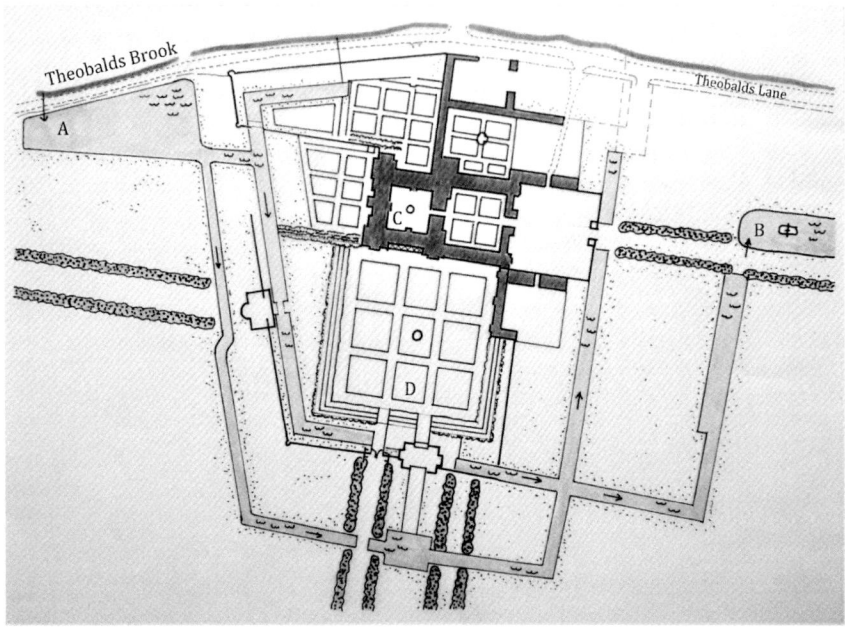

FIGURE 2.10 Layout of the gardens of Theobalds based on the Thorpe survey of 1608. A. Fishpond. B. Horsepond. C. Conduit Court. D. Great Garden. The approach from the London road can be seen on the right of the gardens.

and frequently visited by Elizabeth. In 1598, Paul Hentzner described a 'moat' which surrounded the orchard on three sides:

> one goes into the garden encompassed with water large enough for one to have the pleasure of going in a boat and rowing between the shrubs.[67]

There was a fountain which spouted water from concealed pipes onto unwary passers-by (*giochi d'acqua*), and another which had a little Dutch style ship floating in it, 'complete with canons, flags and sails.'[68] However, it was the way in which Cecil had an overview of the elements in his landscape that was pioneering. He planted a tree-lined way from the south side of his house to link it to the moated garden of Cullings Manor, several fields away, which he had acquired.[69] He linked the gardens and house with Italian type loggias, and brought the wider landscape to the attention with fishponds arranged symmetrically, plus an adjacent lodge, and a large square pond with trees or islands (Figure 2.9).[70] In this he was putting into practice the thinking behind Palladio's design of rural villas, typified by the Villa Rotonda (1552), where Palladio had linked villa and garden through loggia-type porticoes.[71] Although not exact, it can be seen (Figure 2.10) that the proportions of the gardens surrounding the house were related to the proportions of the house itself, and Cecil was the first, or one of the first, to implement this new idea in

England. The design was axial, like that in the Dupérac engraving, which was to become very common in the designed landscapes of the later seventeenth century, but Theobalds was probably the first example in Britain. Incorporating a medieval moat into a designed landscape (as opposed to Wolsey's designed gardens) was new, and was part of the transition from 'manipulated' landscapes to 'designed' landscapes which began in late medieval times and continued through the sixteenth and early seventeenth centuries. Unlike Wolsey, Cecil's example was taken up and moats became popular in landscapes such as Quarrendon, Buckinghamshire and Cope's Castle, Middlesex. This is significant because water was being adopted as part of the design. There is also evidence that it was beginning to become ornamental. Baron Waldstein describes an 'indoor' pool built by Cecil at Theobalds:

> An outstanding feature is a delightful and most beautifully made ornamental pool (at present dry, but previously supplied with water from 2 miles away): it is approached by 24 steps leading up to it. The water was brought up to this height by lead pipes and it flowed into the pool through the mouths of two serpents. In two of the corners of this pool you can see two wooden water-mills built on a rock, just as if they were on the shores of a river. The roof itself was painted in tempura with appropriate episodes from history, and is very finely vaulted. A space beside the pool houses white marble statues of 12 Roman Emperors.[72]

The inclusion of the two wooden watermills in this pool is fascinating. The context of the pool is clearly classical, with its guardian Roman emperors, yet Cecil has introduced the very vernacular element of the watermills. No doubt this was partly because the mills suited the watery scene – perhaps the wheels turned – but this suggests that Cecil was referring directly to the status conferred by owning mills and wanted to convey connotations of lord the manor and long established 'seigneurity'. Like the fishponds he created and the moats he included in his gardens and landscape, the mills were the adjuncts of old, established lordship, and these features appear to be acting as markers of authenticity. As a 'new man', it is quite possible that Cecil, consciously or otherwise, sought to validate his new 'lordship' in this way. The fishponds also retained their important practical value.

The house and landscape at Holdenby (1580s) were made at much the same time as Theobalds, and they also demonstrate a regard for the wider view of the landscape and an 'aestheticisation' of the fishponds in it, as well as the beginnings of a unifying approach to the house and landscape. Sir Christopher Hatton (c. 1540–91), Lord Chancellor to Elizabeth, built a new house on higher ground above the previous medieval manor, overlooking a valley. It is Hatton's positioning of the house and treatment of the fishponds which are of particular interest (Figures 2.11 and 2.12). He terraced the hillside and positioned his house so that it overlooked the fishponds in the valley. By 1587, Hatton had amalgamated the string of ponds in the valley into two large ponds, and added a complex of five rectangular ponds near the church. This tells us two things. Firstly, a 'medieval' fish producing system was installed in a fashionable new landscape at the end of the sixteenth century. Secondly, Hatton

 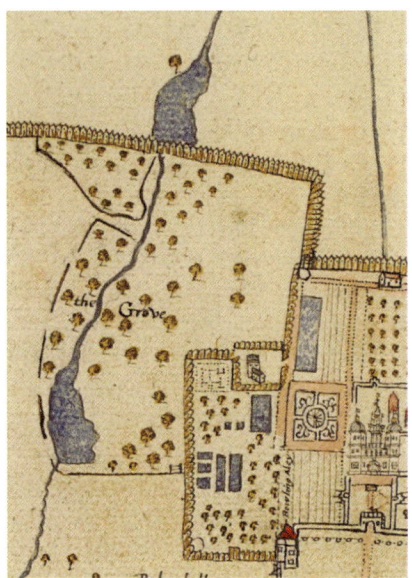

FIGURE 2.11 A detail of Ralph Treswell's 1580 map of Holdenby, Northampton-shire. *Northampton Archives Service. Ref. FH/272/5.*

FIGURE 2.12 A detail of Ralph Treswell's 1587 map of Holdenby, Northampton-shire. *Northampton Archives Service. Ref. FH/272/6.*

was keen to make the *vivaria* prominent features in the landscape. He had sited his house on the hill to look directly down on them, then increased their size to increase their impact. This suggests that as well as displaying them as status symbols, he also wanted to embellish the landscape he overlooked, to aestheticise it, continuing where Knight had left off at Horton. In this English interpretation of an Italian concept (the wider view) the fishponds have a pivotal role, acting as proto-lakes. In the eighteenth century, people wanted houses which looked over large bodies of water, and the prominence of Hatton's fishpond in the valley suggests that some people were beginning to think in those terms in the sixteenth century.

Other 'new men' besides Cecil included these elements in their landscapes, as we shall see. Sir Baptist Hickes at Chipping Camden positioned his new house and garden between a probably medieval fishpond and the mill of Berrington. Both men wanted to cement the implication in the mind that they came from long lineages, and used moats, mills and fishponds to do so.

Water gardens and the aestheticisation of water features

By *c.* 1600, as we have seen, the concept of water as an aesthetic element was beginning to emerge and Sir Francis Bacon (1561–1626) made a significant contribution to this development, in two ways. He created a water-garden at Gorhambury and he wrote a definitive treatise, *On Gardens* in 1597. His water-garden (1600s) is a seminal

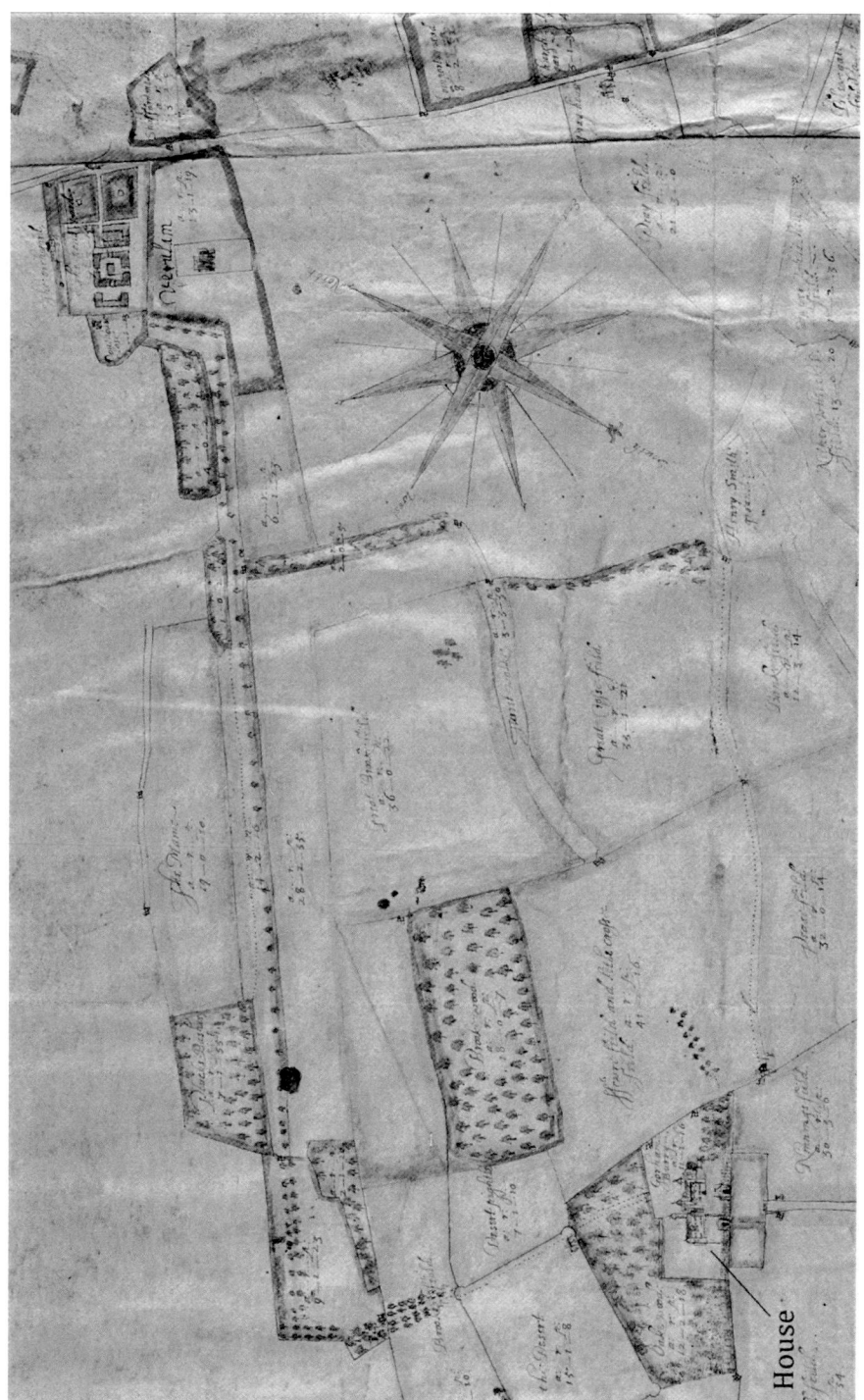

House

FIGURE 2.13 A 1634 map of Gorhambury estate, Hertfordshire, by Benjamin Hare. *Courtesy of the Earl of Verulam.*

example of this adaptation of fishponds into an aesthetic water feature because we have the evidence of Bacon's intentions regarding it. In 1601, Francis inherited Gorhambury from his brother, Anthony. Apart from the intrinsic interest of the gardens themselves, with their ponds and a banqueting house, the most notable aspect is Bacon's initial concept and avowed intention (Figures 2.13 and 2.14). He decided,

> to give directions of a plott [plan] to be made to turn ye pond yard into a place of pleasure, and to speak of them to my Ld. Salisbury.[73]

Whilst other fishponds had possibly undergone similar transformations, Bacon is unambiguously stating what his aim is: the ponds are going to enhance his pleasure in his garden. There is also a sense that he is beginning to design the landscape, not just manipulate it (Figure 2.13). Like Cecil, he is looking at the area as a whole and organising the elements within it coherently. The plan of his estate shows that the water gardens were constructed some way from the main house, presumably because that was where the original fishponds were, and were linked by a line of trees along the way, of eight species, planted in a repeating pattern.[74] Clearly, part of his aim was to have a detached pleasure garden, with its opportunities for solitude and seclusion, and for travelling between the different parts of his estate, as Cecil liked to do.[75] It is this linking, as well as the design of the water-gardens themselves, which shows that Bacon was taking an overview of his landscape – the beginnings of design – and that water was a primary consideration. His 'memorandum' of

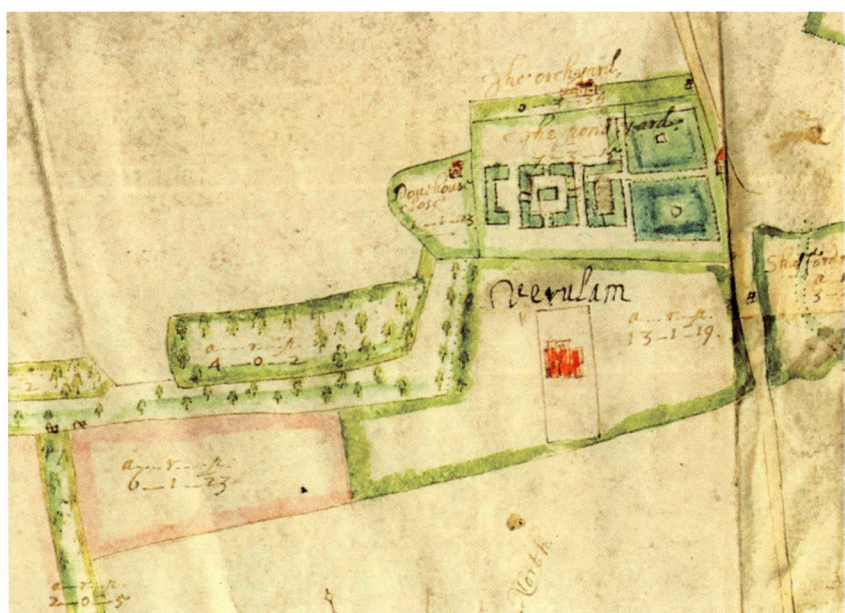

FIGURE 2.14 A detail of the 1634 map of Gorhambury Estate, Hertfordshire. *Courtesy of the Earl of Verulam.*

1608 also describes his intentions for an island surrounded by water with a 'howse' in the middle (Figure 2.14).[76] John Aubrey's 1656 description and sketches evoke a lively and colourful picture of the gardens, the ponds being floored with figures of fishes in coloured pebbles, and a banqueting house paved with black and white marble on the island in the middle.[77] Adjacent to the ponds was Bacon's 'summer house' – Verulam – which afforded good views of them.

It is important here to take in the significance of what Bacon did: he took existing fishponds and made them into a water feature valued primarily for its aesthetic qualities. This had probably happened before – at Theobalds, or Holdenby, or Beddington – but we simply do not have the sources which state that. Hitherto, this role played by fishponds in the development of ornamental water has not been widely recognised. Christopher Currie began to outline the importance of this, suggesting that fishponds could evolve into ornamental ponds,[78] and Christopher Taylor also argued that medieval fishponds had an aesthetic role, but this is the point at which it becomes crystal clear that fishponds could evolve into ornamental water features and that they were the antecedents of the formal water features of the seventeenth century, as well as of the informal lakes of the eighteenth century. In many instances – Petworth, Sussex, Stourhead, Wiltshire, and Burghley, Lincolnshire – an existing fishpond was altered and became a lake. Tracing the evolution of these features fleshes out our understanding of the evolution of lakes.

Given the discussion between Robert Cecil and Bacon, it is not surprising that the water-garden which Cecil made at Hatfield in *c.* 1607 (The Dell), had similar elements to Bacon's, being detached, with a central banqueting house. A probable plan of The Dell, shows a complex design of moats and islands, bisected by a stream, with a central banqueting house astride the river, a mill or pump house, and other structures.[79] The prominence of the mill or pump house in the plan, and the lack of a pond or leat, suggests it was ornamental, rather than functional. Like William Cecil's mills in the pool at Theobalds, this mill seems to have a role as an 'authenticator' in a newly laid out landscape, implying a long-established lordship.

Robert's second water garden was located in The Vineyard (Hatfield), and the noticeable feature here was:

> You have also in those Places where the River enters into and comes out of the Parterre, open sort of Boxes, with Seats round, where you may see a vast Number of Fish pass to and fro in the water, which is exceedingly clear.[80]

Clearly, viewing the river and the fish was important as seats were provided for the purpose. The garden was further embellished with fountains in 1611, by Salmon de Caus, with a cistern to supply them.[81]

Developments at Raglan, Monmouthshire (1549–89), on the Welsh border, illustrate how sophisticated the use of water in elite landscapes was becoming in the late Tudor and early Stuart period. It was the pre-eminent 'court' or palace of Wales,[82] and sophisticated gardens and a water parterre existed, almost certainly created by William and Edward Somerset, the 3rd and 4th Earls of Worcester.[83] These are illustrated on a map by Laurence Smythe in 1652 (Figure 2.15).[84] The most noticeable feature is the

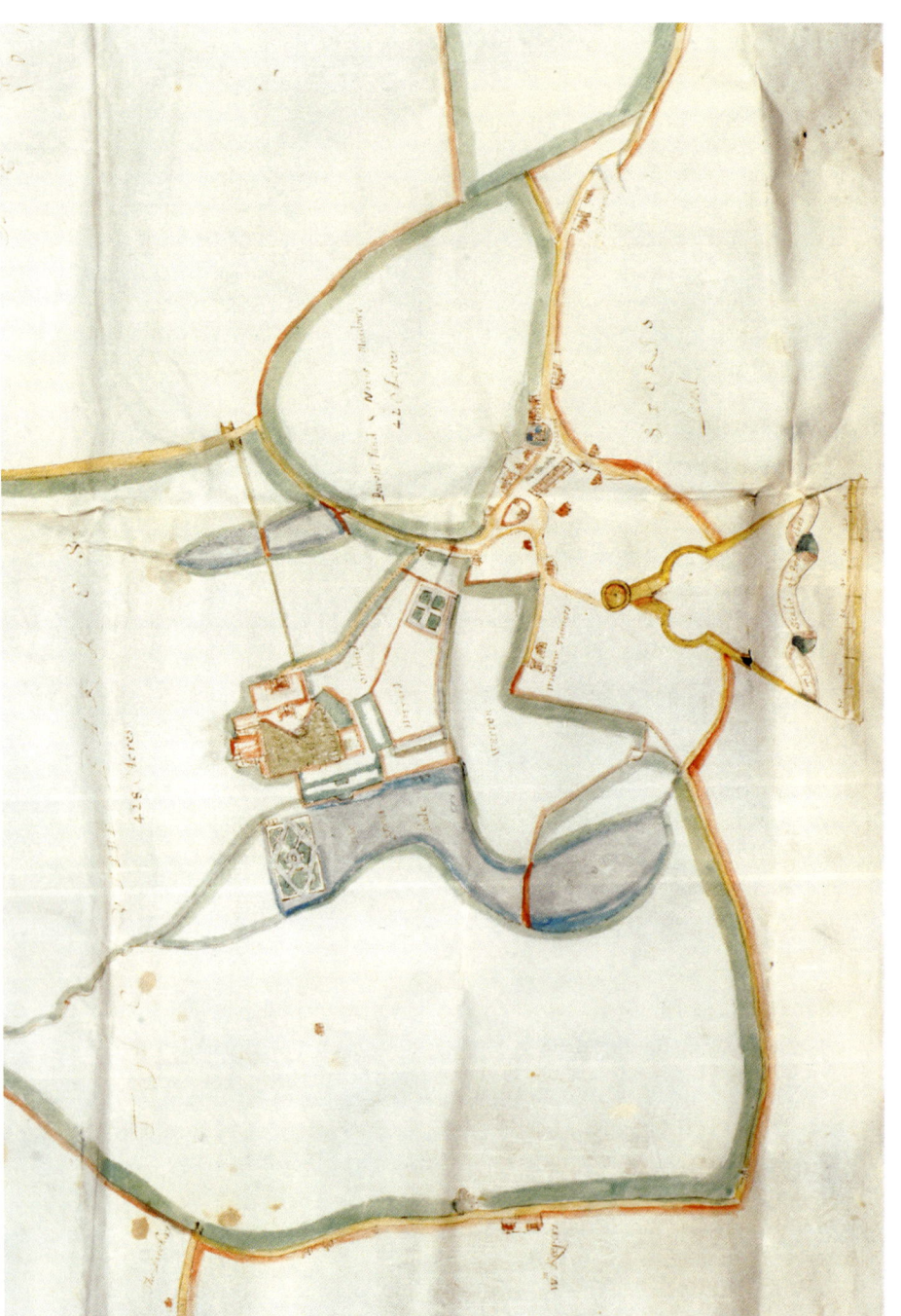

FIGURE 2.15 Raglan Castle, Monmouthshire: The Deplorable Mapp of 1652 by Laurence Smythe. North is towards the top left. *By permission of the National Library of Wales, Cardiff. Ref. Map 1 131/8/3.*

Great Poole (6.2 h), the date of which is uncertain. We cannot even say which century it was made in.[85] The Long Gallery over the chapel, built after 1549 by William Somerset, 3rd Earl of Worcester (1549–89), with a large end window, and extensive garden terraces on the north side of the castle, overlooked the site of the Poole, so it is reasonable to suppose that the 'lake-moat' was there at this time, or about to be constructed.[86] This is corroborated by Thomas Churchyard's poem of 1587:

> The curious knots wrought all with edged toole,
> The stately Tower, that looks ore Pond and Poole:
> The fountain trim that runs both day and night,
> Doth yield in showe, a rare and noble sight.[87]

His mention of pond and pool suggests a distinction between fishponds and the Great Poole. The Poole was much larger than the usual *vivarium*, requiring two large dams to retain it, though it was also stocked with fish.[88] There were also ponds flanking the new approach to the castle (Figures 2.15 and 2.16) which was constructed by Henry, the 5th Earl of Worcester (d. 1646), a marker of status, as we have seen.

William Somerset was Elizabeth's ambassador in Paris in 1570–1. It is tempting to surmise that he was aware of Dudley's efforts at Kenilworth to secure Elizabeth's favour, and set out to emulate him by creating an Elizabethan 'castle-palace' and lake similar to Kenilworth. It appears that the owners of Raglan (though we don't know which ones) felt that a castle of such status should have an impressive body of water adjacent to it, to put it on a footing with places like Caerphilly and Kenilworth.[89] The Deplorable Mapp also shows what appears to be a 300 m bridge across the southern arm of the lake, which seems to link the castle to a clover-leaf shaped pool on the other side of the park, but no more is known about this. Possibly it was created to provide a reason to move across the water, with opportunities for looking back and admiring the castle and the adjacent gardens. This has decided echoes of Leicester's 'bracelet' bridge. As such, it serves as a reminder that competitive display amongst the aristocracy was a powerful force which, like medieval castles, had an impact on the landscape, with 'bigger and better' being a driving sentiment. The sophisticated water parterre at the northern end of the lake was probably made by Edward Somerset, 4th Earl of Worcester (d. 1628). There was a lozenge shaped 'moat' contained within a 'rectangl' of paths, with a possible banqueting house in the north eastern corner (Figure 2.17).[90] It was just under 1 h in size (*c.* 2 acres) and the plan of it on the Deplorable Mapp shows the most sophisticated water parterre known to date. Remains of it can still be seen on the ground today.[91] Aesthetics were important here, whilst the underlying desire to display wealth and status continued to operate. It is also possible that the parterre facilitated fishing. For the Somersets, designing with water in the landscape was a way of adorning it, but also of stating that they were wealthy and fashionable, and on a par with powerful men such as Leicester and the Cecils. Raglan emphasises the continuing importance of water as a status symbol, and of large areas of water being linked with elite residences. This was a thread which extended into the eighteenth

FIGURE 2.16 An artist's recreation of the castle and grounds of Raglan, Monmouthshire, in the early seventeenth century. *Courtesy of CADW Photographic Library.*

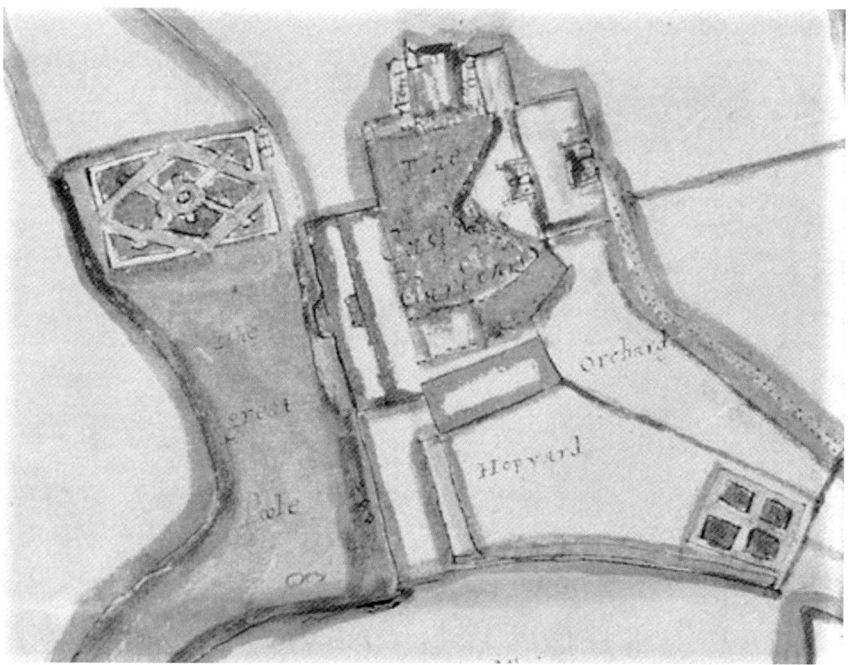

FIGURE 2.17 Raglan Castle, Monmouthshire: detail from the Deplorable Mapp of 1652 by Laurence Smythe. *By permission of the National Library of Wales, Cardiff. Ref. Map 1 131/8/3.*

century, where we see Lancelot Brown routeing approaches to the house across lakes where possible, and placing lakes prominently in view of the house.

Water-gardens in the early seventeenth century were not just the preserve of the highest level of society. Those at Chipping Camden, Gloucestershire (Sir Baptist Hickes), Tackley, Oxfordshire (Sir John Harborne) and Cope's Castle (Sir Walter Cope, d. 1637, Figure 2.18) were made by gentry. Harborne's water garden at Tackley was very geometric in form, and probably constructed with angling as a primary consideration: a plan of very similar gardens appears in *Cheape and Good Husbandry* by Gervaise Markham in 1623, in the section on fish and fishponds (Figure 2.19). He was also known to be a keen angler, and this suggests that he constructed an ornamental garden which also served a utilitarian function.[92] Originally a London merchant, Harborne was a wealthy man who bought manors in Oxfordshire after inheriting from his father in 1609.[93] Despite this, the design remained incomplete, probably because Harborne was unable to purchase one of the quadrants of land from the neighbouring owner.[94]

Both Hickes and Cope adapted existing water features – fishponds, mills and moats. Like those at Tackley, the rationale for Cope's ponds (Figure 2.18) seems to be the sport of angling. A tree-lined way extends eastwards from this area, linking up with further ways to the house. As with William Cecil at Cullings, the moat appears to be highly desirable, if not indispensable, for a 'new man' to authenticate

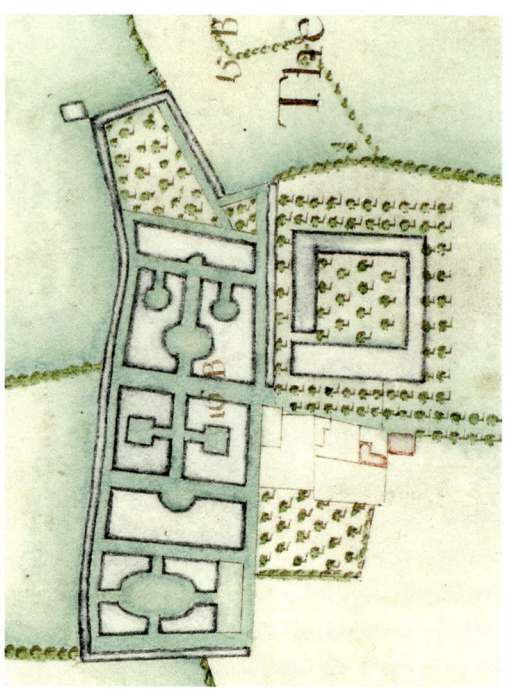

FIGURE 2.18 A 1732 copy of *A Mapp of part of the Manors of Earls Court, Kensington and Abbots 1694/5* by E. Bostock Fuller. North is to the right.

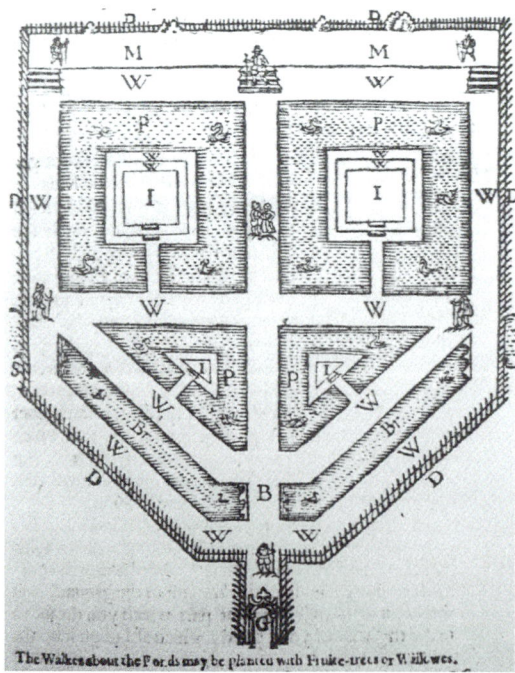

The Walkes about the Ponds may be planted with Frute-trees or Willowes.

FIGURE 2.19 A plan of fishponds in Gervaise Markham's *Cheape and Good Husbandry*, 1623.

the new 'seigneural' landscape which was being created. Possibly, Cope was directly copying the Cecils: in *c.* 1602 Robert sought Cope's advice on water supply in relation to the 'new river' at Theobalds, and Cope would probably have seen that landscape.[95] It is possible that, following Bacon's example, men like Harborne and Cope saw the scope for making, or adapting, a standard fish production system to create a decorative water-garden, as well as using it for angling, in Harborne's case, and to indicate that they were abreast of current fashion.

Although the water-gardens which appeared in the 1600s almost certainly functioned as fishponds, their form was influenced by aesthetic or recreational reasons, which marks out the early seventeenth century from the previous period in terms of water. The intention behind making them was moving closer to the ornamental end of the spectrum. Elite people were also looking for privacy and seclusion when they constructed these gardens. It may be that water-gardens were an evolution of moats in this respect, although, perhaps for the first time, water was being valued specifically for its intrinsic qualities – appearance, sound, changeability. Obviously, it would be unreal to draw a line at 1600 and say that water-gardens appeared after this, but it is difficult to give a conclusive date to any such garden earlier than this. They were largely geometric, often with banqueting houses in which to take refreshments and view them. Water (referred to as a moat) surrounded specific areas of garden, and there was a growing interest in grottoes and 'water plays' – *giochi d'acqua*, as at Enstone, Oxfordshire, which dates from this time. Loggias connecting houses with gardens also became more common, as did terraces or walkways to view gardens from, Bramshill, Hampshire (1600s) being an example (Figure 4.19).[96] There was a change of emphasis as well: landscapes were for experiencing, from the loggias and banqueting houses, boating on moats/canals, walking around to admire their features, and possibly being soaked by the *giochi d'acqua*, rather than primarily viewing them from above, or hunting through them. What is noticeably lacking is the lake. It seems that the concept of a lake simply did not occur to people. If this was because they did not appear in Italian designed landscapes, it illustrates the strength of Italian influence at this time. Instead, the focus, in many instances, was on the geometric possibilities of adapting fishponds and making water parterres.

At this juncture, it is useful to pause and consider the role of islands in fishponds – in some ways, water-gardens could be regarded as collections of small islands arranged aesthetically. The larger medieval fishponds (*vivaria*) did not tend to have islands, but they were common in the medium sized ponds. Thomas Hale, in 1758, considered that they were useful for catching fish by line, as well as with nets.[97] Certainly, they made ponds more accessible to the fisherman, and the convoluted shapes at Tackley and Cope's Castle bear this out. Another consideration was the preference of some fish, carp in particular, for reedy shallows in which to shelter and feed.[98] This question of islands is relevant because they frequently figured in the lakes created in the eighteenth and nineteenth centuries. Why was this? Was it because islands provided a 'destination' for boaters, and an aura of seclusion, or were much more mundane factors at work, such as islands being important for fish and fishing, or being merely useful spoil dumps in the creation of lakes? This discussion will be taken up again later.

In 1597, Sir Francis Bacon produced his essay *On Gardens* (published later in 1625) in which he recommended a garden for every different time of the year – a garden in three parts: a 'green', an area with fountains, and an area of 'natural wildness'.[99] The water must be kept clear (therefore moving) and the bottoms of pools should be decorated to ensure their beauty. This suggests a degree of formality in the form of the ponds. The grass to be kept finely shorn to ensure its beauty, foreshadows Brown's style in the eighteenth century. In his over-arching regard for aesthetics, Bacon signals a sea change in attitudes to landscapes: nowhere does he mention productivity, or utilitarian function. Water was to be primarily for ornament and pleasure. The medieval attitude to landscapes had finally been superseded (in Bacon's mind) and landscapes were beginning to be laid out with aesthetics as a guiding principle, rather than being composed of elements manipulated for function and status. The transition began at the start of the sixteenth century, and was maturing by the end of Elizabeth's reign. Utilitarianism did gain some ascendency in the mid-seventeenth century, but this was a conscious choice, driven by ideology, rather than fashion or aesthetics. To some extent, in his essay, Bacon was reflecting the trends of his time: the Italian influence of ornamental ponds and fountains, terraces and statues. His recommendation for dividing the garden into three parts, as well as trying to have different gardens for the different times of year, was much more forward looking. His influence was far-reaching: tri-partite gardens had become fashionable in Britain by the end of the seventeenth century. He himself may have been influenced by ideas from the Continent as represented in the St. Germain-en-Laye engraving of 1614.[100]

The Commonwealth, tree-planting, and the Restoration

The burgeoning fashion for water-gardens did not survive the travails of the Civil War and the ideological *volte face* of the Interregnum. There was a kind of 'landscape limbo' during the War, and afterwards ostentation of any sort went against the grain of Puritanism. The egalitarian and utilitarian principles of the Puritans resulted in Royalist houses and estates often suffering extensive damage, and estates being sold off, sometimes piecemeal. With no example being set by those governing, and thus no direction of stylistic trends, one result could have been a vibrant flowering of individual tastes and styles among the magnates, the buds of which were appearing in the early seventeenth century. However, there is no evidence of this and the most cogent reason was probably the uncertainty of the times: uncertainty about religion, uncertainty about power and patronage, and therefore wealth. Also, the *mores* of the Puritans did not encompass 'vibrancy' or 'flowering'. What is most noticeable about this mid-century period is how little water featured in the landscapes that were made, as at Wilton, Wiltshire (see below). There is a sense that people maintained existing water features, but were not being innovative in their use of water. This applies to Chatsworth, where the fishponds which existed in 1617 were maintained unchanged until *c.* 1685.[101]

In architecture, a somewhat sterile minimalist classicism, as propounded by Inigo Jones, predominated.[102] Puritan minimalism was linked with economy, and the desire to boost the country's corn production: the Commonwealth government was desperate for the country to become a corn exporter.[103] One response to this was the designs of Samuel Hartlib, in which the underlying principle was to use all the land, including that immediately around the house, in the most productive way possible. The ideas of John Beale and John Evelyn, with their emphasis on orchards and trees, also had an impact in the later part of the century, though water was still a minimal part of the design. At Saye's Court in the 1660s, Evelyn created a garden which owed much to Hartlib's ideas. Men such as Samuel Pepys and Roger North visited it, and it was apparent that Evelyn could practise what he preached in *Sylva* (1664).[104] Evelyn's ideas about landscapes mark the change from a dated enthusiasm for hydraulic toys to a layout of trees extending beyond the garden and far out into the landscape.[105] His deliberate emphasis in *Sylva* on the substantial financial rewards of planting extensive woodlands, allied with the wave of French influence spreading across the country via Le Nôtre, can be seen as fundamental to the development of the extensive geometric gardens with long avenues seen throughout Knyff and Kip's *Britannia Illustrata* 1708, and leading towards an amelioration of the sterile geometry of the French parterre gardens of the court of Louis XIV. The emphasis on tree planting, with plantations becoming increasingly popular towards the end of the century, meant that landscapes tended to increase in size.

Two images highlight developments in the seventeenth century: Wilton in the 1630s, and Longleat in *c.* 1682–1714 (Figures 2.20 and 2.21), both in Wiltshire. In the 1630s engraving of Wilton, the River Nadder can just be seen flowing through the middle compartment, which is a wilderness. Normally furthest

FIGURE 2.20 A 1630s design for the gardens at Wilton House, Wiltshire, by Isaac de Caus. *Courtesy of Metropolitan Museum of Art, New York.*

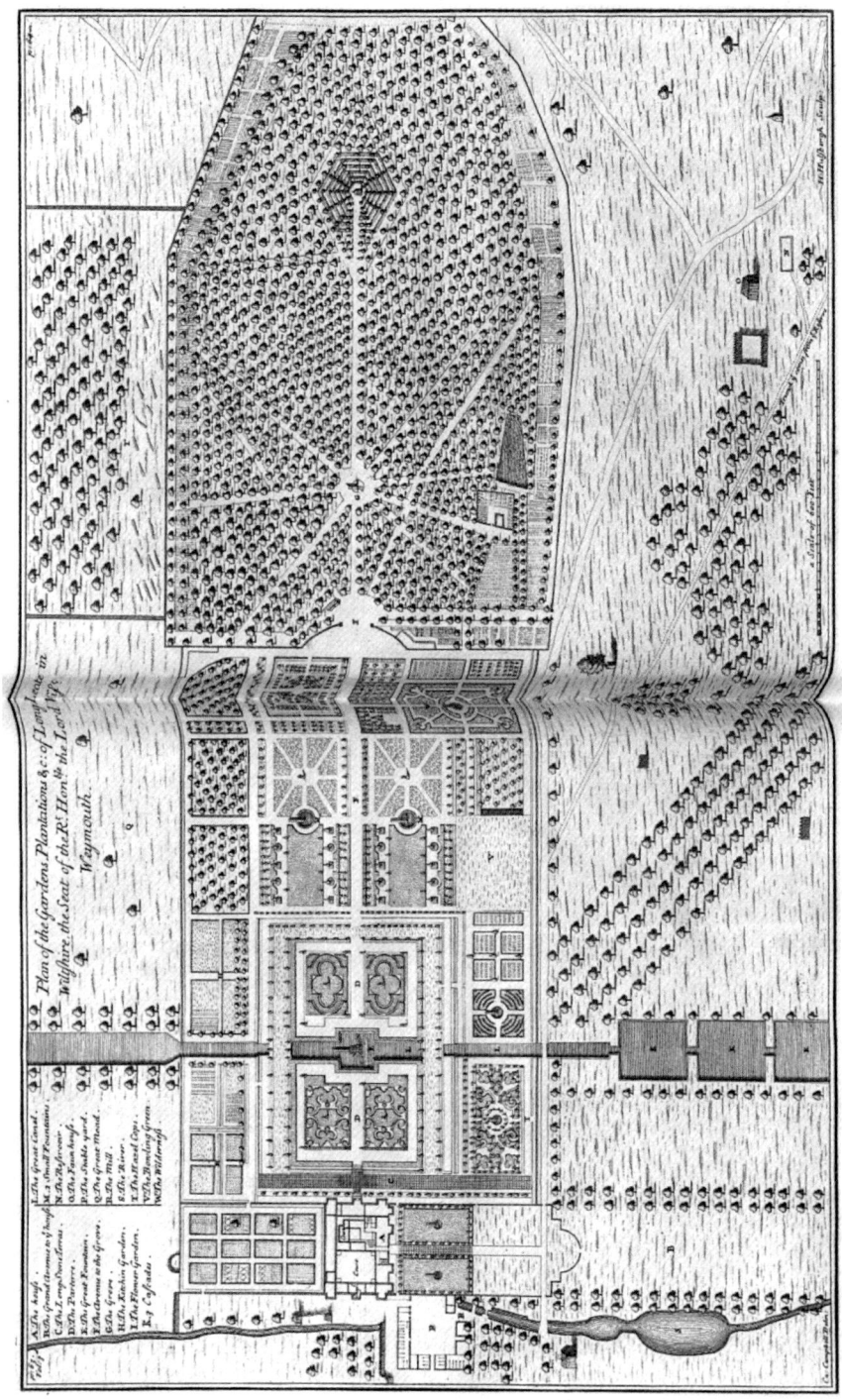

FIGURE 2.21 Colen Campbell's 1725 plan of Longleat, Wiltshire. It is essentially the same as Knyff and Kip's 1708 plate in *Britannia Illustrata*. *Courtesy of Robin Simon.*

from the house, the wilderness seems to be placed here specifically to disguise the river. Isaac de Caus' grotto can be seen at the end of the gardens (top of picture). The significant difference between Wilton and Longleat was that instead of trying to disguise the river flowing through it, the water at Longleat (Figure 2.21) was canalised, with a series of geometric ponds. This treatment of the river marks the decisive change in attitudes to water, and rivers in particular, which occurred in the latter part of the seventeenth century. Rivers began to be put into straitjackets and displayed as 'canals'. Initially, a common design practice in the earlier seventeenth century was to place fountains in ponds at punctuation points in the design, with rivers or ponds adjacent to the gardens to supply the fountains. Later, canals became fashionable features in their own right. The accession of William III in 1688 fostered this fashion; Westbury Court, Gloucestershire, and Grimsthorpe, Lincolnshire, come to mind, and in the early decades of the eighteenth century, moats surrounding residences tended to be altered to give the appearance of canals, Rycote, Oxfordshire, being a good example. Like medieval moats, canals were also used for keeping fish.[106] The Dutch influence of William III saw Britain acquiring many more canals, and these can be seen in the illustrations in *Britannia Illustrata*.[107]

These images of Wilton and Longleat also highlight the increase in tree planting – plantations, groves, wildernesses – mentioned earlier, which started to occur in the intervening half-century. By *c.* 1700, Wilton did have groves or wildernesses just beyond the tri-partite garden. At Longleat, the extended area of tree planting beyond the formal gardens was as large again as the gardens themselves: the designed landscape had doubled in size (Figure 2.21). At this point, attempts were being made to retain a sense of balance in the overall layout, but there were significant areas of other tree planting adjacent to the gardens. This increase in the scale of designed landscapes was to play a pivotal role in the development of lakes.

Water features *c.* 1700

In *c.* 1700, several sources give us a 'snapshot' of what landscapes were like at that point: Henry Chauncy's *The Historical Antiquities of Hertfordshire* 1700 (Chauncy), *Britannia Illustrata* 1708 by Leonard Knyff and Johannes Kip, Robert Atkyns's *The Ancient and Present State of Glostershire* 1712 (Atkyns), and a later work by Colen Campbell, *Vitruvius Britannicus*, of which Volume III, containing landscape plans, was published in 1725. In addition, paintings of landscapes are a valuable source of information, as are the writings of men like Stephen Switzer, Roger North and Thomas Hale, who all wrote in the first half of the eighteenth century.

The images in these sources are the bird's-eye views of landscapes described in the Introduction, or plans like Longleat (Figures 2.21 and 2.22). A considerable amount of information about the water features of the time can be gleaned from these images, such as which were most common, or how many typically occurred in the landscapes of the wealthy.[108] The diversity of these features can be seen in Figure 2.23. Contrary to what we might expect, it appears that water features like

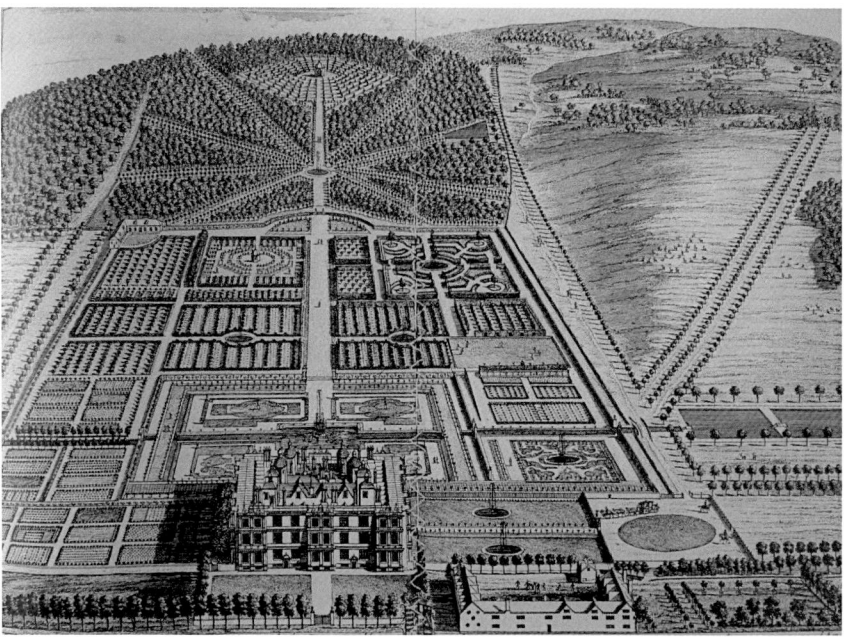

FIGURE 2.22 Longleat, Wiltshire, in *Britannia Illustrata* 1708. *Historic England.*

those shown in *Britannia Illustrata* were actually made much as they were shown by Knyff and Kip – the figure is nearly half, despite several centuries of silting up or erosion.[109] The evidence from the other sources (Chauncy, Atkyns, Campbell) is much the same, but it is still likely that the actual landscapes of the eighteenth century did not look nearly as neat and geometric as their owners seemed to wish: almost certainly the trees, whether in avenues or groves, would not have been growing as uniformly as they were shown and occasionally, views depicted what was planned rather than what existed. As topographical representation and mapping continued to improve in the eighteenth century, it is reasonable to conclude that the later sources, such as *Vitruvius Britannicus* produced by J. Badeslade *et al* in 1739 and 1767–71, can also be considered generally reliable in interpreting the landscape.

Several trends emerge from examining these sources. Firstly, ponds, ornamental or otherwise, were usually connected with fish. Square or rectangular ponds were frequently placed outside the boundaries of the designed landscape, and so were likely to be functional fishponds, rather than ornamental ponds. Conversely, small geometric ponds with fountains, which were the single most common feature in these geometric landscapes, were primarily ornamental. However, this analysis runs the risk of over-simplifying the situation. It appears likely that most ornamental water features were dual purpose and were stocked with fish. This was certainly the case at Dyrham Park, where Switzer states, "In this Canal several Sorts of Fish are

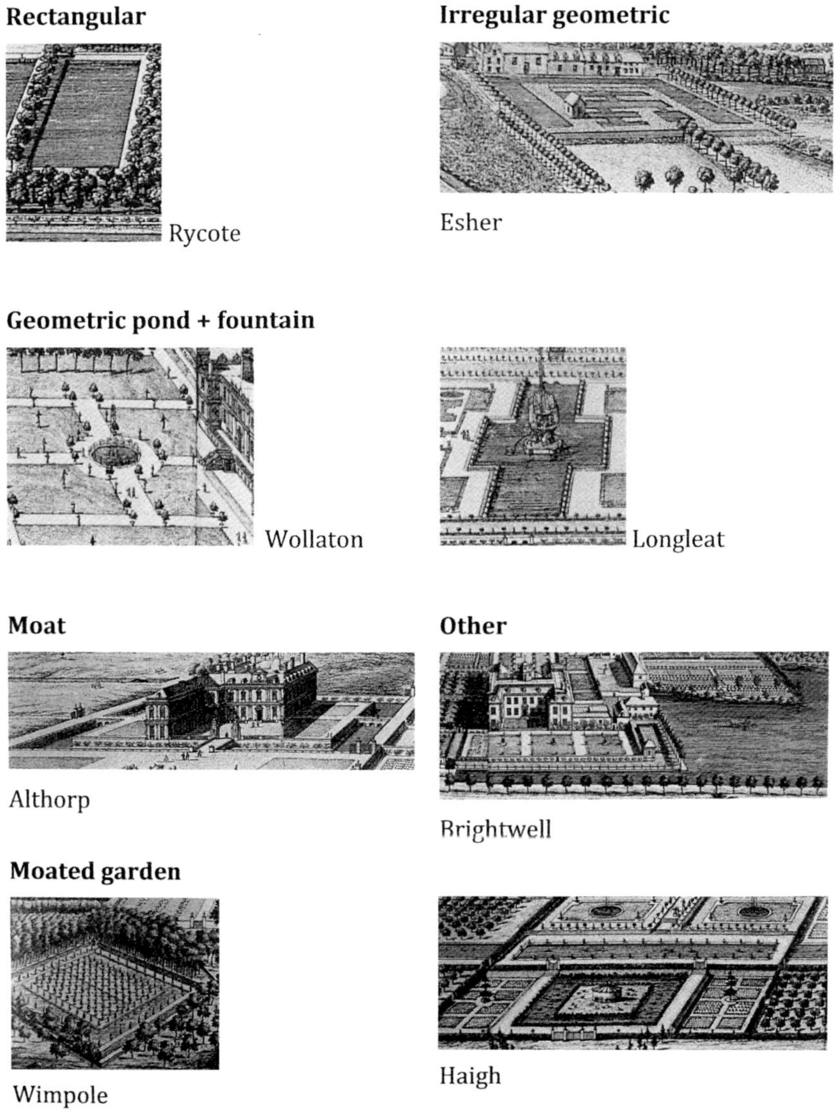

Rectangular

Rycote

Irregular geometric

Esher

Geometric pond + fountain

Wollaton

Longleat

Moat

Althorp

Other

Brightwell

Moated garden

Wimpole

Haigh

FIGURE 2.23 Different types of water features found in *Britannia Illustrata* 1708.

confin'd, as Trout, Perch, Carp, Etc. of a very large Size."[110] Similarly, John Whitney, writing in 1700, specifically mentions fish being kept in fountains:

> I have seen your round Fountain in your delightful best Garden, and the stock of Fish therein kept to be always at hand to pleasure your Friends, which is continually stored with *Trouts* and *Carps* of the largest size; I remember also the Oval Fountain in the Kitchen Garden, which is a good Nursery for the younger fry.[111]

It appears that this also applied to lakes, when they came into being: John Lawrence, writing in 1806, says in his section on fishponds,

> Upon an extensive piece of water, an ISLAND naturally formed, and handsomely wooded and planted with shrubs, is a noble addition to the scenery, and the banks afford the most convenient retiring places for the fish.[112]

At Dyrham Park, the ornamental ponds which were part of the formal design for William Blathwaite in the early eighteenth century, were functional as well as ornamental. A list dated 1710 shows not only what fish were kept in these garden ponds but when they would be ready for the table.[113] By the end of the twentieth century, the ponds had become purely ornamental.

An analysis of the sources mentioned earlier, as well as Badeslade's *Vitruvius Britannicus*, and landscape paintings of the eighteenth century, shows how trends in water features changed over time.[114] By the 1730s, the small round pond with a fountain had declined significantly in popularity, suggesting that it was becoming *passée* by the second and third decades of the eighteenth century. Moated gardens were minority features throughout, and had completely disappeared by the 1720s. Ornamental canals had become more numerous, being popular in the first three decades of the century. Other features also increased in numbers. These included geometric *lakes*, and basins of considerable size, cascades, hybrid lakes and an informal pond (Claremont), features which are discussed in more detail in the next chapter.

The conclusions which can be drawn from this are that firstly, in the decades *c.* 1700, where men had the money they made water features that were as extensive and elaborate as they could, as geometrically as possible. Secondly, any self-respecting gentleman with any aspiration to fashion in his gardens at least had a round pond with a fountain in the early 1700s. Thirdly, ornamental canals were popular, and increased in popularity, throughout the period (*c.* 1680–1730). One fact to note is that a surprising number of residences, when identified on today's OS maps, were sited within, or in, the vicinity of moats. Perhaps a more accurate picture of reality in *c.* 1700 is best illustrated by two images of Bowood, Wiltshire, a relatively modest house and park in 1720 (Figure 2.24). In the painting, the rectangular pond (possibly 0.4 h) is the prominent ornamental water feature and it is shown as completely symmetrical in shape, whereas a pre-1763 estate survey (Figure 2.25) shows it with a wavy edge on the west side. Whilst, in 1763, this could represent a change in fashion and a deliberate softening of the geometry of the water, little else in the landscape backs this up. The conclusion must be that, in *c.* 1725, Sir Orlando Bridgeman, the owner, wanted to construct a completely geometric pond, but failed on the west side, where the land started to rise, and more earth needed to be moved, involving greater expense.

Location of water features

The location of water features in landscapes of *c.* 1700 is of interest because it casts light on how important they were to their owners and, perhaps, what their owners

FIGURE 2.24 A painting of the house and gardens of Bowood Park, Wiltshire, *c.* 1725. *Courtesy of The Trustees of the Bowood Collection.*

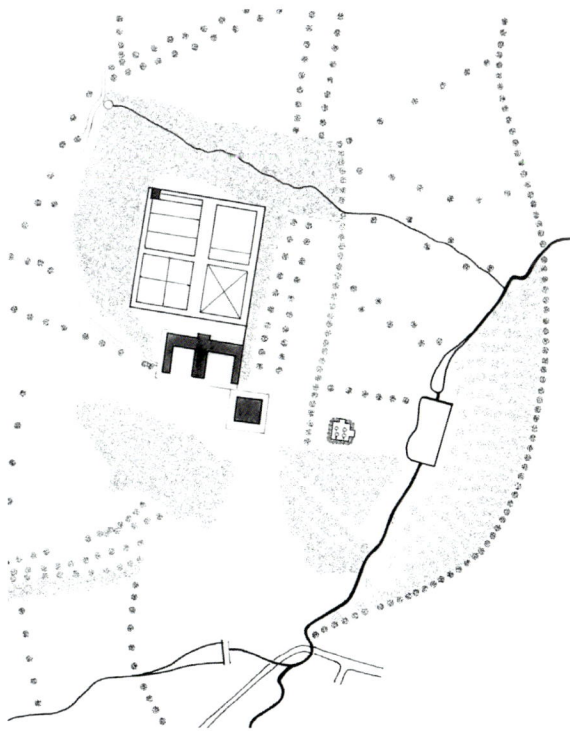

FIGURE 2.25 A diagram of the pre-1763 estate survey of Bowood.

aimed to achieve in making them. The premise here is that owners would be keen to display water features which they valued, and to place them in visible positions, especially in relation to the house. One means of assessing their importance is to determine whether they were visible from the house or not and, if so, which rooms they were visible from. Using Badminton, Gloucestershire, as an example to demonstrate the relationship between house and landscape at this time, Mark Girouard talks of,

> A saloon with apartments to either side, long axial vistas leading up to the saloon or through the apartments to their inner sanctuaries, and the extension of such vistas through the surrounding gardens and countryside.[115]

The significant rooms of a house are usually those which give onto the main façades, either side of the entrance front or the garden front, although often it is not possible to know this precisely. In the late seventeenth and early eighteenth century, the gardens and landscape were viewed from an elevated standpoint, which led to the trend in the 1680s of replacing hipped roofs and dormer windows with half-height windows and a flat roof with a balustrade, to provide a roof walk, as at Thoresby Hall, Nottinghamshire.[116] Factors such as the distance from the house, the topography of the landscape, trees and subsidiary buildings also need to be taken into consideration as they might screen features from the house.

The importance of the view from the house, and of water in particular, is cogently illustrated by Vanbrugh's comments on the houses of Blenheim, Oxfordshire, and Kimbolton, Cambridgeshire, which he built or modified. In a letter about Blenheim he says:

> All the most Valluable parts of the Views, lying to the most Significant Rooms in the Building. . . . The Water (where it will appear to best Advantage, whether Lake or River) is full in View,[117]

Writing about Kimbolton to Lord Manchester, in 1708, he said:

> And the Salon beyond it is Almost as big as the Hall, and looks mightily pleasantly Up the Middle of the Garden and Canall, wch is now brim full of Water, and looks mighty well;[118]

Vanbrugh's comments spell out that seeing the water from the main rooms of the house was most important, presumably to his clients as well as himself.

When considering the location of water features in landscapes some trends do become apparent, though the obstacles mentioned earlier need to be borne in mind. Most of the numerous geometric ponds with fountains were visible from the rooms of a main façade of the house, and perhaps a third of canals shown in the sources mentioned (Chauncy, *Britannia Illustrata*, Atkyns and *Vitruvius Britannicus*), though some would not have been visible from the house at all. Clearly, positioning a large canal in the optimum place was more difficult than with smaller features, especially if an existing landscape was being adapted.

There is some evidence that ornamental water features were being deliberately used to enhance the approaches to houses, which was a significant feature of elite medieval landscapes, as we have seen.[119] Two places where ornamental water did actually symmetrically flank the approach to the house were Temple Newsam, Yorkshire (Figure 2.26) and Burley-on-the-Hill, Rutland. At Temple Newsam, the square ponds have the appearance of fishponds, though with ornamental planting surrounding them, and perhaps dated from when the house was built in the early sixteenth century.[120] This approach route might owe its origins to the status conferred by fishponds at that time. Burley-on-the-Hill (Figure 2.27), shown in a painting by Tillemans of *c.* 1729 also has the main approach impressively flanked by two large geometric ponds. These are now under the reservoir of Rutland Water, although their outline can still be traced in the shoreline of the northwest side of the reservoir. However, water symmetrically flanking the approach to a residence was not a common feature in the late seventeenth and early eighteenth centuries.

Placing water *near* the house, and manipulating the approach to cross over it, was a much more significant trend revealed in *Britannia Illustrata*. At Staunton Harold, Leicestershire (Figure 3.3), for example, where the landscape was laid out in *c.* 1680 by Baron Ferrars, the approach was carried over the large

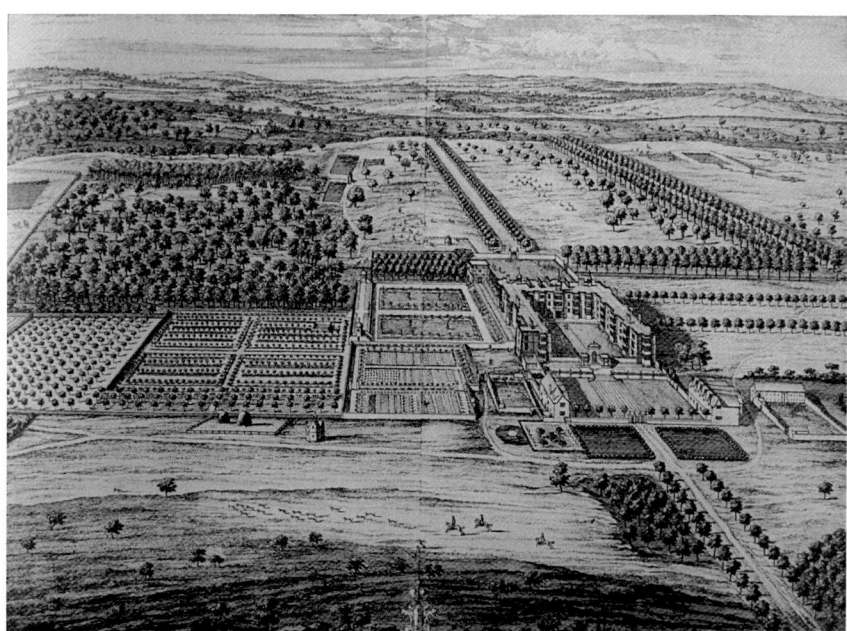

FIGURE 2.26 Temple Newsam, Yorkshire, in *Britannia Illustrata* 1708. *Historic England.*

FIGURE 2.27 Pieter Tillemans's *c.* 1729 painting of Burley-on-the-Hill, Rutland. *Courtesy of the Duke of Roxburghe.*

'canal' by a brick bridge, and was deliberately angled to pass over the impressive stretch of water with a full view of the hall. At Chatsworth (Figure 6.13) and Longleat the approach crossed the river or canal relatively close to the house but the water did not flank the approach, so perhaps crossing water in this way is a modified version of that concept. Another example, this time from Atkyns's *Ancient and Present State of Glostershire* is Old Court at Tortworth, the seat of Matthew Ducie Morton, where the ponds are staggered on either side of the main approach and create an impressive entrance in Atkyns's engraving (Figure 2.28). At a number of other places illustrated in Atkyns and *Britannia Illustrata*, approaches do cross rivers, canalised rivers and moats, or pass alongside canals and ponds.[121]

The conclusion from examining this array of evidence is that, although in this period (*c.* 1700) ornamental water did not often directly flank the approaches to houses, there were often significant bodies of water close by, so that the house and the water would be viewed together. This is important because the desire to see house and water together became a strong factor in the making of lakes in the eighteenth century, not just in where to site them, but in whether to make them in the first place.

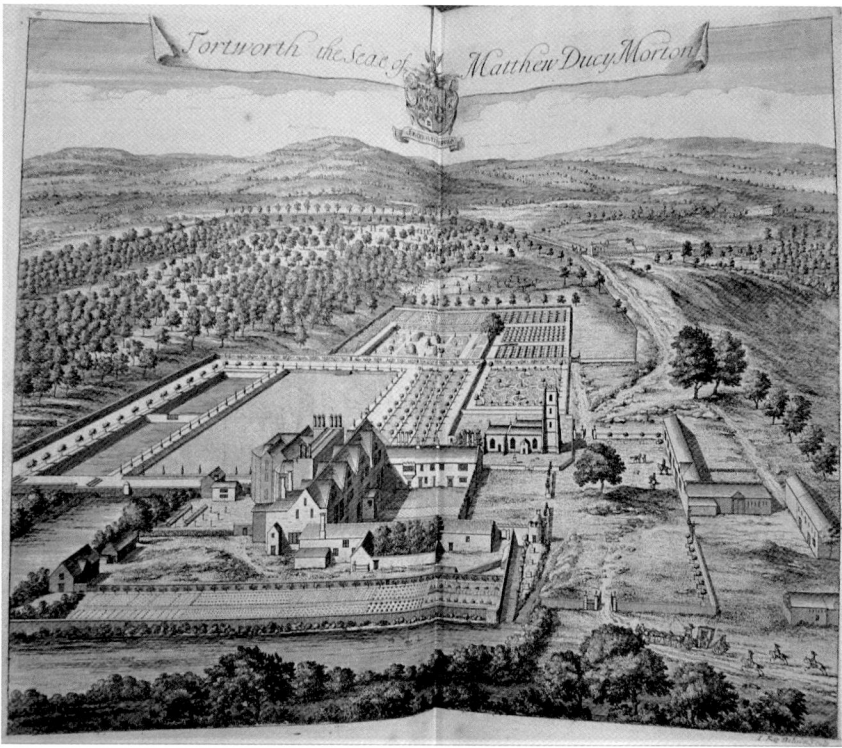

FIGURE 2.28 Tortworth, Gloucestershire, in *The Ancient & Present State of Glostershire* by Sir Robert Atkyns, 1712. *Courtesy of Bristol University Library Special Collections.*

Conclusion

This survey of landscapes prior to 1700 is significant in several ways. Firstly, it sets out the evolution of water prior to the eighteenth century, and reveals that fishponds were an important element in those landscapes. Along with moats and millponds, they led to a high value being placed on water features, which continued into the early modern period. Secondly, it highlights when and how water changed from being primarily functional in landscapes to being primarily ornamental, and this was a prerequisite for the emergence of ornamental lakes in the eighteenth century. Thirdly, it shows that there were no irregular lakes associated with great houses before *c.* 1720. Although there were a few pieces of ornamental water large enough to be considered as lakes (*c.* 8) before 1720, they were all geometric in shape. There was one exception: Thoresby Hall, Nottinghamshire, with a probable date of 1718–21, and this gives us a starting date for informal lakes.[122] It is quite possible, of course, that an earlier example may come to light, but as the next irregular lake to be made was in 1725, at Castle Howard, it is reasonable to point to the 1720s as the decade when irregular lakes first began to be made.

Notes

1 O. Creighton, *Designs Upon the Land: Elite Landscapes of the Middle Ages* (Woodbridge: Boydell Press, 2009) p 78.

2 C. Taylor, 'From Recording to Recognition', *There by Design: Field Archaeology in Parks and Gardens*, ed. Paul Pattison, BAR British Series 267 (Oxford: Archaeopress, 1998); R. Liddiard, *Castles in Context* (Macclesfield: Windgather Press, 2005); P. Everson, '"Delightfully Surrounded with Woods and Ponds": Field Evidence for Medieval Gardens in England', *There by Design*, op. cit.

3 C. Taylor, *Parks and Gardens of Britain: A Landscape History from the Air* (Edinburgh: Edinburgh University Press, 1998) p 30.

4 Currie, op. cit., p 22; R. J. Zeepvat, 'Fishponds in Roman Britain', *Medieval Fish, Fisheries and Fishponds in England*, ed. M. A. Aston, BAR British Series 182 (Oxford: Archaeopress, 1988) pp 17–26.

5 C. Dyer, 'The Consumption of Fresh-water Fish in Medieval England', in Aston, ibid., p 33.

6 J. Carmi Parsons, *Eleanor of Castile: Queen and Society in Thirteenth Century England* (London: Macmillan Press, 1994) p 51.

7 Currie, op. cit., p 22.

8 Ibid.

9 H. Colvin, gen. ed., *The History of the King's Works*, Vol. 1 (HMSO, 1962) p 324, quoted in Creighton, op. cit., p 72.

10 C. Taylor, 'Somersham Palace, Cambridgeshire: A Medieval Landscape for Pleasure?', *From Caithness to Cornwall: Some Aspects of Field Archaeology*, eds. M. Bowden, D. MacKay & P. Topping, BAR British Series 209 (Oxford: Archaeopress, 1989), p 30.

11 S. Switzer, *An Introduction to a General System of Hydrostaticks & Hydraulics* (London, 1729) Vol. I, p 130.

12 C. J. Bond, 'Monastic Fisheries', in Aston, op. cit., p 96.

13 Ibid., p 96.

14 Ibid., p 98.

15 R. North, *A Discourse of Fish and Fishponds* (London: E. Curll, 1714) pp 4–5.

16 The other main type of dam, arch dams, was not widely used until the nineteenth century, and retained water owing to its innate structural strength: N. Smith, *A History of Dams* (London: Peter Davies, 1971) p 33.

17 Ibid., p 164.

18 G. M. Binnie, *Early Dam Builders in Britain* (London: Thomas Telford, 1987) p 33.

19 To avoid onerous repetition, the word 'lake' is taken to mean an ornamental lake unless otherwise specified.

20 North, op. cit.; J. Taverner, *Certaine Experiments Concerning Fish and Fruite Practised by Iohn Taverner, Gentleman* (London: William Ponsonby, 1600) p 24.

21 J. M. Steane, 'The Royal Fishponds of Medieval England', in Aston, op. cit., p 40.

22 Ibid., p 46.

23 Ibid., pp 40–41. Using the currency converter of the National Archives (2020) for the year 1270, £40 would have been approximately equal to £29.000, which gives a rough comparison.

24 Dyer, in Aston, op. cit., p 34.

25 O. Rackham, *A History of the Countryside* (London: Phoenix, ed. 2000) p 362.

26 C. Taylor, 'Medieval Moats in Cambridgeshire', *Archaeology and the Landscape*, ed. P. J. Fowler (London: John Baker, 1972) pp 237–249 quoted in Rackham, op. cit., p 362

27 Not only were many people unable to swim, but moats made undermining the walls much more difficult.

28 North, op. cit., p 25.

29 Creighton, op. cit., p 81.

30 These places are discussed in more detail in Creighton, op. cit.

31 Creighton, op. cit., p 64; Taylor, 'Somersham Palace', op. cit.
32 Bond, op. cit., p 99.
33 Liddiard, op. cit., p 120.
34 Ibid., p 120.
35 Taylor, *Parks and Gardens*, op. cit., p 34.
36 Creighton, op. cit., p 181.
37 Description of Raglan Castle, 1674 version, ref. FmE 4,5,3(d), Badminton Muniments, courtesy of the Duke of Beaufort.
38 J. R. Kenyon, *Raglan Castle* (Cardiff: Cadw, Welsh Historic Monuments, rev. ed. 2003) p 18.
39 North, op. cit., p 27.
40 H. M. Colvin, gen. ed., *The History of the King's Works*, Vol. IV, Part II (London: HMSO, 1982) in M. Andrews, 'Theobalds Palace: The Gardens and Park', *Garden History*, Vol. 21, No. 2 (Winter, 1993) p 144.
41 J. Langdon, *Mills in the Medieval Economy: England 1300–1450* (Oxford: Oxford University Press, 2004) p 16.
42 R. Holt, *The Mills of Medieval England* (Oxford: Basil Blackwell, 1988) p 88.
43 L. Syson, *The Watermills of Britain* (Newton Abbot: David & Charles, 1980) p 54.
44 Watermills are complex systems, often involving one or more leats, a head race, a tail race, sluices and weirs.
45 Creighton, op. cit., p 57.
46 Historic England listing: Stokesay Castle, Shropshire.
47 The tithe apportionment of 1840 itemises this pond as a fishpond.
48 Langdon, op. cit., pp 303–305.
49 T. Williamson, personal communication, March 2017.
50 R. Strong, *The Renaissance Garden in England* (London: Thames & Hudson, 1998) p 23.
51 Ibid., p 23 fn.
52 Ibid., p 37.
53 Standing over 3 m high, it is attributed to a Florentine sculptor, Benedetto da Rovenzzano: J. Whitaker, *Gardens for Gloriana: Wealth, Splendour and Design in the Elizabethan Garden* (London: Bloomsbury Academic, 2019) p 97.
54 Bishop, op. cit.
55 A. Rowe, 'Hertfordshire's Lost Water Gardens', fn 6, p 55 *Hertfordshire Garden History*, Vol II, ed. D. Spring (Hatfield: University of Hertfordshire Press, 2012).
56 T. Mowl, *Historic Gardens of Gloucestershire* (Stroud: Tempus Publishing, 2002) p 24.
57 Ibid., p 25.
58 Ibid., pp 24–25.
59 J. Dixon-Hunt, *Garden and Grove: The Italian Renaissance Garden in the English Imagination: 1600–1750* (London: Dent and Sons, 1986) p 18.
60 Strong, op. cit., pp 29–30.
61 Dixon-Hunt, op. cit., p 85.
62 Strong, op. cit., p 21.
63 Robert Laneham, *Laneham's Letter Describing the Magnificent Pageants Presented Before Kenilworth Castle in 1575* (Philadelphia: Hickman & Hazard, 1822) p 75.
64 Ibid., p 5.
65 Ibid., p 4.
66 T. Mowl, *Gentlemen and Players: Gardeners of the English Landscape* (Stroud: Sutton Publishing, 2000) p 45.
67 Paul Hentzner, *Travels in England During the Reign of Queen Elizabeth*, quoted by M. Andrews op. cit., p 141. Hentzner visited in 1598.
68 Baron Waldstein, trans. G. W. Groos, *The Diary of Baron Waldstein* (London: Thames and Hudson, 1981) p 83. Waldstein visited England in 1600.
69 D. Spring, 'The London Connection', *Hertfordshire Garden History*, Vol. II, ed. D. Spring (Hatfield: University of Hertfordshire Press, 2012) p 39.

70 Paul Hentzner, *Journey into England*, pp 54–55, in P. Henderson, *The Tudor House and Garden* (London: Yale University Press, 2005) p 85.

71 Spring, op. cit., p 15.

72 Waldstein, op. cit., p 87. It is possible that there were 6 steps on each side, assuming it was rectangular.

73 Francis Bacon, quoted in P. Henderson, 'Sir Francis Bacon's Water Gardens at Gorhambury', *Garden History*, Vol. 20, No. 2 (Autumn, 1992), pp 116–131 fn 24. He is addressing Robert Cecil; William Cecil died in 1598.

74 Spring, op. cit., p 13.

75 William Cecil was Francis Bacon's uncle by marriage: Markku Peltonen, *Oxford Dictionary of National Biography* (Oxford University Press, online 2015).

76 Sir Francis Bacon, *The Works of Francis Bacon: Literary and Professional Works*, eds. J. Spedding, J. R. Ellis, & D. D. Heath, Vol. I, 1625 (London: Various, 1861) p 489 quoted in Henderson, 'Sir Francis Bacon's Water Gardens at Gorhambury', op. cit., pp 116–131.

77 J. Aubrey, ref. Bodleian Aubr. MS. 6, fol. 74^r quoted in P. Henderson, 'Sir Francis Bacon's Water Gardens at Gorhambury', op. cit., pp 116–131.

78 Currie, op. cit., p 26.

79 P. Henderson, 'Sir Francis Bacon's Water Gardens at Gorhambury', op. cit., p 134.

80 Sorbiere, *Voyage to England*, p 65 quoted in P. Henderson, *The Tudor House and Garden* (London: Yale University Press, 2005) p 134.

81 Strong, op. cit., p 106.

82 D. Foster-Evans, lecture: *Re-interpreting Welsh Literature of the Middle Ages* (University of Cardiff, Spring 2013).

83 E. H. Whittle, 'The Renaissance Gardens of Raglan Castle', *Garden History*, Vol. 17, No. 1 (1989), pp 83–94.

84 This is complimented by Richard Salter's 1844 copy of a Description of The Castle & Grounds of Raglan 'written in 1674', ref. MS FmE 4,5,3(d), Badminton Muniments. Courtesy of the Duke of Beaufort. Henry, the 5th Earl, was a staunch Royalist, entertaining Charles I at Raglan several times, and holding the castle in the Civil War, until it was taken by Fairfax in 1646.

85 If it existed before the sixteenth century it would surely have been mentioned by the praise poets, such as Guto'r Glyn and Tudur Aled, as they went into considerable detail.

86 Whittle, op. cit., pp 83–94.

87 Thomas Churchyard, quoted in J. R. Kenyon, *Raglan Castle* (Cardiff: Cadw, Welsh Historic Monuments, rev. ed. 2003) p 14.

88 MS FmE 4,5,3(d) in Badminton Muniments: Richard Salter's 1844 copy of a Description of The Castle & Grounds of Raglan 'written in 1674'. Courtesy of the Duke of Beaufort.

89 The lake is no longer there: it was drained when the castle was slighted in 1646 after its capture by Roundhead forces.

90 Whittle, op. cit., p 90.

91 Ibid.

92 E. Whittle & C. Taylor, 'The Early Seventeenth-century Gardens of Tackley, Oxfordshire', *Garden History*, Vol. 22, No. 1 (Summer, 1994) p 53.

93 Ibid., p 38.

94 Ibid., p 45.

95 S. Miller, *The Ponds or Water Maze: An Early Seventeenth Century Water Garden at Cope Castle in Kensington*, forthcoming paper.

96 Discussed in Chapter 4.

97 Thomas Hale, *A Compleat Body of Husbandry*, Vol. II (London: Osborne, Trye & Crowder, 2nd ed. 1758) p 111.

98 Ibid., p 128.

99 Sir Francis Bacon, op. cit., p 489 quoted in P. Henderson, 'Sir Francis Bacon's Water Gardens at Gorhambury', op. cit., pp 116–131.

100 As illustrated in J. Dixon-Hunt, op. cit., p 146.
101 J. Barnett & T. Williamson, *Chatsworth: A Landscape History* (Macclesfield: Windgather Press, 2005) p 59.
102 T. Mowl & B. Earnshaw, *Architecture Without Kings: The Rise of Puritan Classicism Under Cromwell* (Manchester: Manchester University Press, 1995) p 63, p 100.
103 T. Mowl, lecture, Bristol University, 17.11.09.
104 Ibid., p 44.
105 Ibid., p 45.
106 S. Switzer, *Ichnographia Rustica*, Vol. 3 (London, 1718) pp 118–119.
107 D. Jacques & A. J. Van der Horst, *The Gardens of William and Mary* (Bromley: Christopher Helm, 1988) p 99.
108 For further information: Bishop, op. cit.
109 Bishop, op. cit., p 81. Even though labour was relatively cheap, making water features would still cost money, and geometric ones were more expensive to construct than non-geometric features.
110 Switzer, *Ichnographia*, op. cit., Vol. 3, p 120.
111 J. Whitney, *The Genteel Recreation Or, the Pleasure of Angling, A Poem with a Dialogue between Piscator and Corydon* (London, 1700) in 'The Dedication'.
112 J. Lawrence, *The Modern Land Steward in Which the Duties and Functions of Stewardship Are Considered and Explained* . . . (London: 2nd ed. 1806) p 314.
113 E. Dennison & R. Iles, 'Medieval Fishponds in Avon', in Aston, op. cit., pp 205–228.
114 Bishop, op. cit., p 91.
115 M. Girouard, *Life in the English Country House: A Social and Architectural History* (Harmondsworth: Penguin Books, 1980) p 145.
116 G. Worsley, *Classical Architecture in Britain: The Heroic Age* (London: Yale University Press, 1995) p 70.
117 Vanbrugh, in a letter of 18th July, 1709, possibly to Lord Ryalton, in H. Dobrée & G. Webb, eds., *The Complete Works of Sir John Vanbrugh*, Vol. 4 (London: The Nonesuch Press, 1928) p 35.
118 Vanbrugh, op. cit., in a letter of 22nd March, 1708, p 19.
119 Taylor, 'Somersham Palace', op. cit., p 214.
120 Historic England listing: Temple Newsam Park and Garden.
121 Bishop, op. cit., p 95.
122 Ibid., p 117.

3

THE EMERGENCE OF LAKES

When did ornamental lakes appear? The simple answer to that question is in *c.* 1720, as the chronology of lakes in the Landscape Database shows, but that answer needs interpretation, because we also have to bear in mind what is meant by 'lake'. As we have seen, there are two main types of lake: geometric and irregular, with various sub-divisions, and a minimum size of 1 h. Although this is an artificial classification, it enables a statistical approach to be used, which in turn helps with analysing how many lakes were made, and when. This approach enables us to chart the change from the geometrically shaped water which predominated in around 1700 to the informal lakes which were fashionable by 1750. The focus here is on the lakes themselves, not lake designers, though some key individuals are mentioned, such as Sir John Vanbrugh and Lancelot 'Capability' Brown, who both made pivotal contributions to the development of lakes, in different ways.

In the first two decades of the eighteenth century, there was a change in attitudes to ornamental water and the concept evolved that a *large* body of water was desirable, primarily for its visual qualities and leisure possibilities. This was a step change in landscape development. The fishponds which had existed in deer parks for centuries were not generally within the immediate vicinity or in view of the house or castle. Wanting a large piece of *ornamental* water was a new departure. Initially, this development happened within the context of geometric landscapes in the early 1700s, at places such as Welford, Berkshire, Boughton, Northamptonshire, and Stowe, Buckinghamshire. These areas of water still contained fish but their aesthetic role in the design was the primary factor in making them.[1] Unlike the water-gardens of the 1600s, these were large stretches of water – at least 1 h – with a prominent role in the overall design of the landscape. Often called a Broad Water or Great Water, the term 'lake' was not used widely until the 1790s, as we have seen.

Geometric lakes

These larger bodies of water were initially geometric in form, but were 'lakes' in the sense that they extended over an area of 1 h or more, and were clearly ornamental in intention. They were included in the designed landscape, and were 'displayed': they were often positioned to catch the eye, and had ornamental planting around them (Figure 3.1), instead of being in the park. Most geometric lakes were aligned on axes, either relating to the mansion, or other features in the landscape. Occasionally they were not aligned in this way because of factors such as topography or the location of existing features such as fishponds, as perhaps at Staunton Harold. As Table 2 shows, a few geometric lakes were made before 1700, and it is likely that this figure would be higher, as the dates in the 'Dating' column indicate, if we had more accurate data. Welford in Berkshire

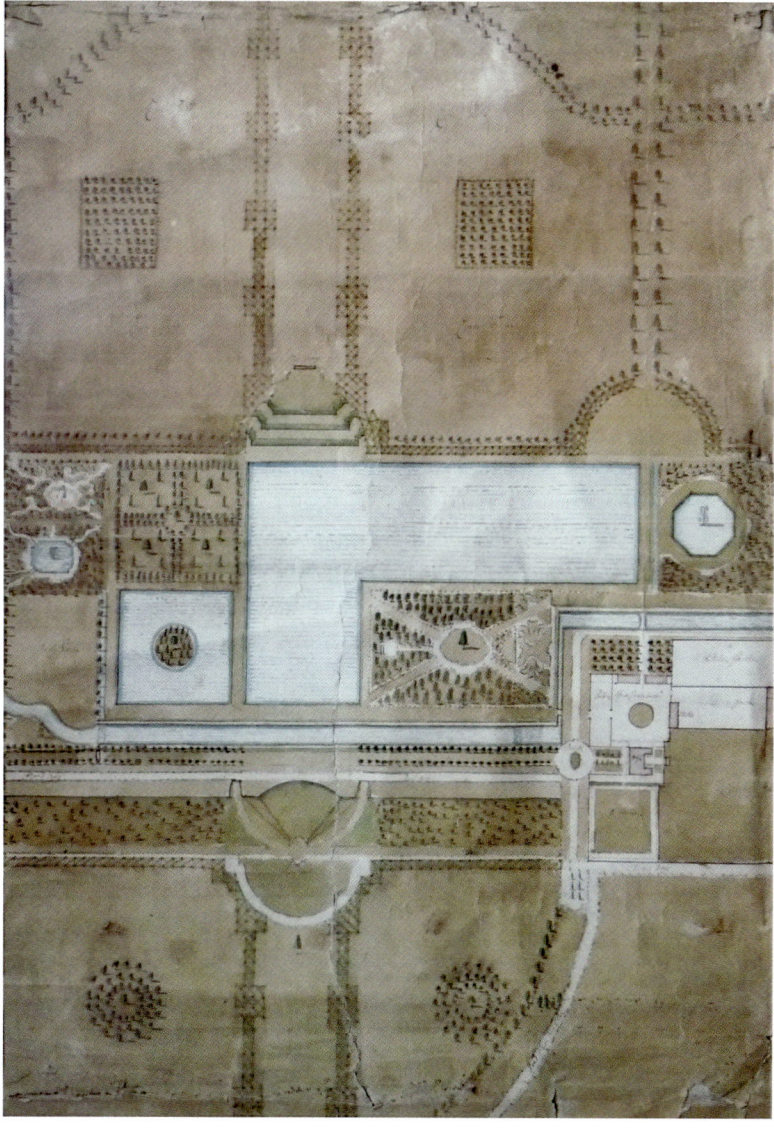

FIGURE 3.1 A plan of Welford Park, Berkshire, *c*. 1700. *Courtesy of J. H. L. Puxley.*

(Figure 3.1) had several geometric pieces of water in the early eighteenth century. It lies along the braided channel of the Lambourn river, so water has always been plentiful. The house is two-thirds down the right-hand side of the plan, and the main water features are the L-shaped lake, and an adjacent square pond with a central island, mirrored by a square garden. (Some of these features can still be discerned, 2020.) This piece of water was flanked by parterres and walks, with other features – the avenue and octagonal pond – aligned on it, and it is clearly ornamental. It is a semi-geometric lake of *c.* 4 h, and the square piece of water is 0.9 h – essentially another geometric lake. The layout at Boughton, Northamptonshire (Figure 3.2), shows similar characteristics, with adjacent parterre gardens. Here, the main body of water is a geometric lake of 2.4 h. What is noticeable about these two landscapes is that the lakes have been inserted into the design by substituting them for parterres: where previously a parterre might have been made, a lake was made instead. An estate survey of Boughton in 1715 illustrates this. There is a parterre where the northern part of the lake was subsequently made in *c.* 1720.[2] This was almost the only way to incorporate large pieces of water, other than canals, into formal, geometric landscapes without disrupting the geometry. It was usually only possible in landscapes where the axes extended at right angles to the house, or each other. Other requirements were a fairly flat terrain, a water supply, the wealth to afford making lakes of this size,

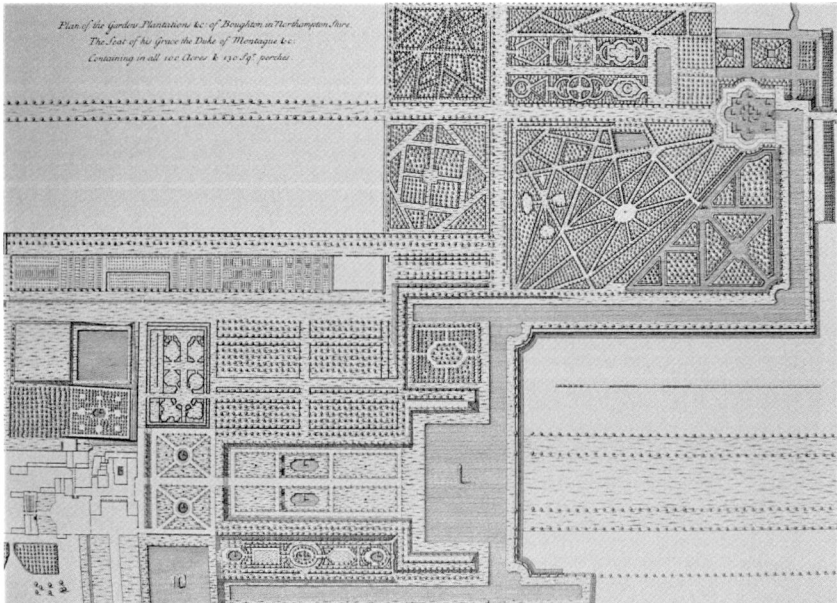

FIGURE 3.2 Colen Campbell's plan of Boughton Park, Northamptonshire, in *Vitruvius Britannicus* 1725. *Courtesy of Historic England.*

and a landscape large enough to be able to devote a 'parterre' to water so that it did not look out of proportion in the design as a whole, or reduce the amount of productive land in the estate too much.

Of the eleven other geometric lakes which were constructed between *c.* 1700 and 1730 (Table 2), two were trapezoidal in shape: Gamlinghay (5.2 h) and Bredby (2 h; see Chapter 4).[3] Whilst it could be argued that Bredby was basically a fish-pond, as it lies outside the main designed landscape, the island and fountain, the decorative avenues on either side, and its axial alignment with the rest of the gardens, belie this. At Gamlinghay, the sophisticated geometric lake (made by *c.* 1715) occupies a prime position in relation to the house, lying across the main axis from the house through the park, and completely interrupting it.[4] Other geometric lakes were octagonal in shape: Stowe, Buckinghamshire (*c.* 1 h), Wimpole, Cambridgeshire (2.5 h) and Wanstead, Essex (4 h). Stowe and Wimpole were at some distance from the house but were also aligned on a direct axis to it. A number of

TABLE 2 A chronological list of geometric lakes

Garden name	County	Approximate date	Date	Lake size (h)
Staunton Harold	Leicestershire	1680	1690	1.1
Bredby	Derbyshire	1684–1702	1698	2
Welford	Berkshire	Late 17 C	1700	4, 0.9
Burley-on-the-Hill	Rutland	By 1701	1701	*c.* 2 each
Seat of William Challoner	Yorkshire	By 1707	1707	*c.* 1
Wanstead	Essex	By 1708	1708	4
Boughton	Northamptonshire	By 1709	1709	1 (octagonal)
Warnford Place	Wiltshire	1710	1710	2
Ditchingham	Norfolk	1713	1713	3
Gamlinghay	Cambridgeshire	1712	1712	5
Wimpole	Cambridgeshire	1721	1721	2 (octagonal)
Boughton	Northamptonshire	1719–20	1721	2.4, 0.6
Stowe	Buckinghamshire	1720s–'30s	1729	1 (0.8–1.8)
Blickling	Norfolk	By 1725. 1680s possibly	1725	0.9
Oatlands	Surrey	1725–35	1732	*c.* 1
Stowe	Buckinghamshire	By 1735	1735	3.9
Castle Hill	Devon	1730s	1738	*c.* 1.4
Althorp	Northamptonshire	By 1730s	1738	*c.* 6
Enville	Staffordshire	1746	1746	1
Hillington	Norfolk	By 1756	1756	1.3
Irmingland	Norfolk	Mid-19th C?	1850	1.2

Source: Data extracted from the Landscape Database.

bodies of water were not quite 1 h, but were nevertheless large and prominent features of the landscape. Those at Combs Hall, Suffolk (0.7 h) and Blickling, Norfolk (0.9 h: Wilderness Pond) were rectangular, whilst the one at Claremont, Surrey (0.6 h) was round (Figure 3.18). Like the rectangular pond at Bowood (*c.* 1725, possibly 0.4 h), the round area of water at Claremont was subsequently the site of a larger body of irregular water.

Of particular note amongst these geometric lakes is Staunton Harold, probably laid out in the 1680s (Figure 3.3).[5] As we can see in *Britannia Illustrata*, there are numerous geometric ponds there, many with fountains, and others which are probably fishponds, though geometric. What is unusual at Staunton Harold is the large, rather 'fat' canal. It dwarfs the landscape and covers *c.* 1.1 h, retaining much of its original shape on today's OS map. Whatever its role in fish production, this piece of water has been constructed alongside the church and various adjacent ponds have been aligned on it. There is ornamental planting around it, and the main approach to the house passes directly across it. Whilst it may well originally have been a medieval fishpond, by 1708 it was primarily ornamental.

It is also interesting to note that the water source which filled geometric lakes is usually evident on today's OS maps. At Wimpole, it is the need for this water source which apparently accounts for the 2 h octagonal basin being placed 1.6 km from the house, as that is where the stream is.

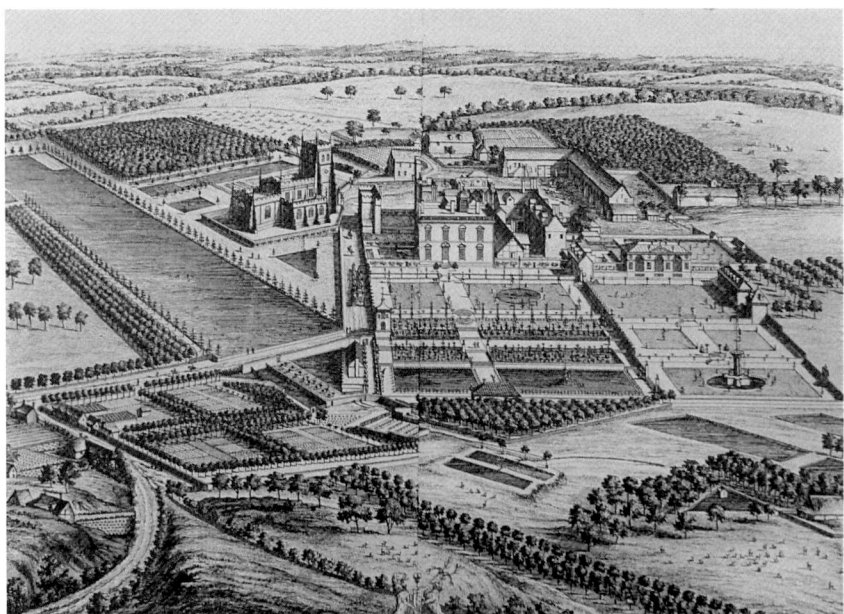

FIGURE 3.3 Staunton Harold, Leicestershire, in *Britannia Illustrata* 1708. *Courtesy of Historic England.*

Semi-geometric lakes

The semi-geometric lakes which emerged from *c.* 1700–30 were developments of the geometric lake, and like them, not many survived the changing fashions of the eighteenth century. The main difference between them and the geometric lakes was that they had straight sides (or arcs) but were not symmetrical. Because geometric landscapes fell out of fashion, few being made after the 1740s, not many semi-geometric lakes are known, and even fewer have survived. They are of interest because they show how landscapes were changing at this time, becoming less symmetrical as well as larger, and the role of lakes in that change is discussed in Chapter 6.

Semi-geometric lakes give us an indication of the process by which lakes changed from being fully geometric to completely irregular in shape. Two such lakes were made at Blenheim (by 1724) and Wolterton (by 1732). A sketch of Blenheim by Pierre Jacques Fougeroux in *c.* 1728 (Figure 3.4) shows a lake of *c.* 3.5 h, with sides composed of straight lines or geometric arcs, giving an overall geometric impression, but an asymmetrical plan view. A sketch of Blenheim in 1724 by Sir

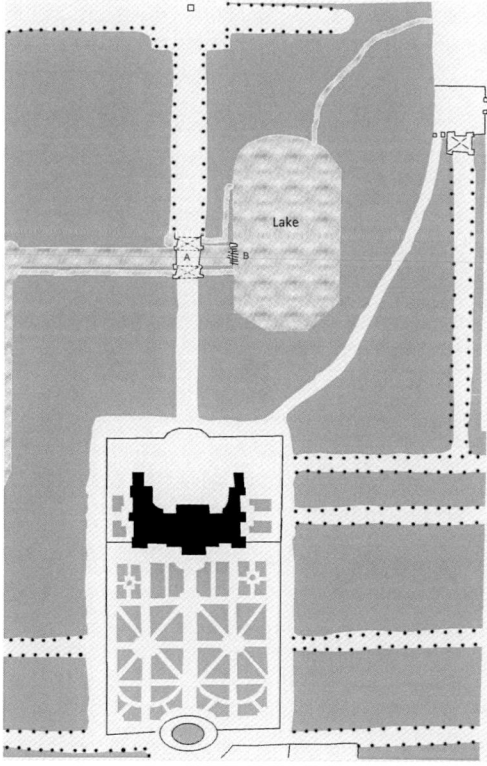

FIGURE 3.4 Blenheim Palace: diagram based on a *c.* 1728 sketch by a French visitor, Fougeroux, showing Armstrong's lake of *c.* 3.5 h. A. is Vanbrugh's bridge. B. is a cascade.

William Stukeley shows the canal leading from this lake to a circular pond, but the lake itself is not clearly visible.[6] It was created under the eyes of Sarah, 1st Duchess of Marlborough, and the house, *c.* 155 m from the lake, looked down a steep slope at it (Figures 5.24 and 5.25) As is well-known, Vanbrugh designed the palace and the bridge over the River Glyme, but the disagreement between Vanbrugh and the Duchess led her to sack him in 1716 and turn to Colonel John Armstrong after the Duke's death in 1722. He was responsible for the creation of the semi-geometric lake and the canals leading from it.[7] These were nearing completion in 1725, as Sarah describes in a letter to a friend. The wording of this letter is particularly interesting:

> the Lake, Cascade, Slopes above the Bridge are all finish'd and as beautifull as can bee imagin'd, the Banks being cover'd with the most delightful Verdure; the Canals are also finish'd the whole length of the Meadow.[8]

This was an early use of the term 'lake', Vanbrugh being possibly the first person to use it in relation to man-made water in his letter about Blenheim in 1709, quoted previously.

The semi-geometric lake at Wolterton, made *c.* 1727 (Figure 1.2), shows how this style of lake could be fitted into a geometric landscape more easily than a symmetrical lake. Just as importantly, they were cheaper to make as they could accommodate the topography to some extent, and less earth had to be moved. It was made for Horatio Walpole, by Charles Bridgeman.[9] There are two interesting points about Wolterton. The first is that it continued the axis extending through the house, although it 'bent' to accommodate the topography. The second is that, as far as we know, it was made in this semi-geometric form, and was not originally a symmetrical piece of water which was altered. The rising ground, indicated by the 30 m contour line on today's OS map (2017), constrained the shape of the lake on the eastern side. A stream entering on the western side also influenced the shape. Apart from The Serpentine (discussed later), the nature and extent of Bridgeman's work with large areas of water is uncertain. The basin in the landscape he created at Claremont covered 0.6 h, so was a sizable piece of water, but not lake sized. Wolterton was angled in a similar way to The Serpentine (begun in 1730), and in both places, Bridgeman worked with existing factors – the terrain and fish-ponds respectively. What is interesting is that he (or his clients) did not choose to impose symmetrical geometry on the water.

Irregular lakes

Irregular lakes began to be made in the 1720s (Table 3).[10] While landscapes were intrinsically linear, with symmetry as the underlying ethos, it was very difficult to make irregular lakes without disrupting those things. Ornamental water basically had to be in the form of canals or rectangular ponds because otherwise it would not fit easily into the overall design. Geometric water was also relatively expensive to make, as more earth had to be moved to make the sides straight, and so generally

TABLE 3 The chronological beginnings of irregular lakes

Garden Name	County	Approx. date	Date	Lake Size (h)
(Welbeck Abbey)	Nottinghamshire		1703	16
Thoresby Hall	Nottinghamshire	1718–23	1719	19.5
Castle Howard	Yorkshire	1724	1724	c. 2.5
(Houghton Hall)	Norfolk	1725	1725	8.6
Holkham Hall	Norfolk		1727	8.3
Londesborough	Yorkshire		1729	4.5
Wentworth Woodhouse	Yorkshire	1730s	1735	8.4, 4.6
Witley Court	Worcestershire	1730s	1735	3.6
Cirencester	Gloucestershire		1736	3.2
Claremont	Surrey		1737	3
Wilton	Wiltshire	By 1746	1738	5.4
Exton Park	Rutland	By 1739	1739	1.2, 1
Painshill	Surrey	c. 1740	1740	1
Fawsley	Northamptonshire		1741	10, 2
Wanstead Park	Essex	Mid 18 C	1745	1.1
Wakefield Lodge	Northamptonshire	c. 1745	1745	c. 8.6
Ditchley	Oxfordshire	1740s	1746	1.9
Newstead Abbey	Nottinghamshire	1740s	1747	9.3
Grimsthorpe	Lincolnshire		1748	16
Newnham Paddox	Warwickshire		1748	1.6, 0.9

Source: Data extracted from the Landscape Database.

it was also relatively small. Exceptional circumstances, such as wealth and topography did mean that some geometric lakes were made, as we have seen at Boughton, Welford and Staunton Harold. However, once irregular ornamental lakes started to become fashionable the geometric straitjacket did not fit, and flexibility, or unbalancing, in landscape design began to increase.

Irregular aspects in garden and landscape features, such as sinuous paths in wildernesses, irregular outlines to plantations, and irregularly shaped ornamental water began to appear in the late 1710s. By the later 1720s irregular elements were becoming increasingly significant, as Badeslade's bird's-eye view of Hamels illustrates, as well as plans published by Switzer (Figure 3.5) and Batty Langley. Although wildernesses and plantations were beginning to change, the first truly irregular element to appear was the water. It changed out of all recognition, in size and shape, during this period (1720s–30s), becoming completely irregular – what we would call a 'lake' today.

An examination of the plans of places such as Castle Howard, Holkham, Londesborough and Thoresby Hall, reveals that irregular lakes were made in otherwise

FIGURE 3.5 A design for an estate by Stephen Switzer in *Ichnographia Rustica* 1742 ed., Vol. II, plate opposite title page. *Copyright W. A. Brogden.*

geometric landscapes (Figure 3.6), which is surprising.[11] Whilst it might seem that these landscapes represented established geometric designs into which forward-thinking patrons inserted informal lakes, this was not the case. In all these examples, the geometric avenues, vistas or parterres were created at broadly the same time as the irregular pieces of water. At Thoresby (Figure 3.8), for example, the formal parterres and wildernesses to the south of the house were planted shortly before, or at the same time, as the lake was made.[12] The same can be said of Holkham – a new landscape which Thomas Coke set about creating in the mid-1720s. At Castle Howard (Figures 3.6 and 3.11), the lake was being made at the same time as the highly geometric and formal parterres to the south of the new house.[13] The lake actually elbows its way into this parterre – wilderness area. The obvious thing to do would have been to line up the western edge of the lake with the eastern edge of the wilderness, thus achieving a much more regular design. However, then the lake would not have been so visible from the house. At Londesborough (Figures 3.6 and 3.17.), Burlington inherited a geometric landscape in 1704, and in the 1720s, he created a formal parterre near the house and extended the southern axis into the

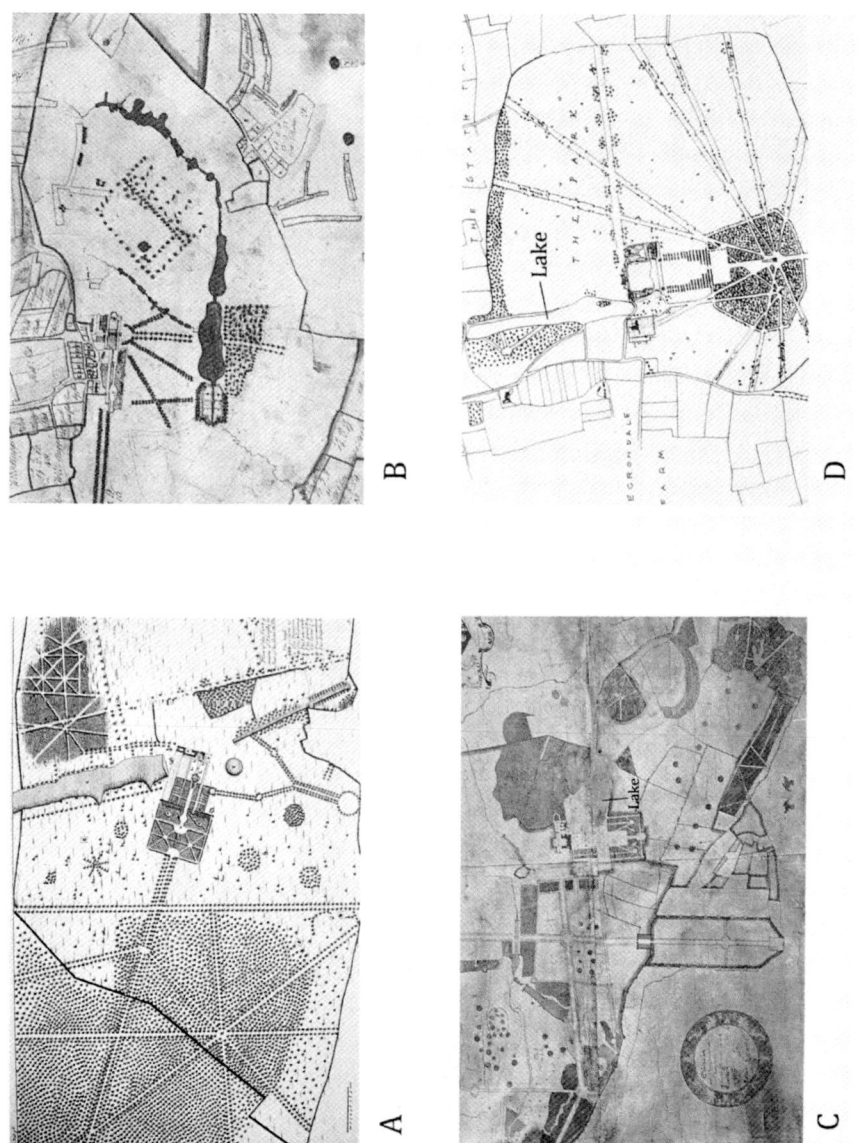

FIGURE 3.6 The earliest irregular lakes. A. Thoresby, Nottinghamshire, *c.* 1719 (20 h) for the 1st Duke of Kingston. B. Londesborough, Yorkshire, 1728–30 (4.5 h) for Lord Burlington. C. Castle Howard, Yorkshire, 1724, (2.5 h) for the 3rd Earl of Carlisle. D. Holkham, Norfolk, by 1729 (8 h) for Thomas Coke.

wider landscape with a *pate d'oie* and kitchen garden flanked by formal plantations.[14] Into this, he inserted a chain of *irregular* lakes, and a chain of small, semi-geometric ponds. Clearly, these men (the Duke of Kingston, the Earl of Carlisle, Thomas Coke and Lord Burlington) were happy to put large areas of irregular ornamental water into their largely formal landscapes. In contrast, in 1708, it was not acceptable, as Staunton Harold and the landscapes in *Britannia Illustrata* clearly illustrate. Attitudes had changed significantly in two decades.

It is also worth noting that in these four landscapes it was acceptable for large areas of woodland to be irregular in outline. Perhaps this was because such areas tended to be very large indeed, as at Thoresby and Castle Howard, and so the lack of geometry was less easily perceived. (Badminton in *c.* 1700 also shows this.) Geometry was imposed on woodland by cutting straight rides and vistas through it. Obviously, this could not be done with water, so it is all the more surprising that these lakes were not geometric in shape. It might be expected that the estate owner would restrict the size of the lake to one which he could afford to make geometric, which appears to have been the case at Boughton in the 1720s, so the conclusion must be that the irregularity was deliberate.

Perhaps the most pivotal factor in this development in design was the increasing scale of landscapes: they were becoming bigger and bigger. Thoresby Park in *c.* 1719 (Figure 3.8) admirably demonstrates this, and the effects it had. It also represents a pivotal point in the story of lakes: the first known irregular lake was made there. Evelyn Pierrepont, the 5th Earl of Kingston-upon-Hull (1665–1726), inherited the estate from his father in 1690 (Figure 3.7). The formal gardens occupied an area of *c.* 7 h, with the River Meden to the west in the park, among trees. A plantation with geometric rides lay to the south. By the time Campbell published his plan in 1725, the park had been considerably extended to the north and significant plantations with geometric rides had been made to the south (Figure 3.8). A new house had been built, designed by Campbell, and the formal gardens extended to cover *c.* 18 h. In these gardens, a formal cascade fed into an octagonal pond facing the house, and the canalised Meden in the gardens was covered over, but reappeared to the east. Into this geometric but unbalanced landscape the large irregular lake was inserted. These changes were carried out very much according to the Campbell plan, as we can see on an estate map of 1738 (Figure 3.9). To the north-west, the estates of Clumber and Welbeck adjoined Thoresby. The Thoresby lake extended just beyond the park pale, but was enlarged by *c.* 1750 after more land had been acquired, and it became 25 h (62 acres), very close to the 65 acres mentioned by Campbell on his plan.[15] By 1738, three detached areas with sinuous paths had been created, reminiscent of Batty Langley's plans, and suggestive of Rococo gardens. The area covered by the lake was low-lying, and adjacent areas appear to be poor-quality, infertile land, which may account for the willingness of the Duke of Kingston to make such a large lake.[16] The scale of this landscape is not easy to grasp from the plans: the distance from the house to the southern end of the formal gardens in 1738 was around half a mile (0.8 km). Subsequently, Brown drew a plan for Thoresby, and

FIGURE 3.7 A 1690 estate map of Thoresby Park, Nottinghamshire. *Courtesy of Gregor Pierrepont and University of Nottingham Manuscript and Special Collections. Ref. Ma 4 P19.*

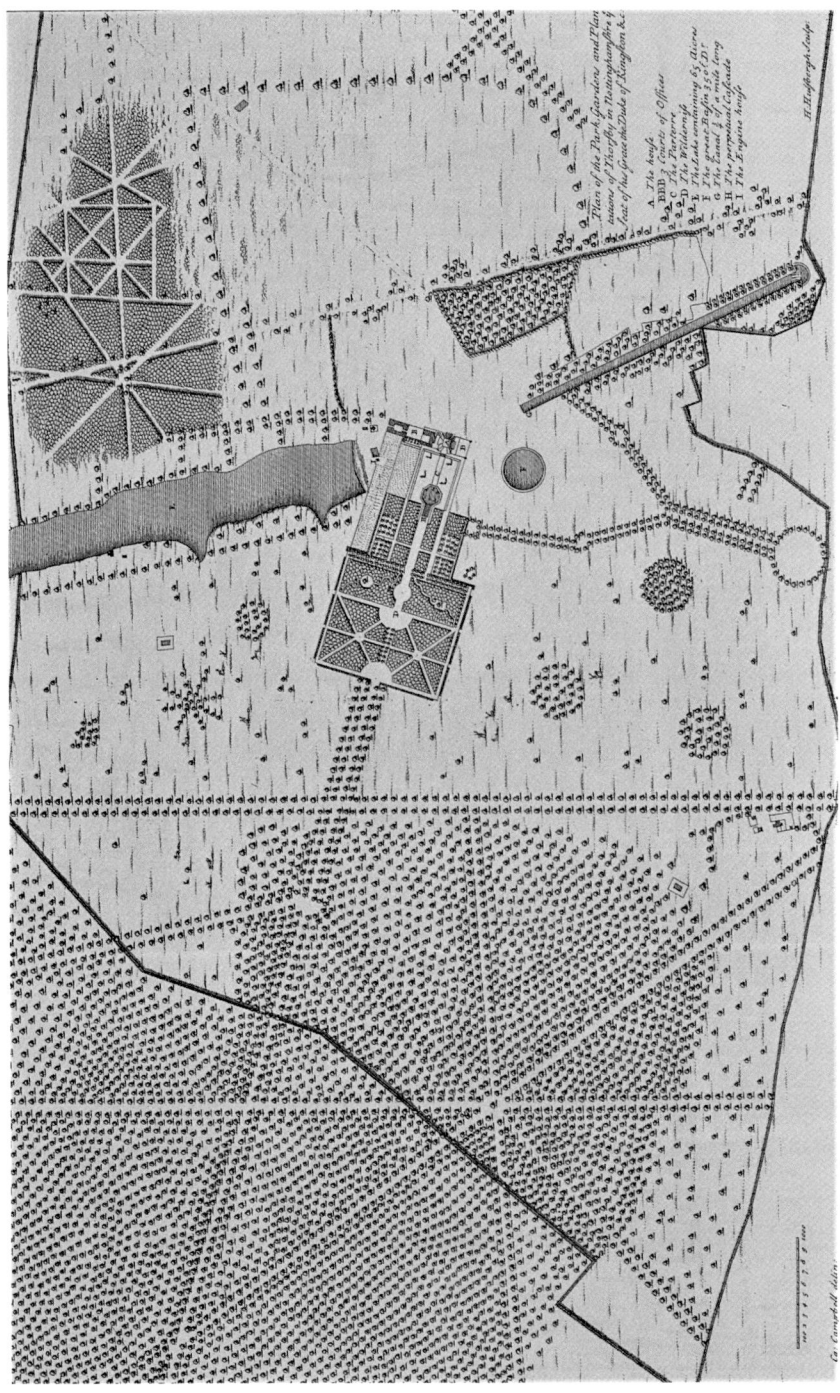

FIGURE 3.8 A plan of Thoresby Park in *Vitruvius Britannicus* 1725. *Courtesy of Gregor Pierrepont.*

FIGURE 3.9 A 1738 estate map of Thoresby Park, Nottinghamshire. *Courtesy of Gregor Pierrepont and the University of Nottingham. Ref. Ma 4 P 20.*

Repton produced a Red Book (Figure 7.6), with many of his suggestions being implemented.

Thoresby is significant for two reasons. It clearly demonstrates the increasing scale of designed landscapes, and the unbalancing which was occurring in them. Secondly, in *c.* 1719–21, it was the very first irregular lake to be made.[17] The dating is reasonably certain as a Peter Tillemans painting (Figure 6.7) showing the lake, new house and new stables, was delivered to Kingston a week before he died in March, 1726, and estate accounts show a corresponding amount of labour for 'levelling' (surveying) for the period 1718–19.[18]

Kingston is not a familiar figure today, but he was one of the foremost men of his time, politically, culturally and socially, and he ranks alongside Burlington and Coke in his influence on the development of lakes. A prominent government member (Privy Councillor and Lord Privy Seal), he was a Whig and supported George I, receiving a dukedom in 1715, and the Garter in 1719.[19] The contents of his library show that he was abreast of current thinking about landscapes. It included copies of Francis Bacon's works, Joseph Addison's *Remarks*, John Evelyn's *Sylva*, d'Argenville's *Theory and Practice of Gardening*, Le Pautre's *Ornaments for Architecture*, and *Designs for Fountains*, Palladio's I *Quattri Libri*, as well as works by the Earl of Shaftesbury and William Temple.[20] It is not an exaggeration to say that he, along with Carlisle (his close friend) and Coke, set the trend in landscape fashion which led to the emergence of irregular lakes.[21] It is not known who designed his landscape, but it is possible that Vanbrugh had a hand.[22] Kingston, Carlisle and Vanbrugh were all members of the Kit Cat Club, and Vanbrugh mentions Kingston several times in his letters.[23] As we have seen, he (Vanbrugh) also planned a 40 acre lake at neighbouring Welbeck for the 1st Duke of Newcastle-upon-Tyne (John Holles, d. 1711), who was a cousin of Kingston.[24] Though that lake was not made, Vanbrugh as an inspiration, or even the designer, at Thoresby is a reasonable possibility, as we shall see.[25]

Holkham was also laid out in the 1720s. Like Carlisle, Thomas Coke had made the Grand Tour, and met William Kent in Italy.[26] He returned to Norfolk in 1718, and in 1727 began making an irregular lake by damming a small river which ran through the park (Figure 3.10).[27] This was well before construction of a new house began in 1734. Coke laid out a largely geometric landscape which followed the general line of the north – south axis of the irregular lake. Like Kingston, he did not seem to have any qualms about having a large, completely irregular lake in his otherwise geometric landscape. South of the house, a lawn and basin gave onto a large grassed area bordered by a stepped plantation, leading to Kent's Obelisk Plantation, cut through with rides and vistas, begun in 1729. These axes were carried out into the wider parkland. What Coke created was a transitional geometric landscape, which had a Palladian mansion at its centre, a good sprinkling of classical buildings, and an irregular lake.

The man who was prominently connected with some of the irregular lakes which appeared in the 1720s was Sir John Vanbrugh. He was the creator of the irregular South Lake at Castle Howard in 1725, as his letter to Carlisle in February attests:

> I have the new piece of Water much at heart; I hope 'twill do well but I doubt there's no certain proof till the dry comes. A vast deal of Rain has been this way.[28]

He was obviously relieved to hear, in March, that the lake was filling satisfactorily: "The rising of the Water in so hopeful a manner, is indeed a Cordial

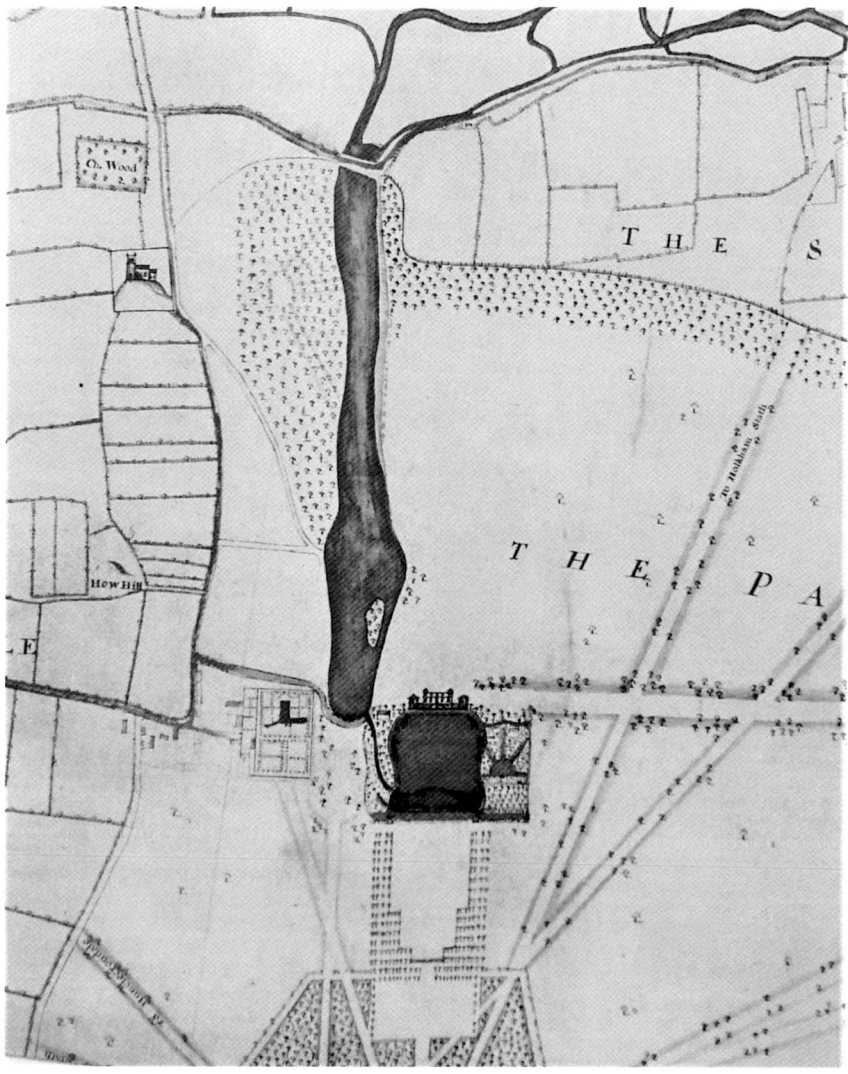

FIGURE 3.10 An estate map of Holkham Park, Norfolk, *c.* 1745. *Courtesy of the Earl of Leicester and the Trustees of the Holkham Estate.*

to me;"[29] Clearly, Vanbrugh was not as confident about lake building as about house building as, on 16th December, 1725, he wrote to Carlisle again: "I am glad I hear of no mischief about your Water under Ray Wood So I hope your Dam holds firm."[30] This lake of Vanbrugh's, which was *c.* 2.5 h, completely broke with the geometric tradition in being a most unusual shape. It was generally rectangular, but with a 'frilly' outline (Figure 3.11). Viewed from the terrace bounding Ray Wood, or from the Temple of the Four Winds, it

Lake

FIGURE 3.11 A detail of a 1727 estate map of Castle Howard, Yorkshire, showing South Lake. *Reproduced by kind permission of the Howard Family. Ref. CH P1.4.*

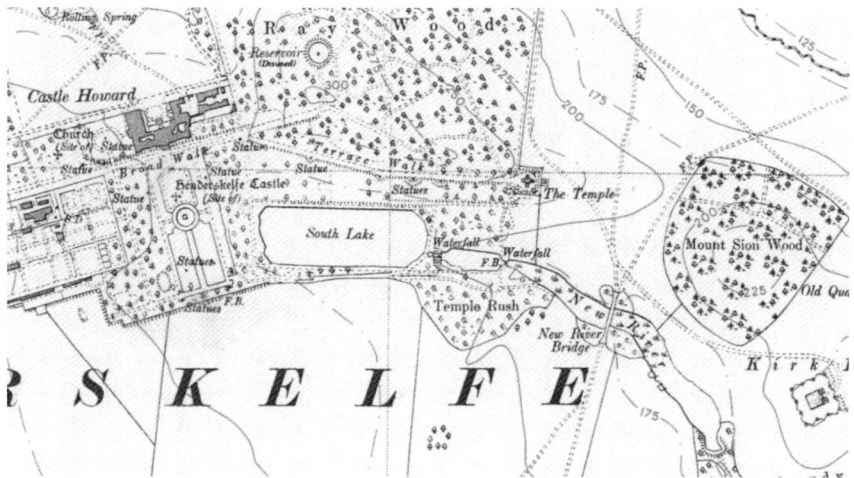

FIGURE 3.12 South Lake, Castle Howard, Yorkshire, OS 6" map, 1950. *Reproduced with permission of the National Library of Scotland.*

may well have seemed to have a pleasing irregularity. Certainly, those features provided vantage points from which to look down on the lake. The outline or margins of this lake have been altered more than once since the 1720s, being made more formal in the late 1840s (Figure 3.12).[31] The ground of Ray Wood rises fairly steeply and probably would have afforded views over the lake from some of the paths which wound through it. While the design of Ray Wood was particularly remarked on by contemporaries such as Daniel Defoe, the lake just below it had a pivotal role.[32] It provided a visual and physical link between the formal wilderness and the perceived informality of Ray Wood. As at Thoresby, Daniel Defoe does not see fit to comment on the lake, so his visit probably pre-dated the making of it.

At Castle Howard, the Earl of Carlisle was one of the first people to include irregular features in his landscape. Only Moor Park, Surrey (1697), shows similar irregular paths in a garden of an earlier date (Figure 3.13). Sir William Temple published his essay *Upon the Gardens of Epicurus* in 1697 and it may be relevant that Carlisle visited him at Moor Park in that year.[33] Temple's actual gardens and his opinions on gardens were predominantly regular and ordered, but both contained noticeably irregular elements: the engagement with Chinese gardens and *sharawadgi* in the *Essay*, and the very sinuous paths in a small garden at Moor Park lying alongside the river, between the formal gardens and the park.[34] Clearly, Carlisle was interested in the irregular aspects of Temple's work. Looking at the date of Carlisle's visit, and the construction of Ray Wood (*c.* 1705–8), we can

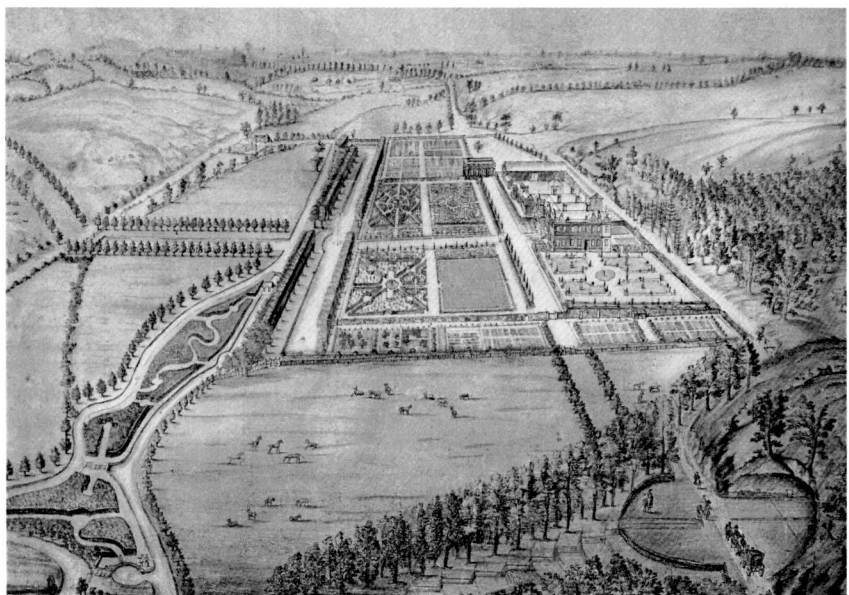

FIGURE 3.13 A 1697 painting of Moor Park, Surrey. *Reproduced by permission of the Surrey History Centre. Ref. PX/64/148.*

see a time lag of about ten years between a possible source of inspiration, and accomplishment.

Apart from the lake at Castle Howard, what other evidence is there for Vanbrugh designing irregular lakes? He tried to design one in the Glyme valley for the Duke of Marlborough, as Figures 3.14 and 3.15 show, but was thwarted by Sarah, Duchess of Marlborough.

Figure 3.14 shows geometric canals and an irregular lake, possibly with a cascade, pencilled in on the plan. Vanbrugh's bridge is merely indicated in pencil. The later plan (Figure 3.15) seems to have been drawn in order to show properly the planned canals and lake in Figure 3.14, and was more formal. This indicates that they must have been part of Vanbrugh's original conception, and not devised by Colonel Armstrong in the 1720s.[35] The outline of the bridge can also be seen, the foundation stone having been laid in June, 1705.[36] Remains of the Palace of Woodstock are also shown (demolished by *c.* 1719).

Figure 3.15 has more of the appearance of a hybrid lake, suggesting that Vanbrugh modified his original design to make it more palatable to the Marlboroughs. This second plan shows the more regular lake as part of the canalised River Glyme. It is almost as though two hands are at work in this design, as the 'canal' immediately north of the lake is quite informal, whereas the other canals are very formal, and hints at a 'tussle' occurring between patron and designer. It is possible that

Lake

FIGURE 3.14 Vanbrugh's planned lake at Blenheim, *c.* 1705 (early), with an elaborate cascade. The lake was *c.* 3 h. Emphasis added by author. *The Bodleian Libraries, The University of Oxford. Ref. MS. Top. Oxon. A.37★, fols. 1-2.*

something similar happened at Castle Howard, with the 'frilly' lake there being a compromise between what Vanbrugh proposed, and what the Earl of Carlisle would accept. Interestingly, in Figure 3.15, at a later date, someone has roughly sketched in a lake to the east of the bridge, approximately where the Duchess had

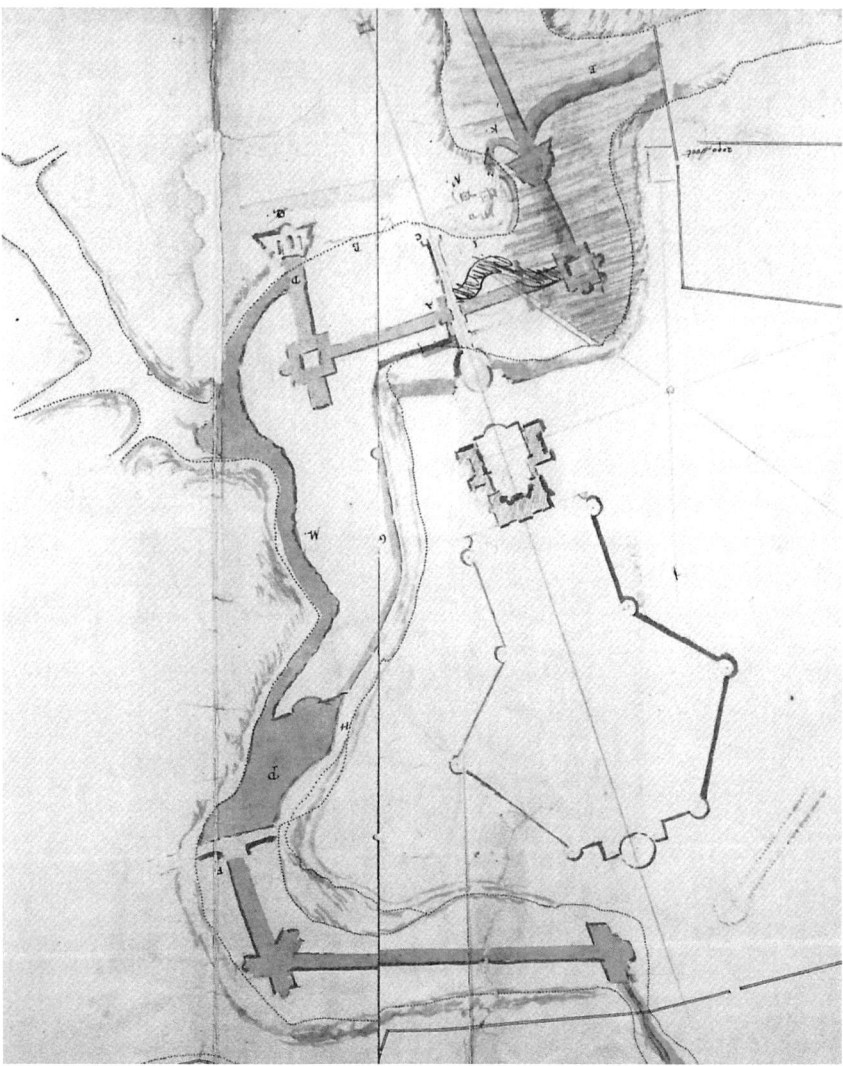

FIGURE 3.15 Vanbrugh's plan for a lake at Blenheim, *c.* 1705 (late). *The Bodleian Libraries, The University of Oxford. Ref. MS. Top. Oxon. A.37*, fols. 1-2.*

one made in 1724. Presumably, this was Vanbrugh as it is a similar shape to the eastern lake in Figure 3.16.

Of more significance is the plan of Blenheim (Figure 3.16) which Colen Campbell attributed to Vanbrugh. In his description of the plates of Blenheim in Vol. 1 (1715), Campbell emphasises his debt to Vanbrugh, saying:

> I present the Curious with all the Plans and Elevations, by the particular direction of *Sir John Vanbrugh*, who gave the Designs of this Magnificent

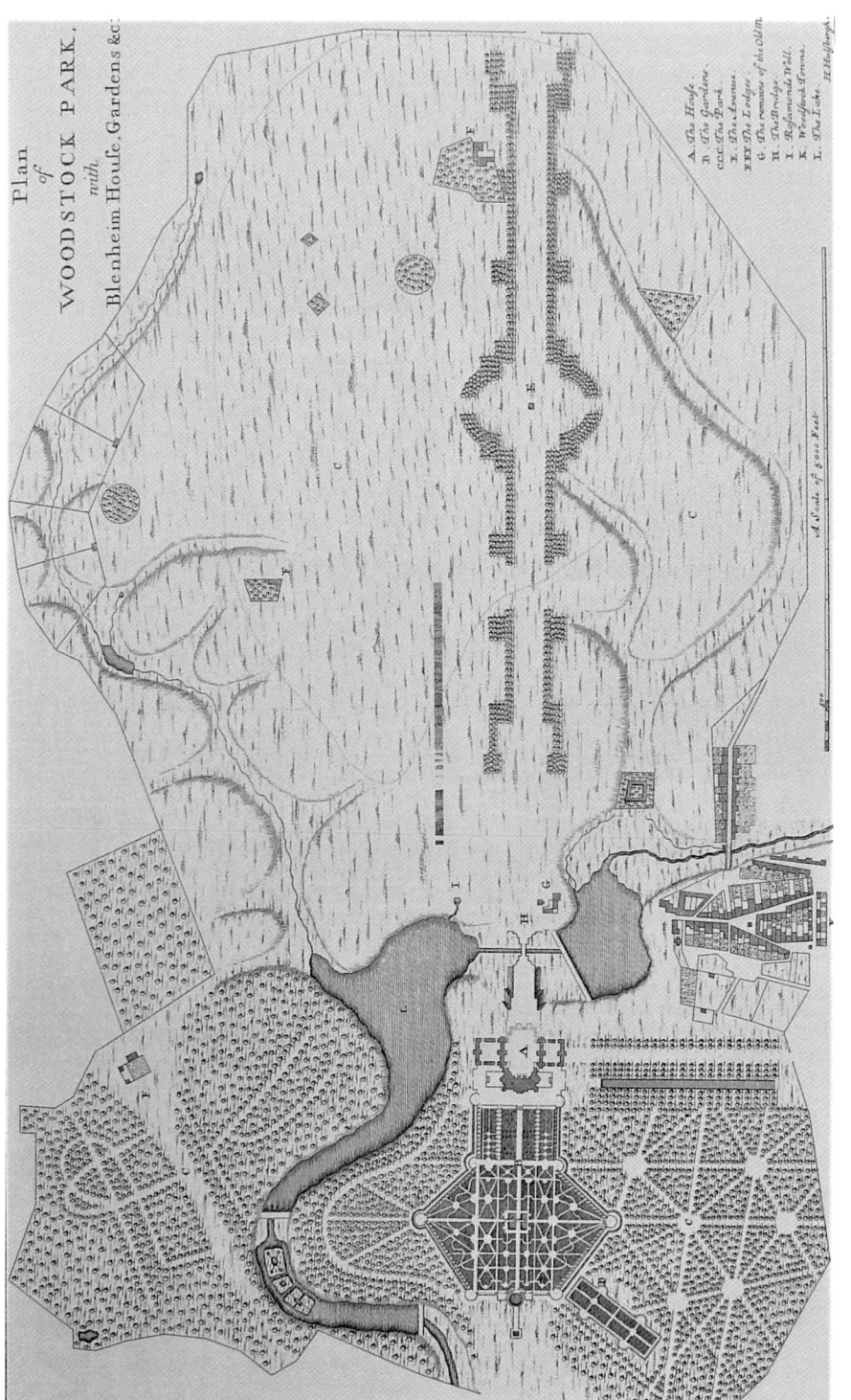

FIGURE 3.16 Vanbrugh's plan for Blenheim in *Vitruvius Britannicus*, dating from 1715 or earlier. *Historic England.*

Palace . . . most generously assisting me with his Original Drawings, and most carefully correcting all the plates as they advanced. . . . Here are noble Gardens; a stately Bridge,[37]

and in Vol. III, 1725, he says,

Having given the Plans and Elevations of this stupendious Fabrick in my First Volume, I have here given an exact Geometrical Plan of the Park, Gardens, and Plantations, which are of very great Extent.[38]

This means that Campbell had received the landscape plan from Vanbrugh by 1715, although he did not publish it until 1725, along with the other garden plans.[39] The two irregular lakes are most striking, pre-dating Brown's lake (47 h) by almost 50 years. The plan testifies to Vanbrugh's precocious vision, and demonstrates that he was a pioneer of the concept of a lake as we think of it today: an irregular lake. The planned lake at Welbeck (16 h) and his possible influence at Thoresby (20 h) seem much more plausible in this light. In the designs for lakes at Blenheim and his intention to retain the ruins of Woodstock Manor, he anticipated concepts which did not become widely popular in landscape design for half a century: irregular lakes, and the Picturesque.[40] Usually, Vanbrugh is respected for his architecture, but he sowed the seed of irregular ornamental water in the top echelons of society, and it began to bear fruit in the 1720s and '30s.

It was Vanbrugh's early use of 'lake' in his plans and letters to patrons – the word appears in the legend to the plan in Figure 3.16 – that helped to introduce the word and, more importantly, the concept. As we have seen, use of the term 'lake' to mean man-made ornamental water was unusual until the 1750s, and then it was not widely used, although some such as Switzer, Stukeley, the Duchess of Marlborough (all connected with Blenheim), and Pope, began to use the term from the 1720s.[41] It is relevant to note that Defoe (d. 1731), in his *A Tour Through the Whole Island of Great Britain by a Gentleman*, written before 1724 but published in 1742, consistently referred to ornamental water as 'a great piece of water' or 'a noble piece of water' except when he was describing the water at Stowe, Grimsthorpe, Newstead Abbey and Blenheim, and at these places he refers to 'lakes'. It is surely no coincidence that these were places belonging to great men of the time – the *cognoscenti* – and it is likely that it was the owners who used these terms rather than being Defoe's own inspiration.

Londesborough was Lord Burlington's property in East Yorkshire which he inherited in 1704. Perhaps this place reflects the ambitions, in terms of water, which he did not have the space or topography to attempt to realise at his house at Chiswick, Middlesex. He did not start to develop this landscape until the 1720s, probably creating the lake and string of informal ponds in 1728–30.[42] Not a great deal is known about his work with water there apart from the features shown on the estate survey (Figure 3.17). Like the previous three examples, it is noticeable that

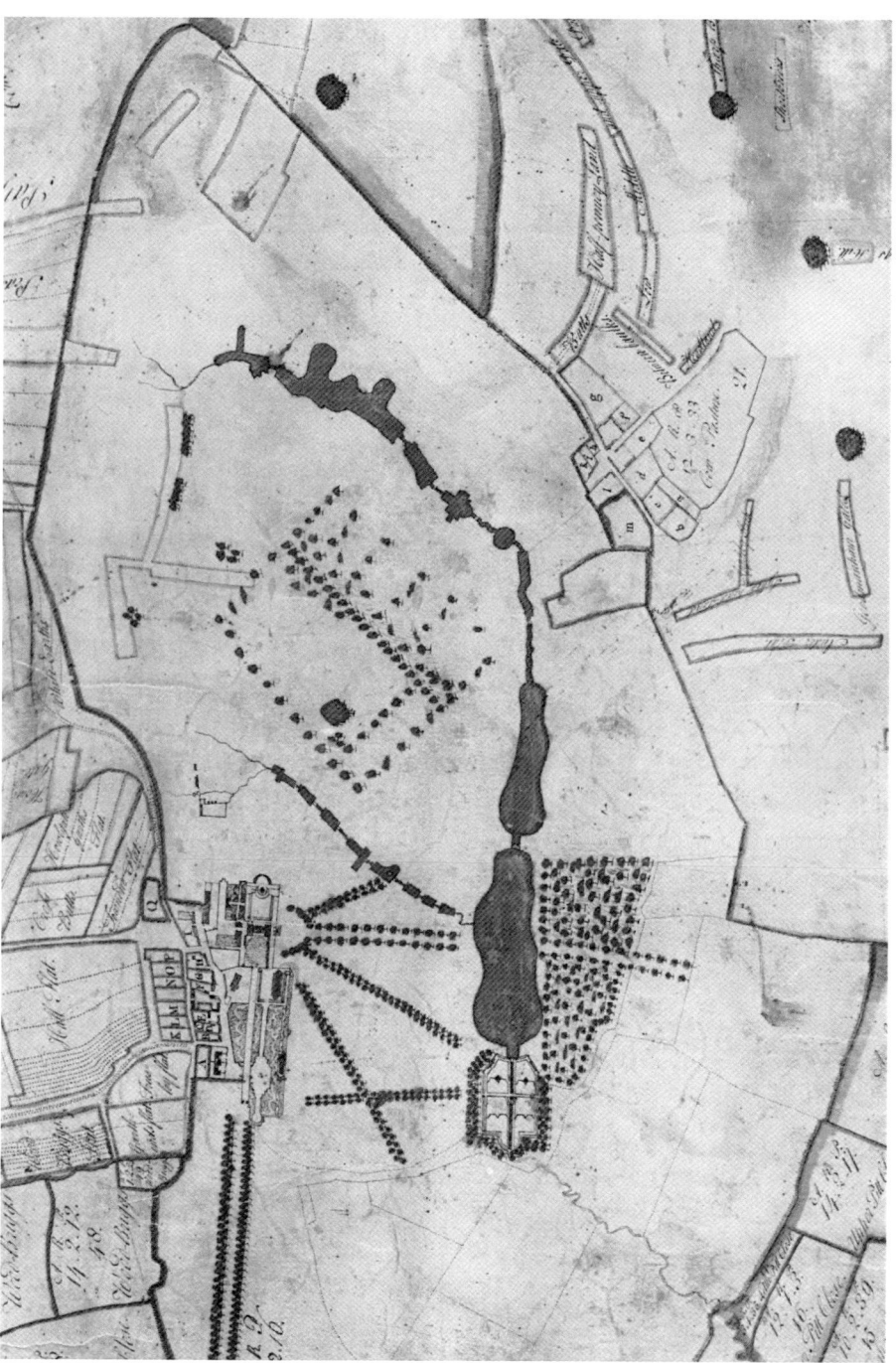

FIGURE 3.17 A 1739 estate survey of Londesborough Park, East Yorkshire. *Courtesy of East Riding Archives. Ref. DDX 31 173.*

formal and informal elements are juxtaposed. The formal gardens adjacent to the house contained sinuous paths which are not present in the illustration in *Britannia Illustrata*, and it may be assumed that Burlington introduced this form at Londesborough, as he used it extensively at Chiswick.[43] He appears to have retained the avenues and *pate d'oie*, and created the irregular lake almost to complement them.

It is notable that the irregular lakes which began to appear in the 1720s had irregular though straight-sided plantations adjacent to them, and both were increasing in size, as we have seen. The water was in the park, where the constraints of 'garden' did not apply but, as the positions of the lakes at Thoresby, Castle Howard and Holkham show, it was not isolated in the park, as fishponds usually were, but was relatively close to the house and directly overlooked by it. It is as if the lake was forming a link between the formality of the house and gardens and the informality of the park – a concept which may have applied to the activities in them as well as their physical organisation. It may well be that women were beginning to use parks more extensively than they had before, though information on this subject is very scanty before the middle of the eighteenth century. If women were beginning to participate in activities in parks, a closer relationship between gardens and parks would have been an advantage, a theme which is discussed in Chapter 6. However, cause and effect are difficult to determine precisely.

The juxtaposition of geometric landscape elements with irregular lakes continued in the 1730s. Chiswick, a landscape laid out by Lord Burlington in the 1720s and '30s, latterly with the aid of Kent, captures this phase of transitional geometric landscapes admirably. It was dominated by axes radiating in various directions, with triangular and rectangular gardens dissected by twining, irregular paths. The various buildings – the house and temple, the obelisk – were classical in style, tending to emphasise the geometry, as did the geometric ponds, but the treatment of the river shows Burlington's attempts to create irregularity and informality, which perhaps appear stilted to modern eyes in Roque's engraving of 1736. Like The Serpentine in Hyde Park (Figure 1.6), it was viewed by contemporaries as a radical departure from the straight-sided canals of the first two decades of the eighteenth century.[44] The cascade which Burlington and Kent created after 1736 is in line with the move towards greater informality which gained hold in the middle of the century.

Claremont in the 1730s likewise encapsulates the transitional geometric phase (Figure 3.18). Originally fashioned by Vanbrugh for Thomas Pelham-Holles, Earl of Clare, later Duke of Newcastle, a predominantly geometric layout was imposed on the topography, as illustrated by Campbell in *Vitruvius Britannicus*. The water is geometric, including the round basin, as is the tree planting. Only the serpentine paths ascending the hill behind the house break the geometric mould. By the time Rocque made his engraving in 1736, the most noticeable change was that the basin had been made into an irregular lake (*c.* 3 h) by Kent, and graced with an island (see Chapter 6). Kent, unlike Brown, did not make any large bodies of water, although he modified several.[45] A comparison with Rocque's plan shows a strong resemblance to the upper right quadrant of Robert Castell's plan (see Chapter 6), although the water is not as irregular.

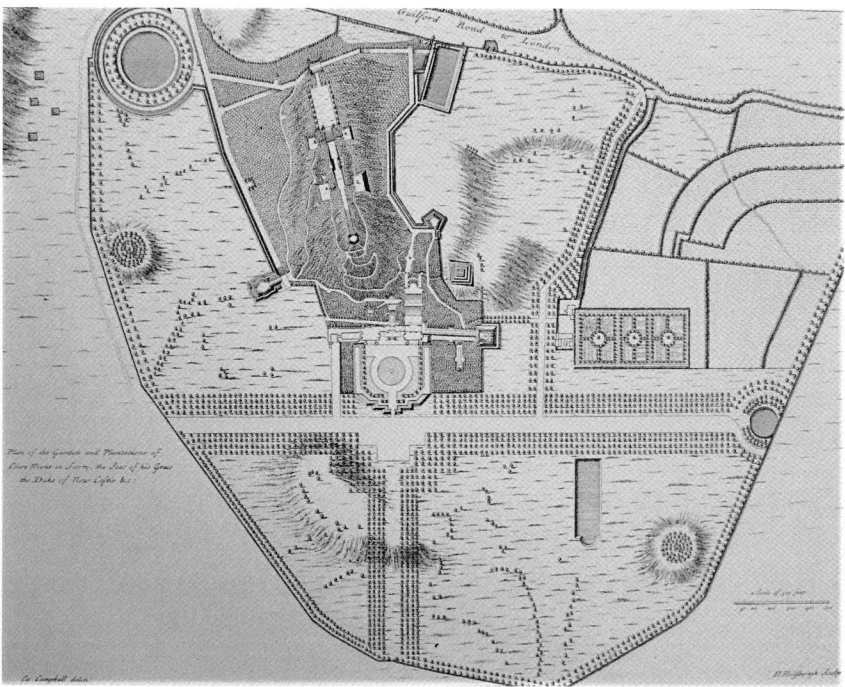

FIGURE 3.18 A plan of Claremont, Surrey, in *Vitruvius Britannicus* 1725. *Historic England.*

Serpentines

A parallel thread in the development of ornamental water was a feature which could be described as a serpentine canal. The word 'serpentine' has been used in the twentieth century to denote a narrow, winding irregular lake, or one with a 'hooked tail' like the one produced at Holkham by William Emes in *c.* 1784, but this was not what contemporaries meant by the term. In the eighteenth century, 'serpentine' was usually followed by 'river' and often meant a canal composed of geometric arcs: a serpentine canal, which was essentially a geometric construction. That Bridgeman had this idea as early as 1729, is illustrated in his plan for Sherborne, showing a serpentine canal, which may have been made although possibly not until 1743.[46]

Queen Caroline's lakes – the Long Water and The Serpentine – which together are referred to as 'The Serpentine' today (*c.* 16 h), were made by Bridgeman in 1730–31.[47] They were novel because of their lack of symmetry, the slight informality of The Serpentine itself, and the angles which they incorporated. These angles derived from the original string of ponds from which the Long Water was formed, shown in Rocque's 1762 engraving of Kensington Gardens and Hyde Park (Figure 1.6). Within Kensington Gardens, the Long Water was geometric and Bridgeman maintained the formality of the gardens in that area. Where the water was in the park, it appears to be slightly informal, mirroring the less formal character of that part of the park, although a plan of *c.* 1738 attributed to him shows a more formal arc to The Serpentine. Two factors made it very unusual: the angle in the canal and its considerable size (*c.* 16 h altogether). The impact of this design was that it gave royal approval to what was seen as a signal departure from the norm for formal water.

Rocque's map of *The Environs of London* 1745, shows a very similar configuration of water at Wanstead, Essex (Figure 3.19). It is labelled 'Serpentine Ponds', and was also made from existing geometric ponds, as shown on a Rocque engraving of 1735. This suggests the angle of the canal was carefully constructed, and that it was this angle which was being showcased. Note that the northern edge has been lined up along an axis of the gardens.

Whilst examples of serpentine canals are not numerous, more are coming to light. A slightly curved canal made at Holkham in *c.* 1743, almost certainly by Kent, linking the basin south of the house with the lake, may be an example, but was not typical of this phenomenon.[48] Badminton and Longleat show another manifestation of the serpentine canal. The water at Longleat appears on an estate survey of 1747 (Figure 3.20), the two lakes being composed of strictly parallel geometric arcs. In fact, they are geometric lakes, and the larger one was on a similar scale to The Serpentine itself. The 2nd Viscount Weymouth was almost certainly inspired by

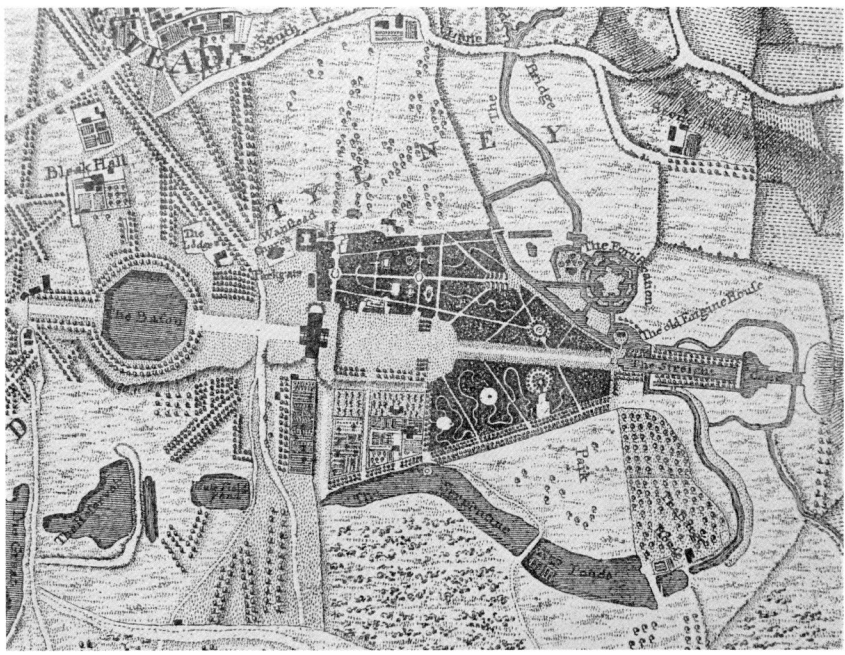

FIGURE 3.19 Rocque's 1745 map of Wanstead, Essex. *By kind permission of Cambridge University Library. Ref. Atlas 2.74.2.*

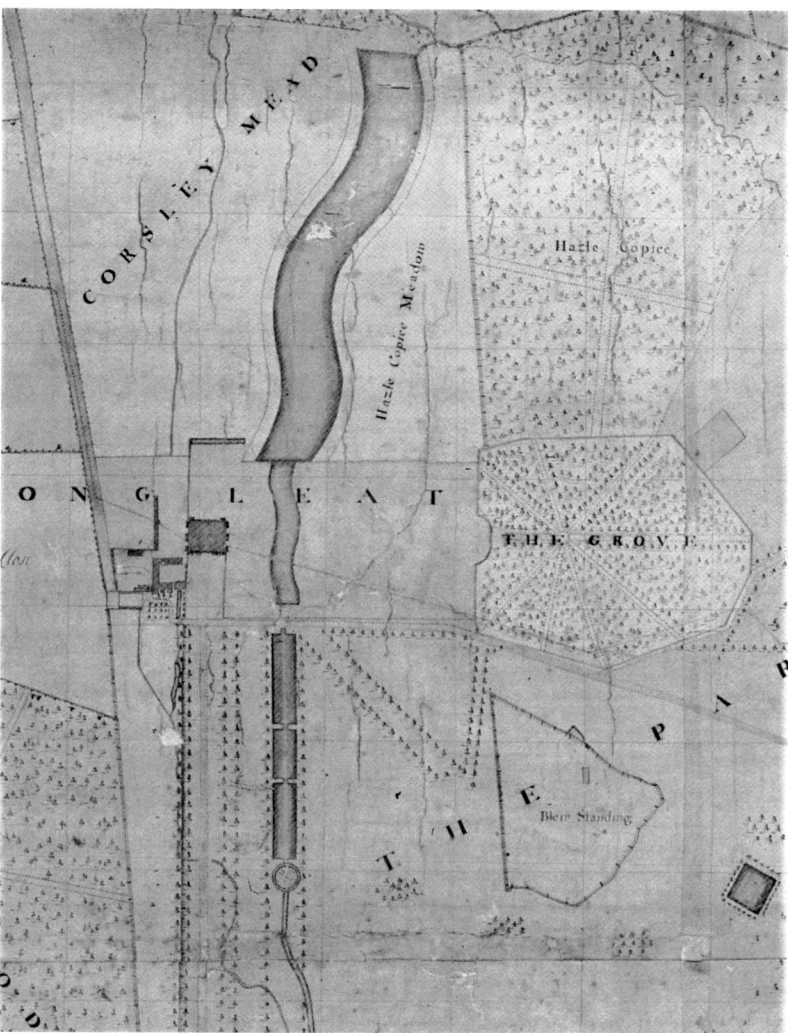

FIGURE 3.20 A 1747 estate map of Longleat, Wiltshire, by John Ladd. *Reproduced by permission of the Marquess of Bath (copyright), Longleat House, Wiltshire. Ref. Map 68.*

Queen Caroline's. His lakes were nearing completion in 1736, as part of a two-year transformation of the landscape there. He was appointed as Ranger of Hyde Park in 1739, suggesting a familiarity with developments there, and a yacht was bought for Longleat in 1736; two had been bought in 1731 for the Royal Family to use on the Serpentine.[49] Similar in form to the one at Longleat, the canal at Badminton appears on an estate survey of 1750 by Robert Whittlesay (Figure 3.21).

There is little evidence that the term 'serpentine' was much used by contemporaries to describe ornamental water, other than to signify The Serpentine itself:

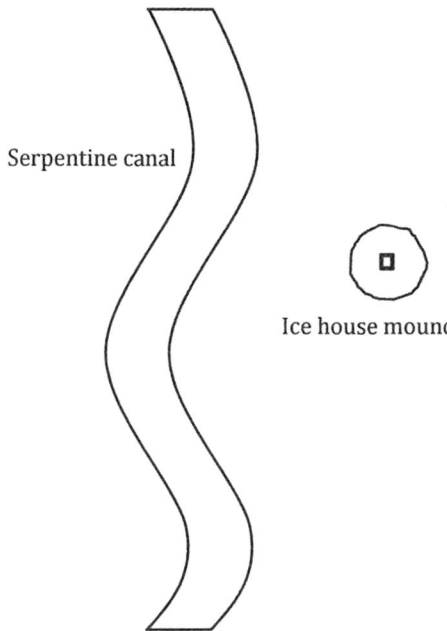

Serpentine canal

Ice house mound

FIGURE 3.21 A diagram of the serpentine canal on an estate survey of Badminton, 1750, by Robert Whittlesey.

"Next Monday they begin upon The Serpentine River and Royal Mansion in Hyde Park", as described in *The London Journal* of 26th September, 1730.[50] One early use was by Francis Blomefield, who used 'serpentine' in relation to Kimberley in 1739 (Figure 1.3):

> the piece of water which . . . is there said to contain 12 or 14 acres, is now extended into a noble lake of about 28 acres [11 h], which seems to environ a large wood or carr on its west side; rendering its appearance to the house much more grand and delightful; the rivulet that ran on its east side is now made a *serpentine river*, laid out in a neat manner.[51]

His use of 'lake' is also a relatively early use of the term. Another early use of 'serpentine river' is in a letter from Mrs. Elizabeth Robinson of December, 1743, about Mr. Haytley's painting of the Brockman's new landscape at Beachborough Manor, Kent (Figures 6.9 and 6.10) She says,

> all that is to be put into a Picture painted by Mr. Haytley. he [Mr. Brockman] wants some better name than that of a pond for his water which puzzles him very much for he fears it is too small to call a Serpentine River;[52]

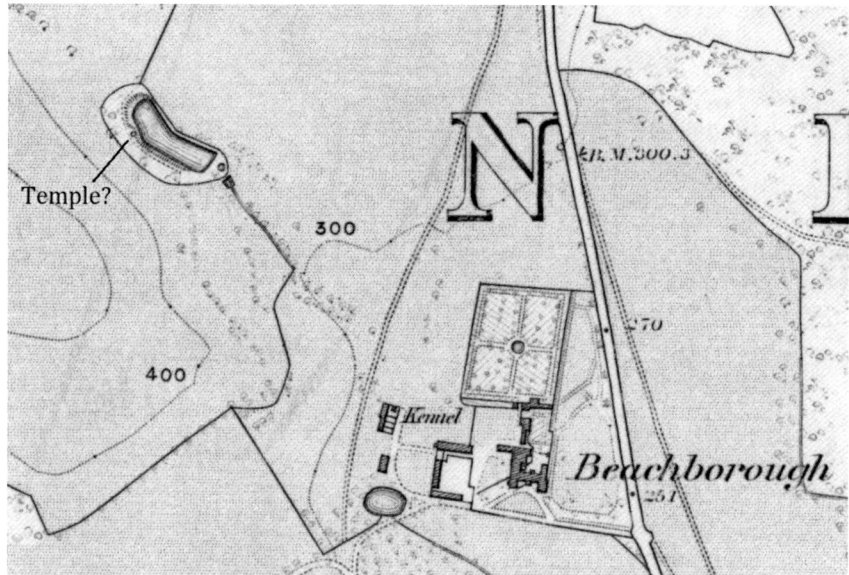

FIGURE 3.22 The water at Beachborough, Kent, OS 6" map, 1872. Note the round structure on the western side, which is probably the temple shown in Edward Haytley's painting. *Reproduced by permission of the National Library of Scotland.*

It was only about 0.17 h, or under half an acre in size. The Brockman quandary about a 'Serpentine River' suggests that informed and fashionable people were well aware of the lake created by Queen Caroline even before Rocque published his map (1746). The actual shape of the Beachborough pond is also reminiscent of the Serpentine, albeit in miniature, as Mr. Brockman realised (Figure 3.22). What is typical of both is the angle, and this suggests that some people were still thinking very much in terms of geometric water, the novelty being that the straight canal had been given a 'bend'. According to the Oxford English Dictionary (OED), it is 1824 before anyone uses 'serpentine' in a conceptual sense to describe ornamental water: J. C. Loudon, in his *Encyclopaedia of Gardening* says "Those wavy serpentine canals . . . are never mistaken for natural scenes".[53] The word does not appear on any maps, other than Rocque's, as far as can be ascertained.

Despite the popularity today of the term 'serpentine lake', this was not a term used in the eighteenth century. People talked of 'serpentine ponds' and 'serpentine rivers' which as we have seen, usually meant something 'canal-like'. Arthur Young was one of the few people to refer to a 'serpentine lake', at Ditchley, Oxfordshire, in 1774 (Figure 3.23).[54] The hook at the west end of this irregular lake presumably reminded him of the bend in the Serpentine itself, or possibly Wanstead. The OED entry for 'serpentine' gives an interesting quotation from George Eliot in 1853: "I am hoping for a row . . . on the Serpentine, which is

FIGURE 3.23 Ditchley, Oxfordshire, OS 6" map, 1900. *Reproduced by permission of the National Library of Scotland.*

really almost as good as a lake".[55] This suggests that she regards The Serpentine as too circumscribed, or too geometric, to be regarded as a real lake. The term only seems to gain currency after 1948, when Christopher Hussey talked of "The most famous Serpentine Lake, that in Hyde Park", tying it again to The Serpentine.[56] It is probable that the emergence of serpentine canals was a phenomenon which was parallel with the sinuous walks in wildernesses that were becoming popular in the 1730s, and which can be seen at Claremont (Figure 3.18) and Thoresby (Figure 3.8). Switzer had published plans incorporating these in *Ichnographia Rustica* 1718 (Figure 3.5), as did Batty Langley in *New Principles of Gardening* in 1728. Although these walks are characterised by their sinuous informality, perhaps contemporaries felt that the serpentine canal, though it has quite a geometric air, was the watery equivalent. Although the fashion for these canals does not appear to have outlived the 1740s, they are significant because, despite being essentially geometric, they indicate that people were looking at ornamental water in a different way, reinforcing the concept that it had become acceptable to be innovative with water. However, there does not appear to be any direct link between serpentine canals and irregular lakes. In fact, these angled canals were a dead end in terms of lake evolution.

Chronology of lake numbers

By the 1740s, the concept of an ornamental lake was beginning to become established, and an increasing number of lakes were being made. Numbers began to increase in the 1730s, with irregular lakes starting to become popular in the 1740s (Figure 3.24) and numbers peaked in the 1760s–70s.[57] It is interesting to note that even though the period 1780–99 includes the county maps which were generally made in the last two decades of the century, and more information could be expected from these sources, there is still a significant dip in the numbers of lakes being made in the 1780s before they began to rise again. There was also a similar fall in Acts of Enclosure in that decade.[58] The costs of the war in America and of the Napoleonic Wars may have been a factor, with the implicit impact on the economy. The general trend shown by this graph is clear: from a very low point in the first two decades of the eighteenth century, lake numbers peaked in the 1760s and '70s, and then started to fall. In the period 1750–99 most of the lakes which were made were irregular, following an almost identical pattern to numbers for all lakes in the second half of the century, and Table 3 gives an indication of when they started to be made, and their varying sizes.[59] In the eighteenth century, sizes generally varied from 1 h to 25 h or more.[60] Two things are immediately noticeable about irregular lakes. Firstly, many lakes were made by unknown people, and we can only conjecture who made them. Perhaps owners employed their own men, or local experts, to make these lakes, and no doubt further attributions will be made. Secondly, of the known lake-makers, Lancelot 'Capability' Brown's name is predominant.[61]

Much has been made in earlier chapters of the link between irregular lakes and the large fishponds (*vivaria*) of Roman and medieval times, and the point has

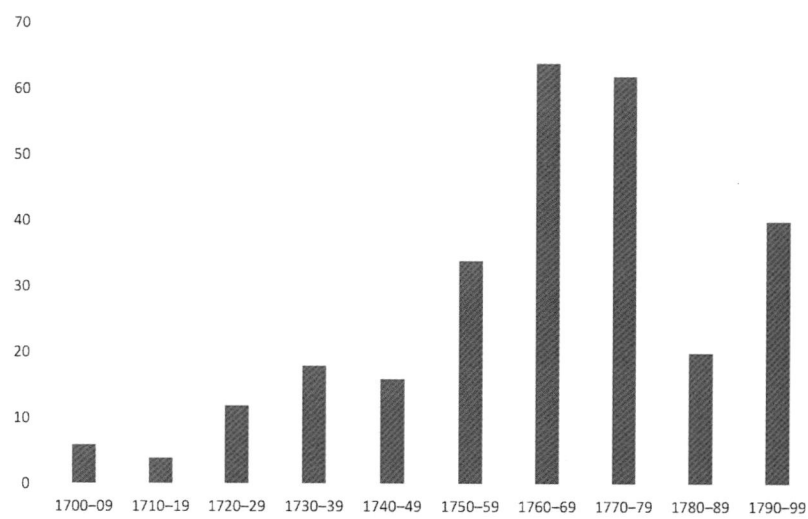

FIGURE 3.24 A chart of the numbers of all types of lakes, 1700–99.

arrived where it is possible to see this link clearly. As we have seen, the irregular lakes which had become popular ornamental landscape features by the 1750s often looked like their *vivaria* forbears, and were constructed in a very similar way. Indeed, it was quite common for an existing fishpond to be extended and modified into a lake. This is clear in the changes which can be seen in the water features at Stourhead, Wiltshire, which serves as a seminal example. Figures 3.25 and 3.26 show the fishponds in the valley when the first Henry Hoare bought the property in 1717, and the shape of the lake today. It is quite clear, from these two illustrations, that the earlier fishponds formed the basis for the ornamental lakes made by Henry the Magnificent (Henry Hoare II). The 1792 painting (Figure 3.27) illustrates the stages which the water features went through before the final lake was made. It shows the rectangular pond he made (just below Flora's Temple), and the original grotto and pond, both of which were flooded when the lake was made in 1754. This kind of progression can be seen in numerous cases in the eighteenth century in places where fashionable ornamental lakes were made by extending existing fishponds. It was very good practice, in fact, as the presence of the fishponds showed that a lake would probably be feasible. Other landscapes where this occurred are Petworth, Surrey, Blickling, Norfolk, Burghley, Lincolnshire, and Compton Verney, Warwickshire, to name some of the most well-known, and there are many others where this is likely to have been the case.

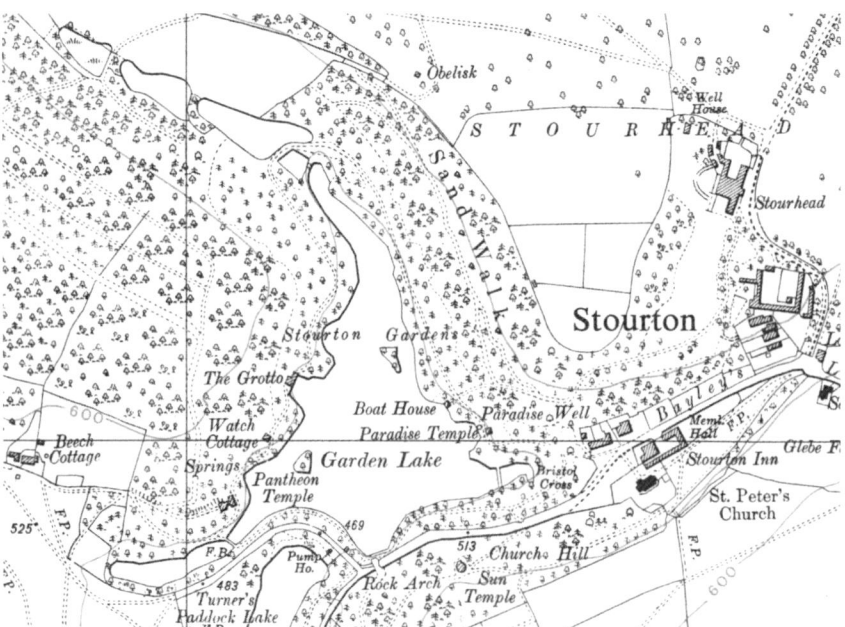

FIGURE 3.25 The lake at Stourhead, Wiltshire, OS 6" map, 1962. *Reproduced by permission of the National Library of Scotland.*

FIGURE 3.26 A 1722 estate map of Stourhead, Wiltshire. *Wiltshire and Swindon Archives. Ref. 383/316.*

FIGURE 3.27 A 1792 painting of the bottom of the lake at Stourhead, Wiltshire. *Courtesy of Wiltshire Museum, Devizes. Ref: DZSWS: Map. 564.*

Conclusion

The main purpose here has been to establish a chronology of when lakes began to be made, and how they evolved. At the beginning of the eighteenth century, the concept of a man-made *lake* did not exist. However, large areas of ornamental water were starting to become popular, though they were geometric in shape. Generally, they were not very large, because of the expense of making them, and of fitting basically square or rectangular elements into linear designs. As we have seen, the first irregular lake was made at Thoresby in *c.* 1719, and lake numbers peaked in the 1760s and '70s. These included different forms such as hybrid lakes and river-lakes, which will be discussed in the next chapter.

The accepted thinking about lakes was that their evolution was driven by landscape style – the 'naturalistic' style became popular under Kent and Brown *et al*, and that water then became informal – but this was not the case. Instead, lakes developed first, and landscapes changed around them. Men like the Duke of Kingston, the Earl of Carlisle, Thomas Coke, and Lord Burlington were happy to plan geometric landscapes and put irregular lakes into them. However, their desire for large, irregular bodies of ornamental water had a significant impact on the geometric landscape style, and eventually led to its disintegration, a theme which is discussed in Chapter 6.

Notes

1 'Aesthetic' here embraces the concepts of 'ornamental' and 'leisure'.
2 1715 survey of Boughton in John Cornforth, 'The History of the Boughton Landscape', *Country Life* 11th March, 1971, Figure 3.
3 1801 survey of Gamlinghay Park, Cambridgeshire, in T. Mowl & L. Mayer, *Historic Gardens of Cambridgeshire & the Isle of Ely* (Bristol: Redcliffe Press, 2013) p 83.
4 Mowl & Mayer, op. cit., p 83.
5 Historic England listing: Staunton Harold Park and Garden.
6 J. Bapasola, *The Finest View in England: The Landscape and Gardens at Blenheim Palace* (Oxford: Blenheim Palace, 2009) p 37.
7 Bapasola, op. cit., pp 35–36.
8 From *Letters of Sarah Duchess of Marlborough at Madresfield Court*, 1875, ref. 166, quoted in Bapasola, op. cit., p 37.
9 T. Williamson, 'The Archaeology of the Landscape Park', BAR British Series 268 (Oxford: Archaeopress, 1998), pp 72–80. It is supplied by two small streams.
10 The full list can be found in the Landscape Database.
11 Larger versions of these maps are given later.
12 Information extrapolated from accounts for Thoresby Hall, ref. Ma 4287 1719–20, Manvers Archive, University of Nottingham, Manuscripts.
13 J. Vanbrugh: letter to Lord Carlisle, 20th February, 1721, in Dobrée & Webb, op. cit., p 129.
14 D. Neave & D. Turnbull, *Landscaped Parks and Gardens of East Yorkshire* (Unknown: Georgian Society for East Yorkshire, 1992) p 50.
15 Bishop Richard Pococke, *Travels Through England*, ed. James Joel Cartwright (London: Camden Society, 1888), Vol. 42, p 73. It is possible that Francis Richardson, the 2nd Duke of Kingston's gardener, may have been associated with this work.
16 As observed by the author in 2015. The Earl of Kingston received a dukedom in 1715.
17 There is always the caveat that subsequent research may uncover an earlier lake.
18 University of Nottingham, ref. Ma 4283 1718–19.

19 G. F. R. Barker, *Evelyn Pierrepont, 1st Duke of Kingston*, revised by M. E. Clayton, ODNB online (2015).

20 University Library, Cambridge, ref. Munby.bb.2. Published in 1726, the catalogue of his library was the first one printed in English of a personal library.

21 The author's theory is that Kingston, Jamaica was named after him, although concrete evidence has not surfaced to date.

22 Vanbrugh was working at Castle Howard from 1699, and Thoresby would have been virtually *en route* for him if he was travelling from London.

23 Letter in 1703, Dobrée & Webb, op. cit., p 8.

24 K. Downes, *Sir John Vanbrugh, A Biography* (London, 1987) p 461.

25 There is even the possibility that Copley's informant mistook Welbeck for Thoresby: when initially constructed, Thoresby lake was *c*. 48 acres.

26 C. W. James, *Chief Justice Coke: His Family and Descendants at Holkham* (London: Country Life, 1929) pp 187, 202.

27 The lake became more irregular in the 1730s, with the addition of an island in the southern part: T. Williamson, personal communication, November 2016.

28 Dobrée & Webb, op. cit., p 157. The letter is dated 11th February, 1724, but the common practice before the calendar changed in 1752 was for the new year to begin after Lady Day (25th March).

29 Dobrée & Webb, op. cit., pp 157, 159.

30 Ibid., pp 156, 171. This is also one of the earliest uses of the word 'dam'. The usual word was 'head'.

31 As illustrated by the First Edition 6'' OS map, surveyed in 1852–3.

32 Daniel Defoe, *Tour Through the Whole Island of Great Britain*, Vol. II, 1748, p 181. Defoe died in 1731, and made his tour in 1724.

33 C. Saumarez Smith, *The Building of Castle Howard* (London: Random House, 1997) p 128.

34 J. Swift, ed., *The Works of Sir William Temple in Two Volumes*, including *'Upon the Gardens of Epicurus'* (London: Round et al, 1731) p 170.

35 C. Dalton, *Sir John Vanbrugh and the Vitruvian Landscape* (Routledge: Abingdon, 2012) p 107.

36 Bapasola, op. cit., p 16.

37 C. Campbell, *Vitruvius Britannicus*, Vol. I (London, 1715) p 5.

38 Ibid., Vol. III (London, 1725) p 10.

39 This was probably the case with Thoresby as well, and no doubt all the other places which appear in Vol. III. It also implies that Campbell's Thoresby plan was a depiction of planned landscape works, some of which were probably in progress in 1715.

40 Vanbrugh was very keen to retain the ruins of Woodstock Palace, which was a significant cause of acrimony between him and the Duchess of Marlborough. In 1713, referring to his wooing of the young Mrs. Yarburgh in York, Lady Mary Wortley Montagu (Kingston's daughter) wrote in a letter the amusing comment, *our Dullness has inspir'd him with a Passion that makes us all ready to die with laughing. 'Tis credibly reported that he is endeavouring at the Honourable state of matrimony and vows to lead a sinfull life no more . . . but you know Van's taste was allways odd. His Inclination to Ruins has given him a fancy for Mrs. Yarborrough.* Dobrée & Webb, op. cit., p 201.

41 Switzer was employed at Blenheim on the construction of the bridge in 1707–8. Bapasola, op. cit., p 28.

42 Neave & Turnbull, op. cit., p 50.

43 Ibid.

44 Defoe described it as a 'Serpentine River': *Tour Through the Whole Island of Great Britain*, Vol. II, 1748, p 168.

45 Opinions are divided about what Kent was responsible for in connection with water at Holkham.

46 T. Mowl, *Historic Gardens of Gloucestershire* (Stroud: Tempus Publishing, 2002) p 64.

47 Willis op. cit., p 96.

48 T. Williamson, personal communication, March 2017.
49 T. Mowl, 'Rococo and Later Landscaping at Longleat', *Garden History*, Vol. 21, No. 1 (Summer, 1995) p 59; the *London Journal* 1st May, 1731, quoted in H. Colvin, gen. ed., *The History of the King's Works*, Vol. 5 1660–1782 (London: HMSO, 1976).
50 OED, "serpentine adj. 3a", *London Journl.* 26 September 2/3 (Oxford University Press).
51 F. Blomefield, *An Essay Towards A Topographical History of the County of Norfolk*, Vol. 2 (London: W. Miller, 1805) 539.
52 Letter from Mrs. Robinson to her daughter, Elizabeth Montagu, 28th December, 1743, quoted in Harris, op. cit., p 220.
53 J. C. Loudon, *Encyclopaedia of Gardening* (ed. 2) in OED, "serpentine iii. iv".
54 Arthur Young, *A Six Month Tour Through the North of England*, Vol. 3, 1774 (London: Various, 1774) p 327.
55 OED, "serpentine n. 10c".
56 OED, "serpentine adj. 3a".
57 Bishop, op. cit. For the purposes of this survey, if a geometric lake was converted into an irregular lake it was considered to be new but irregular lakes which were altered in shape or enlarged were not considered to be new, because the basic style remained the same. Thus, a lake which was made by Brown, but was altered by Repton, was counted once.
58 M. Turner, *English Parliamentary Enclosure* (Folkstone: Wm. Dawson and Sons, 1980) p 142.
59 Refer to the Landscape Database for a complete list.
60 Bishop, op. cit., p 297.
61 Based on information drawn from the Landscape Database.

4

THE MAKING OF LAKES

Making lakes is a complex business, not least because large amounts of water are involved. This is cogently illustrated by what the 9th Marquis of Lansdowne said about his 16 h lake at Bowood, Wiltshire (2015). The dam has to pass a 10-yearly inspection for the Environment Agency, and he pointed out that if it failed, the resulting deluge would flood the centre of Chippenham, 7 miles away. He was not exaggerating.

Construction

Building the dam retaining the water is the most important and complex part of lake-making, but the construction of a lake also has to take account of a number of other factors. One of the essential components is a source of water such as a stream, river or spring, which constantly fills up the lake. This is what really distinguishes a lake from a pond, as well as the size qualification of 1 h. This constant replenishment means that lakes can usually be much bigger than ponds. The basic construction criteria which were discussed in the Introduction apply to all lakes, both geometric and irregular, though the different types of lake were constructed slightly differently, which affected where they could be made, as well as the costs of making them.

As we have seen, the ornamental lakes which evolved in the eighteenth century were very similar to the large fishponds (*vivaria*) of medieval fish production systems, in size, shape and construction, with the technology used in making *vivaria* being transferred to the making of lakes.[1] The general principles governing the construction of both remained largely the same until the nineteenth century, although by the eighteenth century, construction standards had declined somewhat.[2] Making the dam to hold back a large amount of water has always been the most difficult part of making a lake, though evidence about how those dams were actually made is not plentiful. There is some evidence of people who had the expertise to make large fishponds of 1 h or more. Men such as William de Chester and Brother John of Waverley (1247–51) have been mentioned, repairing royal *vivaria* and their dams for Henry III.[3] Also, a tombstone in Gunton churchyard, Norfolk, to James Briggs 'Head Pondmaker', 1709, is further evidence that such highly skilled men existed, and were considered important enough to merit a headstone.[4]

One example where there is concrete evidence about the dam is Bowood, Wiltshire, where Lancelot Brown made a lake for the Earl of Shelburne in 1766. From the accounts, it can be seen that a John Case and his assistant, Richard Darch, travelled for two days to Bowood to 'survey and level' a suitable site for the dam. They also designed the dam itself, calculating the 'necessary angles and resistance to the water', and drew the plan and section of it. For this they charged £9. 1s, which did not include any of the work of building it.[5] A plan for a sluice in a dam by 'Mr. Swindon' in the Bowood archives gives us another name, as well as information about how the sluice and part of the dam were made. Men like these almost certainly appear in estate accounts, but without specific descriptors, and the detail of their function is hidden.

The story of the lake at Bowood continues, however, and provides valuable information about the process of lake-making. It was filled in 1766, but by the spring of 1767, there were problems with the dam. It had already been overtopped in the July of 1766, because of heavy rains that summer. In May, 1768, it was decided that the lake had to be drained, as one side of the dam was starting to collapse, and possibly the whole dam was slipping on its foundations. There was also a problem with the siting of the sluice – it was the wrong (downstream) side of the internal clay wall. We do not know whether John Case, as the designer, played any part in this process or not. Letters between Brown and Lord Shelburne show that finding a solution was definitely Brown's responsibility. There is what could be called a structural report, with plans, analysing the problem and indicating a solution, and this suggests that another expert was called in by Brown, unless he produced the report himself. The problem seems to have taken most of a year to solve as it is not until March, 1769, that we find that the lake was being filled again.[6]

Another early source, Roger North, gives substance to the theory that there were men with specific skills in making large fishponds. Talking about how to make dams, he says,

> The Advantage of Trades, is, that by continual Experience, they find nearer Ways of doing Things, spending fewer Strokes, and less Time, than others can. And in the Conduct of this Work, there is much to be sav'd.[7]

This tells us that 'tradesmen' existed who were skilled in making dams, and that it was cheaper to employ them because they could do things more quickly and efficiently.

North, in *A Discourse of Fish and Fish-ponds*, 1714, describes exactly how to make a dam for these large fishponds. This is important because it is basically a blueprint for the making of dams for ornamental lakes.[8] He makes the proviso that he is talking about areas of the country with clay soils, not sandy soils, and his method is this (Figures 4.1, 4.2 and 1.7):

1   the plan of the pond should be a half oval;
2   the dam should be across the valley to hold back the water [along the straight edge of the oval];
3   a trench should be dug along this straight edge, a foot or two deep [15–30 cm] and rammed full of clay, to stop the water of the pond seeping under the dam [a cut-off trench];
4   on this rammed clay in the trench, a "Wall of Clay" well rammed, must be built up right across the valley, and covered with packed earth ("dug out of the ground where the pond will be") to protect the clay from drying out and cracking, as the dam would then leak;
5   the dam wall must be built three feet higher than the required depth of the pond because the earth will sink however hard you ram it to make it solid.[9]

He also gives the dimensions for dams: a medium fish 'pond' (1.6–2 h/4–5 acres) would need a dam 4.3 m high/14 feet at the centre and at least 15 m/50 feet wide at the base, and the sides must be sloped to give a top of 4.9 m/16 feet wide. This would ensure the dam was strong enough, and could also take carriages across the top, or be planted with trees.[10] Sluices have to be built into dams to control water levels and prevent flood water pouring over the top of the dam and breaching it, and North gives detailed instructions for making them, preferably from one whole

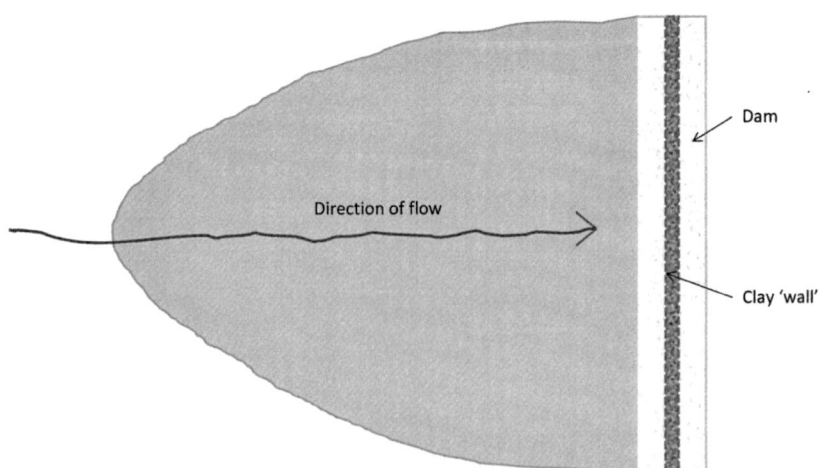

FIGURE 4.1 A plan of a fish 'pond' (*vivarium*), based on Roger North's instructions in *A Discourse on Fish and Fishponds*.

piece of wood.[11] A sluice of this kind, made from elm, was found at Burghley in the 1980s, during repair works to the dam.[12] Sluices also enable the ponds (or lakes) to be drained for maintenance. North's account also illustrates that the of construction fishponds in the early 1700s was basically the same as the one John Grundy illustrated in a 1747 plan of a dam for an ornamental lake at Grimsthorpe, Lincolnshire (Figure 4.2), and clearly shows the link between the construction of fishponds and lakes. Grundy's dam corresponds to a common profile, having a shallower upstream slope (1 foot in 3½ feet) and a steeper downstream slope (1 foot in 2 feet). Sometimes, instead of a clay core inside the dam, a layer of clay was laid on the upstream side of the dam, with the area near the crest of the dam being protected by a layer of stone on top of the clay ('pitched' stone).[13]

Making dams in areas with less stable soils such as sand, was more difficult and in 1600 John Taverner mentions this problem. He says that if your earth "is a light sand, or onely chalke", stakes should be driven into the ground and packed with fine soil [clay], which should be watered as it is rammed, to bind it solidly.[14] He also recommends that sluices should be made of one piece of wood if possible, and any joints packed with tar and [horse?] hair, going into considerable detail about their construction.[15] John Lawrence, who is still talking about this problem in 1806, also gives details of the process of reinforcing dams with stakes.[16] Clearly, the construction of the dam is a complex process, and one with the potential to go wrong, as the danger of leaking is often mentioned. Events at Bowood bear this out.

As we have seen, a major difference between ponds and lakes is that lakes have some form of water replenishing them – usually a stream or river – whereas ponds do not. This points to a second vital difference: you have to put a lake where the water source is, as springs or rivers are necessary for filling lakes, and keeping them full. Ponds, ornamental or functional, are sited where they are *desired* or *needed*, so an ornamental pond might be desired as part of a garden design, as at Longleat (Figure 2.21), or a reservoir tank might be needed to supply ornamental ponds. In the latter case, it might be necessary to channel water to the reservoir from a suitable source, as Stephen Switzer did at Hampton Court, Herefordshire.[17] This kind of channel was known as a water carriage (see below). It would be quite possible that the bottoms of such ponds would need to be 'clayed' all over (lined with clay) to

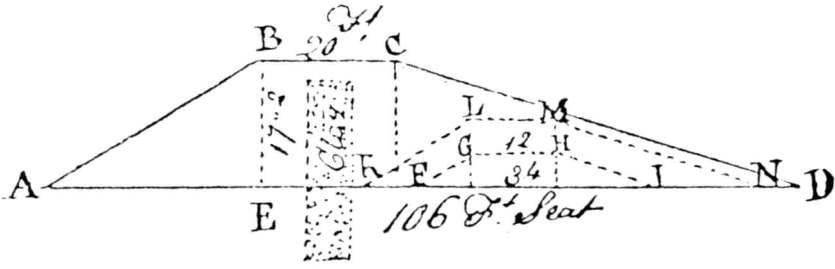

FIGURE 4.2 A cross-section for a dam at Grimsthorpe, Lincolnshire, by John Grundy, 1747.

keep the water in, as Switzer says in *Ichnographia Rustica* 1718. Water piped from a spring could be used to fill ponds, as he explains, and small ponds could be filled from rain-water and run-off (dewponds). Switzer was one of the earliest authorities to write about how to make 'ponds', and the reservoirs to supply them. However, it must be remembered that Switzer, like North, uses the word 'pond' to mean simply an area of water. The word 'lake' was not in common use when he was writing. His information about how to make dams is useful though, and basically the same as North's. He makes the point that accounts he had read of the ways to make such ponds and dams varied in all the books he had consulted, and working practices varied in their details.[18] He also says that "in the West (whence this is wrote,) every Ploughman and Shepherd is able to make good Reservoirs and Ponds for holding of Water"[19] and complains that there is no *written* record of exactly how to make ponds, implying that there was a body of specialist knowledge among workmen which was unrecorded.[20]

A common component of the types of ornamental water systems which Switzer designed was the water-carriage. These were relatively small conduits made of stone or elm wood to convey water from a source to where it was required, to fill ornamental ponds, or reservoirs, or supply other water-works such as cascades or fountains, as at Hampton Court, Herefordshire. He may also have advised John Kyrle-Ernly on the water-carriage for his cascade at Whetham, Wiltshire, in 1712.[21] Later, Enville, Staffordshire, boasted a water-carriage in the 1760s, known as 'the Navigation', which was a water top-up system for the ornamental pools.[22]

Puddling

It is often assumed that the ornamental lakes of the eighteenth century were lined with puddled clay to stop them leaking, but was this actually the case? The question of whether lakes were usually lined is important because of the implications for how big they could reasonably be, how expensive they would be to make, and where they could be made. A point to note is that when clay is being used as a waterproofing agent, it is 'watered' or 'puddled' (mixed with water and kneaded), so that 'puddling' is often used as a shorthand term for lining a pond with puddled clay. 'Clay lining' is a similar shorthand term. The lining of a 6 h lake, for example, with clay would add very considerably to the expense of making it, to the extent that a smaller lake would probably be considered.[23] However, if a lake was lined, there would be much greater flexibility about where to site it – it could be put almost anywhere.

One of the main objections to 'unlined lakes' is that the water would leak out. However, as explained later, lakes are made at the lowest point of a drainage system and water cannot move anywhere but towards the lake. The critical factor is that water is flowing in, as well as out, and though water may be lost through percolation into the adjacent rocks, water levels can generally be maintained by closing sluices.

FIGURE 4.3 At Stourhead, Wiltshire, the course of the stream can be seen winding along the lake bed, 2016.

Because lakes have a stream or river running through them, currents are present in the lake which would erode a clay lining. A picture of the drained lake at Stourhead (Figure 4.3) illustrates this. Whilst these currents would not have such a direct scouring effect when the lake was full, they would still have an effect over time.

There is significant confusion about the use of clay (puddling) in the making of ornamental pieces of water. In particular, the 'claying' of the dam has been confused with lining the lakes themselves with clay. As we have seen, North recommended using puddled clay in the construction of a dam as a central vertical core – a 'clay wall' – inside the dam itself.[24] Switzer's account in *Ichnographia Rustica* is another source of confusion, and what he says is pertinent to the question of whether lakes were usually puddled or not. Talking of reservoirs or ponds, he says:

> If they are cut out of the whole Ground, they are commonly circular, and ought to be well clayed, except [unless] the Hill abound with Water. But it may be possible there is some Hollow or Valley in the Hill; then a Head [dam] made with the sinking, widening, and clearing of it will do, and save a great deal of Money; but there should be a Trench dug down in the middle of the Head [dam], about a foot wide or wider, and some strong Clay well ramm'd down, or else the Water will soak away thro' the Head; and this Trench ought to be cut down lower than the Bottom of the Reservoir or Pond.[25]

What Switzer is saying here is that where you have a valley, or hollow, supplied by a stream or river, you can get away with clearing the ground and building a dam across the valley. As long as you put a clay core wall inside the dam, with a cut off trench, it will hold water without the pond being lined with clay. In fact, though Switzer does not specifically say so, this method would be appropriate for large or small pieces of water. However, if there is no stream or river to refill the 'pond', then the best practice is to 'Clay all Over' the bottom and sides of the area of water. As a caution, he says:

> There be Some who affirm, that there is no need of Claying all Over, but only the Sides, and this doubtless may do where-ever there is any Layers of Clay, or Clayey Gravel under the bottom of your Pond, which often-times naturally happens, or if the Spring lies near; but if it be a deep, loose Sand or Gravel, or if it be towards the Brow of a hill, or toward the Ground, I doubt [think] it ought to be Clayed all over, even if the sides were Brick, [as] has been commonly used: Yet 'tis certainly best to Clay the Bottom, and that with extraordinary good Clay, such as has been prov'd; and if it were twelve or fourteen Inches thick, still the better.[26]

It must be remembered that Switzer is talking about relatively small areas of water, and that by 'reservoir' he means a tank, not anything like the canal or public water-supply reservoirs of the late eighteenth century. In *Hydrostaticks* 1729, for example, he gives dimensions for reservoirs or basins as 2–2.4 m/7–8′ deep and 0.1–0.3 h or 100–200 feet square, so he is talking about pond-sized pieces of water, not lakes.[27] To summarise: Switzer says, if a water supply is abundant in a valley, you can simply put a dam across the valley and make a piece of water, as long as you include a clay wall inside the dam, plus a cut off trench, but where there is no such water supply, the best practice is to line the intended site of the water completely with clay. These points are important because Switzer's instructions have been interpreted by historians in the past as instructions for lining lakes with clay, to waterproof them, which is not what he meant, and this is taken up in the following.

The use of the word 'pond' for any piece of water also adds considerably to the confusion. North and Taverner are talking about fish 'ponds' of several hectares which are being made on *feasible* sites, with a water source running through them, or a spring, constantly replenishing them. Generally, Switzer's 'ponds' were ornamental garden ponds and would have been significantly smaller than the 'ponds' which North and Taverner were discussing, and this is where confusion has arisen.

Two other original sources, Richard Woods's work at Cusworth, Yorkshire, and John Grundy's reports on Grimsthorpe, Yorkshire, in the 1760s, give us more insight into the question of puddling. Woods' detailed plans, sections and notes give more information about the use of clay in dams, and show that he was thoroughly conversant with best practice for making dams: sinking a cut off trench for the internal clay wall into the bedrock, or an equally firm foundation. Woods himself said the cut-off trench must be:

> sunk until you come to a solid and firm bottam either in a close gravel or sand or clay, and I case you should be obliged to sink 3: 4: 5: or 6 feet [1–2m] below the bed of the water before you come to such a bottam you must have patience, and persue it till you are sure you are safe. . . . Let the clay be put in thin courses not more than 6 or 7 inches [15–18cm] at each course, and well ramed.[28]

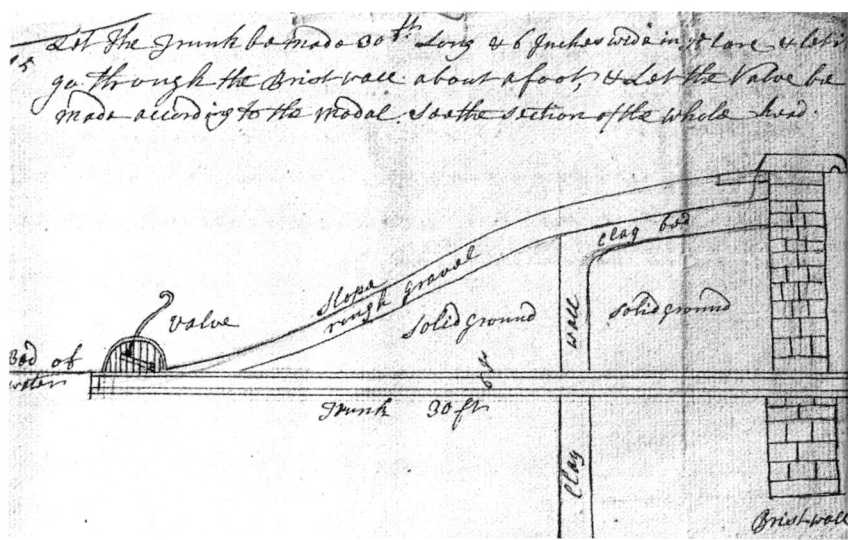

FIGURE 4.4 A cross-section of the dam at Cusworth, Yorkshire, 1764, by Richard Woods. The water is flowing from left to right.

He also made the customary arrangements for coping with flood water or draining the lake by specifying a barrel sluice under the dam made from timber planks, as the cross section shows (Figure 4.4), although the design is more like a weir than the usual dam. How the 'valve' was to be accessed is a mystery, as it would have been under the water. Perhaps it had to be fished for, as did Brown's at Burghley.[29] What is significant is that the clay is being used as a 'wall' inside the dam, and extends into a trench below it – a cut off trench – to prevent the water passing through or under the dam.

John Grundy, junior, who had worked as an engineer on Deeping Fen, Lincolnshire, in the 1730s, made the Great Lake (16 h) for the Duke of Ancaster at Grimsthorpe, Lincolnshire, in 1748.[30] He used the clay core method for his dam, like Woods did, but when the Duke of Ancaster subsequently asked him to extend this lake, his report of 1766 shows that he was very concerned about whether the site would allow this as the estate is mainly on limestone, and local workers told him about 'chasms' (swallow holes) in the area.[31] In *Another Scheme for Enlarging the great Piece of Water at Grimsthorpe* Grundy wrote:

> the present great Piece of Water, will be enlarged more than 20 Acres [8 h] with this most ornamental and valuable advantage that no termination thereof will appear from the House but it will have the beautiful effect of a very large river, running quite through the Park . . . [but the biggest problems] are that the greatest part of the Ground on which it is proposed to be executed is Chasmny and full of Swallows so that without some very careful

and effectual means are used to stop them the Ground cannot be made to hold Water.[32]

The implication here is that "careful and effectual means" were not normally necessary to make lakes hold water. He initially proposed an enormous cut-off trench across the valley but finally favoured a full clay lining for the lake, which he referred to as an 'artificial bottom', as the only fail-safe measure for retaining the water. However, he wrote, "I am apprehensive that there is sufficient Quantity of such loomey Clay to be got on the sides of [adjacent to] the Work which will greatly facilitate this Business".[33] He then quoted a price of £1733. 10s, or *c*. £177,000 today (2017 conversion figure) for the work, which was never carried out, presumably because it was too expensive.[34] Grundy's use of the term 'artificial bottom' is very interesting. He gives detailed instructions for making it, which suggests that this was not at all a routine procedure – there would have been no need for such detail if it was routine.

The lakes which Lancelot 'Capability' Brown made at Petworth, Sussex, for the 2nd Earl of Egremont in *c*. 1757 are a further source of confusion over the use of puddled clay. The main lake is known today (2020) as Upper Lake, and the smaller lake is known as Lower Lake. Brown made four contracts with Lord Egremont between 1753 and 1756 and several pieces of water are mentioned:

> the **1753** contract *includes* making "the Horse Pond";
> the **1754** contract includes a receipt for "a Plan for the Lake in the Park near the Half Moon Wood – £3. 3s";
> the **1755** contract mentions various works, including enlarging a pond, to make it 2,460 feet round, for cattle, with 'Clay Walls' and pitched sides;
> the **1756** contract is solely about making "the intended Piece of Water which is to be made in the flat part of the park".

Terminology is important here. 'Pitching' meant to cover a surface with stone, to protect it, often from stock, in the case of ponds or lakes. 'Clay Walls', in the context of making lakes, means the clay core wall inside the dam. Roger North throws light on this term:

> Now first, for making the Bank or Head [dam], you must be sure it is tight, and that it do not sew or leak . . . therefore a Bed or Wall of Clay the whole length of the Bank, must be carry'd up with good Ramming, from a Foot or Two below the Surface of the Ground.[35]

'Piece of Water' is a term which usually refers to a significant body of water, for example the lake in a park. Similar terms were 'Great Water' and 'Great Pool'.[36]

The article in the 1753 contract says "To now make the Horse pond in all its parts, the Leaden Work excepted". This cannot relate to the making of the main lake (Upper Lake) as there is so little detail for so large a project. Secondly, the half-moon shaped pond (3 h, Figure 4.5) in the north of the park appears to be

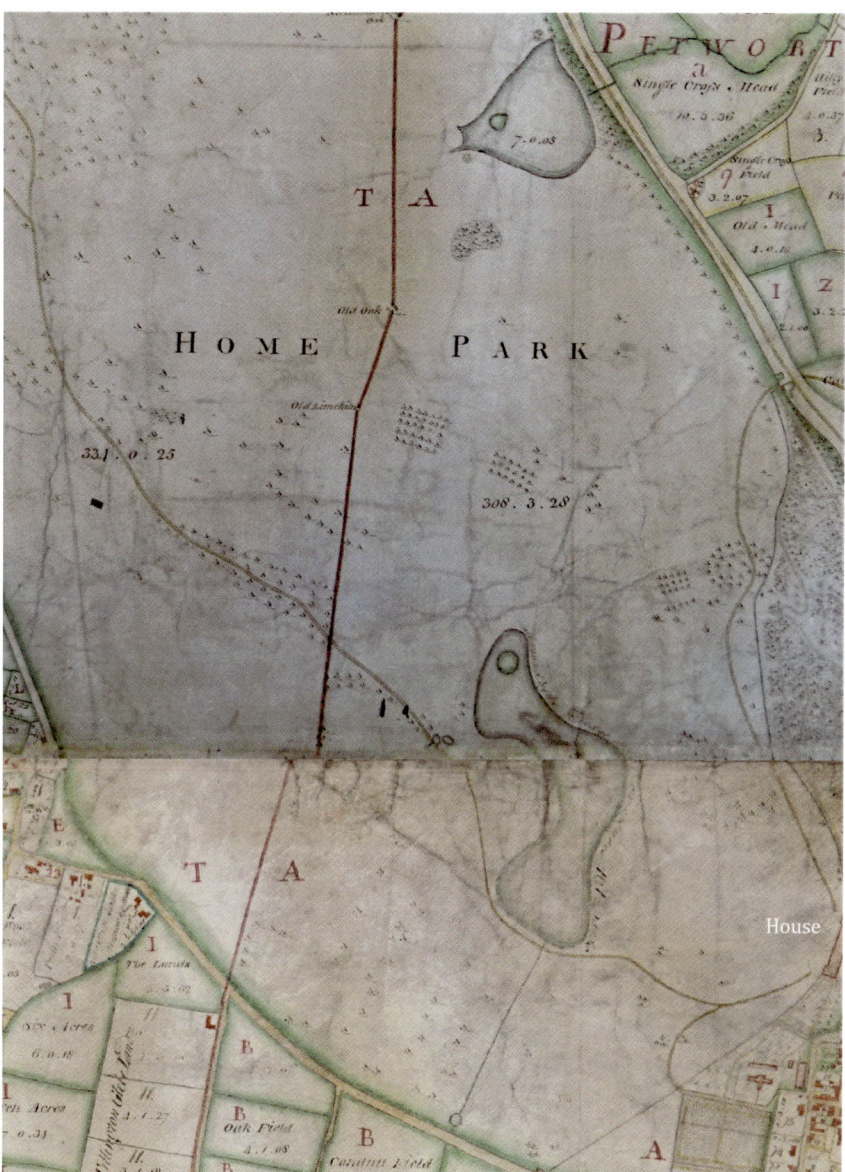

FIGURE 4.5 Part of the 1779 Crow survey of Petworth Park, Surrey, West Sussex Record Office. *Courtesy of Lord Egremont. Ref. PHA 3606.*

the subject of the 1755 contract as it has a perimeter *c.* 2,417 feet today. The first article, which deals with this pond, is as follows:

> To enlarging the Pond according to ye Stakes put in for ye Purpose & digging out all such Parts as are not deep enough, (making the shallow Places three feet & a half) & making all the necessary Clay Walls, & levelling the Bottom of it, and pitching the Sides which are 2460 feet round to prevent the Cattle from Damaging it, as likewise to turf the Edges of the Water & to lay in the Plug.[37]

A small pond had existed on this site since at least 1610 and Brown was commissioned to enlarge it considerably, constructing a dam (clay walls) where necessary, making a sluice (plug) as well as levelling the bottom.[38] From this it is clear that Brown was not lining the bottom with clay. The 1756 contract, unlike the others, does not deal with anything else but only the construction of "the intended Piece of Water which is to be made in the Flat Part of Petworth Park" and the first article is "To make a secure Head" (dam).[39] The perimeter of this piece of water was 4,854 feet (1618 yds.) in 1779.[40] This, with all its detail, almost certainly relates to the making of the main lake (Figure 4.5, bottom), and no clay is mentioned in it. Also, this half-moon shaped lake does not appear on Brown's 1752 design plan for Petworth.

Brown's contracts (not just at Petworth) are quite specific, giving numbers of carts and horses, for example, as well as harness and fodder for them. What emerges is that there is no mention of clay, other than the 'Clay Walls', and very large amounts of clay would have been necessary to line the 6 h lake. Even if clay from the lake bottom was spread up the sides of the lake to line it, where is the labour and equipment for doing this? Brown was careful to itemise anything which 'My Lord' was providing, wherever he was working. At Grimsthorpe, Grundy doubted if there would be enough clay adjacent to the planned lake (8 h) to line it.

The situation at Stowe, Buckinghamshire, is somewhat different but equally illuminating. The Copper Bottom Lake (*c.* 0.7 h) has recently been repaired with Bentonite matting to stop it leaking (2016).[41] Adjacent to the Eleven Acre Lake, it crosses a limestone seam and has given trouble in the past: the name supposedly derives from attempts to plug previous leaks with copper.[42] There are similar problems with New Water (*c.* 1 h). The fact that these lakes have repeatedly leaked suggests that they were not lined with clay. The slim shapes and small sizes of these lakes, compared with the Eleven Acre Lake (4.5 h) at Stowe, suggest that the makers were aware that larger lakes would not be feasible in this location. If they were lined with clay, presumably problems with currents scouring and/or the limestone seam led to that lining failing.

The importance of the water source – the replenishing mentioned earlier – is illustrated by Brown's unsuccessful attempt to make a lake or pond in the Grecian Valley at Stowe for Lord Cobham. Nattes' view (Figure 4.6) appears to show the shape of the lake bed which Brown made. He moved some 23,500 cubic yards of

FIGURE 4.6 The Grecian Valley, Stowe, Buckinghamshire, 1805, by J. C. Nattes. *Courtesy of Buckinghamshire County Museums Collections.*

earth to make it[43] and in 1746/7 was trying to make an oval lake or pond, according to Cobham's instructions:

My lord,

　As to finishing the Head [dam] of the Oval . . . indeed I think it would be better not finishing this season, I thinking that a sumer's talks and Tryels about it may make it a very fine thing. The Springs fill the Oval much about a barleycorns head a Day. I can only add that my hope is still biger than my fear that your Lordship will see it full.[44]

(A 'barleycorns head' signified one third of an inch, or *c.* 8.5 mm.)[45]

It would appear that an oval pond or lake was planned. Clearly, Brown had doubts about the sufficiency of the water source and wanted more time to address the problem, though he says that he hopes, on balance, to do so successfully: the site was further up the valley of the Elysian Fields where the Worthy River had been dammed to create irregular ponds. This demonstrates the importance of the water source for making lakes. If Brown had lined it with clay, he would not have been experiencing such difficulties retaining the water. Perhaps this was one of a number of places where achieving a satisfactory lake would take several years, or may not have been feasible, as the bedrock is limestone with superficial deposits of diamicton.[46]

Because lakes are made by damming a stream or river in a valley, by definition the stream or river is at the bottom of the local drainage area. Figure 4.7 illustrates this. The significance of this is that if any water should leak out of the lake, it would not be a large amount because the direction of flow of sub-surface water would be towards the lake bottom.[47] Some percolation into the rocks and soil surrounding the lake will occur, but significant amounts of water cannot pass out of the lake, because all the ground water is draining towards the lake.[48] The water level in the lake would remain largely constant after some initial loss. The rate of percolation depends on variables such as the depth of the lake above the water table, and the permeability of the rocks of the valley sides. In times of drought, lake levels will fall, as surrounding strata become drier, and more water from the lake percolates into them. Lake levels do then tend to drop, but the river or stream constantly replenishes the lake, although this may not keep pace with the loss in times of severe drought.

The construction of industrial reservoirs is also relevant to the discussion of puddling, as knowing how they are made throws light on how lakes in the eighteenth century were made. In the same way that the construction of medieval fishponds informed the construction of eighteenth-century lakes, in turn those lakes influenced the construction of the reservoirs which were beginning to be built in the second half of the century, to supply canals, and also water for towns. These reservoirs were built by men like Robert Thom, John Rennie, Thomas Telford and James Jardine and were generally bigger than the ornamental lakes of the time but, to date, the evidence suggests the construction methods were the same. Seeswood Pool, Warwickshire, was one of the earliest reservoirs, begun in 1764 by Sir Roger Newdigate, to supply canals for the coal industry on his Arbury estate, and is very much like an ornamental lake or fishpond, on the northern edge of Arbury Park.[49] It was enlarged to 7.8 h in 1777, so a similar size to the lakes Brown was making at that time.

Another early industrial reservoir was the Glencorse Reservoir, built in 1818–23 by Thomas Telford and James Jardine to supply Edinburgh with water, with a 23.5 m high dam. What Norman Smith says about it is particularly relevant:

> The type of earth dam built in Britain for canal reservoirs was very much the one adopted for water-supply works. . . . During its working life the dam has not, it seems, experienced any problems or required much maintenance. . . . Several anxious months passed during 1821 before bed-rock was reached and a solid base located for the dam's puddled clay core wall.[50]

FIGURE 4.7 Drainage patterns. The river is at the bottom of the valley, and all the subsurface water in the drainage area is moving in the direction of the arrows. The dotted line is the planned lake surface and the solid line is the original river.

Glencorse Reservoir was 21 h, so comparable to a large eighteenth-century lake, and was built with the same type of dam, but nowhere does Smith (or G. M. Binnie) mention a lining for lakes or reservoirs. In this context, a statement by Andy Hughes, Chairman of the British Dam Society, is important:

> It is very rare to line a lake with clay; the site is chosen because we have had glaciated conditions and generally the water table bends down into the valley bottom so the water does not escape unless the geology is sloping away from the lake to the next valley – very rare.[51]

No evidence has come to light that reservoirs, then or now, were lined, and this is strong evidence that ornamental lakes were not usually lined with clay.[52] Indeed, this highlights the progression from the *vivaria* described by North, who also does not mention any clay lining, to the lakes made by Brown *et al*, ending with the reservoirs of the late eighteenth and nineteenth centuries, which were often vast. Although dam building techniques changed in the nineteenth and twentieth centuries, in terms of the materials used and the shapes of dams, the basic principles remain constant: raising an impervious wall across a valley, which also extends down into the bed-rock, to retain the water supplied by a river.

In summary, we can see that the idea that ornamental lakes were lined with clay in the eighteenth century is largely inaccurate, and it has arisen because there has been much confusion over the use of clay. Clay was almost always used in the construction of dams, but sometimes this idea has been 'transferred' to the lakes themselves, or primary sources have been misread, and it has been assumed that lakes were lined with clay. Ornamental lakes in the eighteenth century were made where the geology and topography were suitable, and they were not usually lined with clay. It is possible that a few lakes were lined – at great expense – but little evidence of them has come to light so far.[53] Furthermore, the realisation, based on the knowledge of how to make fishponds, that it was unnecessary to line large bodies of water with clay, may have been one of the key factors in the evolution of irregular lakes: men could afford to make pieces of ornamental water much larger than before.[54] A similar increase in size of geometric lakes would have been unfeasible owing to the expense of moving lots of earth to create the straight sides, and so large bodies of ornamental water – lakes – had to be basically irregular.

Construction of the different lake types

The basic construction criteria we have looked at apply to all lakes, both geometric and irregular, but the different types of lake were suitable for different types of site, and this influenced where they could feasibly be made, as well as the costs of making them. As we have seen, lakes fall into two basic categories: geometric and irregular. Geometric lakes (Figure 1.1) are fully geometric and symmetrical in plan view, whilst semi-geometric lakes are made with straight sides or arcs, but are asymmetric (Figure 1.2). Irregular lakes are completely irregular, and constructed with dams which tend to be straight or gently curved (Figure 1.4), whilst river-lakes are sinuous and constructed with weirs (Figure 1.5). Hybrid lakes fall between the two main categories, having at least two straight sides, plus irregular edges (Figure 1.3).[55]

Geometric lakes, such as those depicted at Bretby, Derbyshire, (Figure 4.8) in 1708, had straight sides (or an arc) and one or more of those sides might be acting as a dam, especially if the land was fairly flat. At Bretby, the rhomboidal 'pond' was in a slight valley falling towards the top of the picture, so the further, 'top' edge was a dam, and quite possibly the two adjoining sides were as well, albeit low, bank-like dams. In fact, this 'pond' at Bretby was a geometric lake of 2 h, the site of which is still visible on today's OS maps, and the rhomboidal shape reflects the opening out of the valley it was in. The square piece of water in the foreground was also large – probably at least 1 h. The fishponds which occupy the valley to the east of the house today, are just off the right-hand side of the engraving, but were not deemed worthy of inclusion in the picture. Because geometric lakes often required more than one dam and a considerable amount of earth-moving on anything other than virtually flat sites, they were more expensive to construct, hence their position at Bretby, in the shallower valley to the left (west) of the house, rather than in the deeper valley to the east. If an irregular lake had been made at Bretby in the later eighteenth century, it would have been in the valley to the right of the house, where the fishponds were. Because geometric lakes were relatively expensive their sizes and numbers were usually limited, and smaller areas of water were made, as Table 2 shows. By the end of the 1730s, it was unusual to make a geometric lake, Temple Lake at Enville, Shropshire, in the 1740s being an 'outlier'. Evidence about geometric lakes is not plentiful as the new fashion for irregularity later in the century often led to them being altered, or completely over-written by irregular lakes.

Semi-geometric lakes were constructed along the same lines, but not being symmetrical, were likely to be less costly to make because they could be adapted so that the

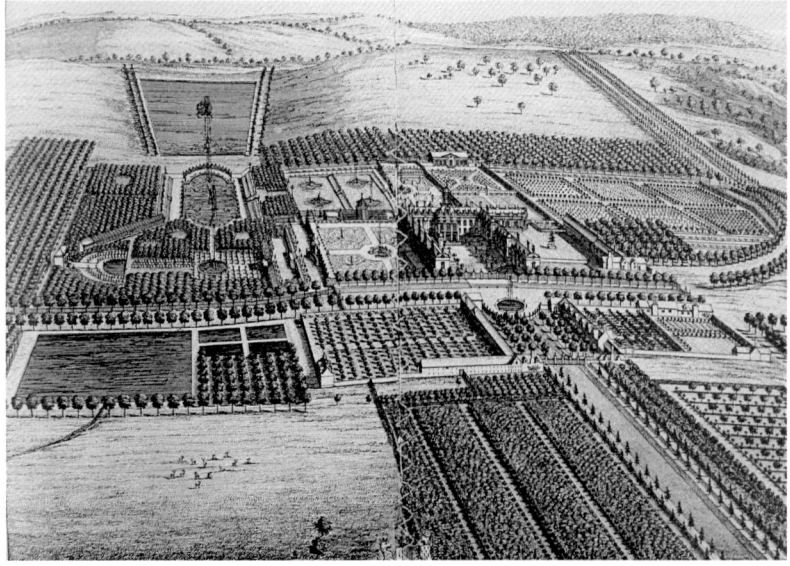

FIGURE 4.8 Bretby, Derbyshire, depicted in *Britannia Illustrata* 1708. *Historic England.*

natural terrain acted as a dam on various sides. One dam at least was required to retain the water though, and Wolterton (Figure 1.2) is a good example. As they were potentially slightly less expensive, these lakes were likely to be bigger (see Appendix I). By the 1720s–'30s, irregularity was beginning to appear in lake forms, and the hybrid lake became more common, remaining a feasible alternative into the late nineteenth century.

The **hybrid lake**, with some irregular sides and two or more straight sides, was one of the most adaptable lake forms both in terms of topography and fashion. Wollaton, made *c*. 1774–85, is a good example (Figure 4.9). A hybrid lake could be made reasonably easily on flattish sites because two low dams could be used, instead of one large dam, and low dams are easier to construct. However, on undulating sites, as at Bramshill (Figure 4.19), larger dams would be required. Hybrid lakes did not necessarily have more than one dam because sometimes it was possible to create a straight side where the lake abutted a slight rise in the ground, and this appears to be the case at Wollaton. The dam is at the south-west end, whilst the ground rises slightly on the north-west and north-east sides, acting as a dam. On the south-east side, the lake abuts against an outcrop of the Lenton Sandstone Formation.[56] It was this possibility of leaving the remaining 'sides' irregular which generally made hybrid lakes less expensive to construct than geometric lakes.[57]

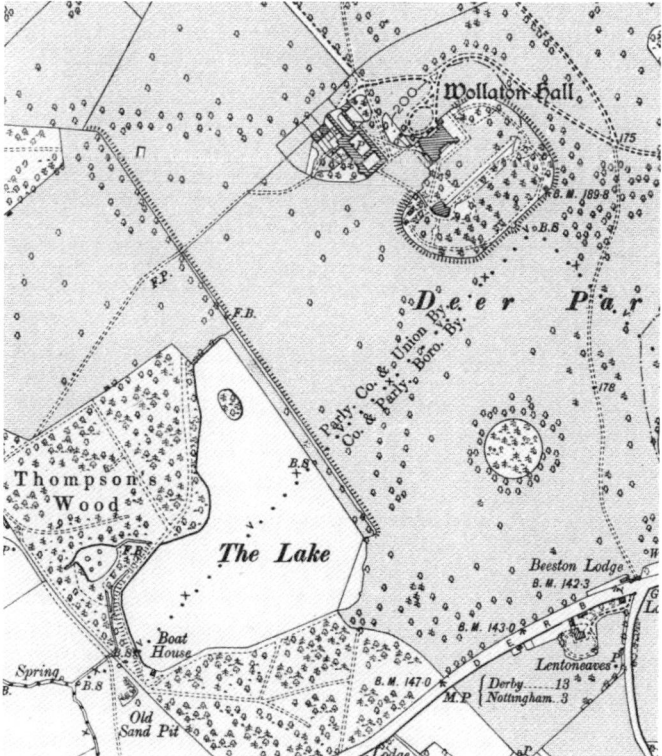

FIGURE 4.9 The hybrid lake at Wollaton Park, Nottinghamshire, OS 6" map, 1899.
Reproduced by permission of the National Library of Scotland.

Stylistically, hybrid lakes were a combination of a geometric and an irregular lake, with the potential for a geometric lake to have two sides deformalized to make it more fashionable. The relative scarcity of maps from this period (1700s) limits our knowledge, and we do not know how often this was done. The lake at Raynham Park, Norfolk, appears to be an example of this (Figure 4.10). The sides

FIGURE 4.10 A 1758 estate map of Raynham Park, Norfolk. The house is at the top and north is to the left of the plan. The water shown is the same as that on a *c.* 1730 map of the estate. *By kind permission of Viscount Townshend.*

of the lake are straight (the dam is on the left side), except for the south-western edge (bottom of the plan), which is shown as a series of symmetrical arcs, suggesting a semi-geometric intention. This lake dates from at least the 1730s, as all the water features – the lake, the canal supplying the pump engine, and the moated garden – appear in the same form on an estate map of that period. By 1838, when the tithe map was produced, the south-western shore of the lake is depicted as less regularly geometric (Figure 4.11). This is also the side on which the land begins to rise gently, and begs the question: was the lake ever as geometric as it was depicted, or was it more difficult, and therefore more expensive, to make it geometric on that side? Or, was the lake shore deliberately made more irregular at a later date? Unfortunately, we will probably never know, and this kind of ambiguity also highlights the lack of information and the subjectivity involved in classifying lakes in this way; nevertheless, the classification is useful in identifying commonalities. The lake at Trentham may be another example of this process as in *c.* 1700 the sides of the two pieces of water are shown as straight though asymmetrical (the dams are on the left side, Figure 4.12). By 1727, when the Coppy Map was produced (Figure 5.23), the western shore of the western lake was depicted as a series of irregular curves. This is also the side on which the land gets higher, and again the question arises: was the lake ever as geometric as the *c.* 1700 map suggests? Another factor, which applies to all lakes, is that they tend to silt up over time, particularly if they are on fairly flat sites, unless they are maintained regularly, so there may be a tendency for

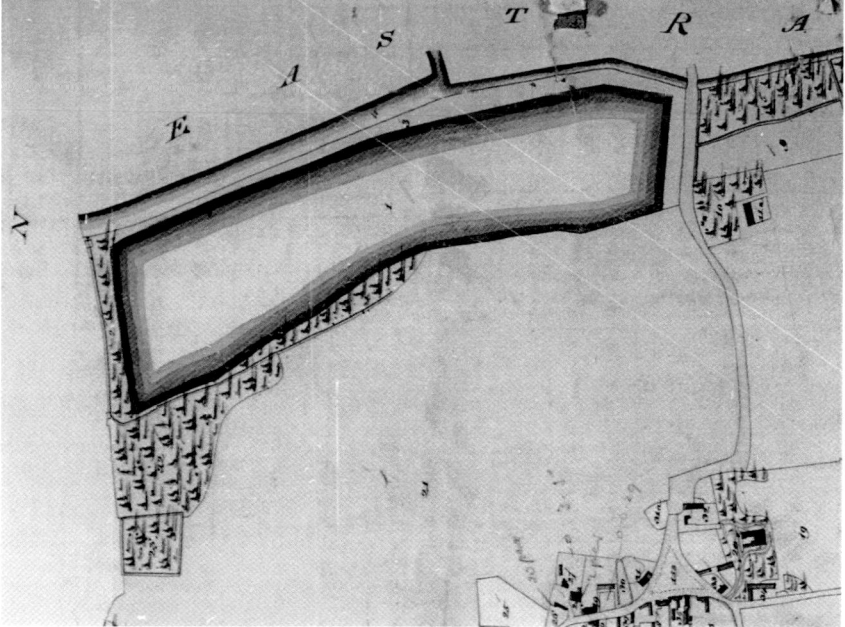

FIGURE 4.11 The tithe map for West Raynham, 1838. © *Crown Copyright Images reproduced by courtesy of The National Archives, London, England. www.NationalArchives.gov.uk and www.TheGenealogist.co.uk.*

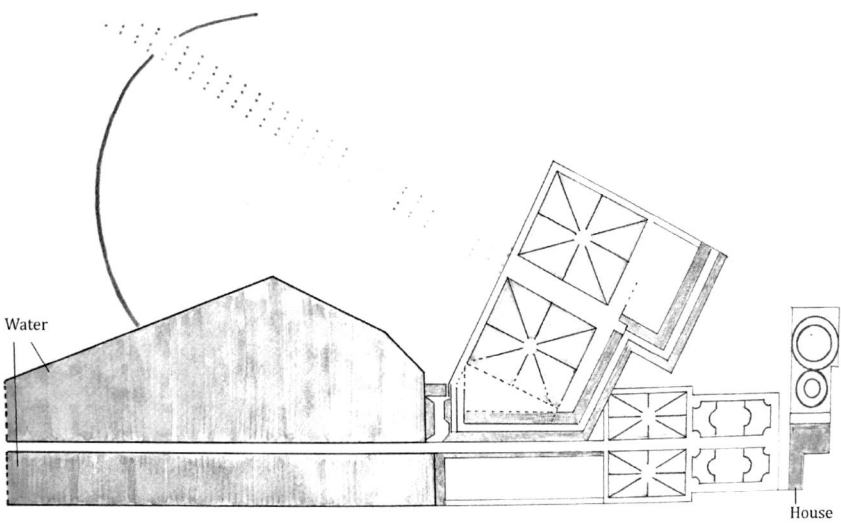

Water

House

FIGURE 4.12 A diagram of pieces of water at Trentham Park, Staffordshire, from a plan dating from the 1700s. North is to the right.

straight sides to become less geometric. Silting also means that lakes may become significantly smaller, which is the case at Raynham. The First Edition 6" OS maps are particularly good at indicating this silting up.

Irregular lakes were usually constructed in a valley by putting a dam across the direction of flow of the river or stream, and allowing the water to pond back behind the dam to produce a contour lake. This method produces a completely irregular shoreline, and is marked by its complete lack of geometry. The only straight part might be the dam itself, and this was often disguised – with planting or an adjacent island. Dams might also be slightly curved, though a marked degree of curving would lead to an earth dam of this kind being weaker. It was quite common to redesign the shoreline: Brown often graded it very carefully, in parts or completely, and Edward Kemp, whose lakes were often on flatter sites, excavated parts of the shoreline to achieve a totally different shape, one which might be characterised as spreading (see Chapter 7). On virtually flat sites, with a high water table, it was sufficient to scoop out the earth to form an irregular lake, using it to landscape the adjacent areas, or make islands, as Kemp did at Birkenhead Park in the 1840s (Figure 7.15). However, irregular lakes were often made in river valleys which were deeper and as a result significant dams were required, often 3–8 m high, unlike fishponds, which tended to be relatively shallow (it was easier to monitor and catch the fish). As we have seen, these higher dams were more difficult to build successfully, as they were largely made from earth, and relied on the weight of that earth to hold back the water.

Having built a dam across a valley, the shape of the resulting lake largely depended on the profile of the valley. A shallow valley would produce a spreading, wide lake, like the one at Kimberley, Norfolk (Figure 1.3). The deeper the valley, the narrower would the lake tend to be, as at Fonthill, Wiltshire. These lakes had one dam; if there was more than one dam, a hybrid lake would be formed. Various terms are associated with irregular lakes, such as 'long water', 'broad water' and 'serpentine', but these are descriptors, not definitions. 'Irregular lake' is something of an umbrella term: it incorporates lakes of very varied shapes, but the common concept of an 'informal' lake – a lake with sloping banks, made by damming a stream or streams and following the course of the valley like a natural lake – turns out to be largely accurate.

River-lakes were narrow and sinuous, and very river-like. Hitherto, this type of lake, with its special characteristics, has gone unrecognised, perhaps because the method of their construction has not been appreciated. They were made by building one or more weirs across a river to pond it back, creating a lake rather like a thickened river. The most important characteristic of the river-lake can be seen by examining the beginning and end of the river-lake, and comparing it with the river above and below those points (Figure 1.5): it is wider but not dramatically so. It is also much the same general width from start to finish, and again, does not seem markedly different from the original river. Horton, Northamptonshire, is a good example. The rectangular pieces of water at the west end of the lake are probably the remains of a moat, or of two ponds in a formal garden made by the Montagus in the seventeenth century, and re-used to form part of the lake made by George Montagu (2nd Earl of Halifax, 1739) in the mid-eighteenth century.[58]

At first glance, river-lakes and narrow irregular lakes may seem very similar, but river-lakes are made using weirs to pond back water, whereas irregular lakes use dams. Both types of lake rely, to differing extents, on utilising an existing water course. However, a person making an irregular lake expends greater effort and money building dams. They are also more complex to make, involving sluice gates, or other spillways. Weirs are cheaper and easier to build than dams (see later) because they are usually much lower. They do not usually retain a great amount of water, and so the shape of the lake they make is similar to the course of the original river, as we have seen at Horton. River-lakes can be constructed in valleys where the fall of the river is gradual. They are not suitable for areas where there is any significant fall as a dam would be necessary to retain the water, and a different kind of lake would be produced, such as a hybrid or irregular lake. There are two great advantages in constructing a river-lake: the water supply is ensured and proven and, as a result, the cost is relatively low.[59]

River-lakes did not appear until the 1750s (see Appendix II). It is not known who the 'inventor' was but Brown was certainly an early practitioner, with the possibility that he produced a plan for one at Wallington in *c.* 1750.[60] Generally, river-lakes were not large, 1–2 h being common, as they were usually narrow. However, size did depend on length (often determined by park size) and where the river was sizeable, the result could be a large lake, as with the River Derwent at Chatsworth: 10 h.

The intention in making a river-lake was to create a body of water which was sufficiently large to pass as a lake, and to act as an ornament, perhaps reflecting the house, whilst not going to the considerable expense and labour of building dams. In some instances, some earth moving to enhance the river-lake may have been undertaken, as at Chatsworth, where hydrographic factors also had to be taken into account (see below). There, Brown put in a weir to service a new mill in 1760–1.[61] He then embellished this scheme in the early 1760s, putting in a second weir, at a cost of £239 12 s 1 d, further upstream to pond back the river and make it wider, as part of a scheme of landscape improvements commissioned by the 4th Duke of Devonshire.[62] The eastern river bank was dug away immediately above the weir (*c.* 700 m south of the house), to widen the river more and enhance the impression of being a lake. Horatio Walpole mentioned this in 1760: "The Duke is widening it [the river] and is making it the middle of his park".[63] Clearly, the river-lake was the focus of the newly improved landscape, as an engraving of 1779 by Watts shows (Figure 4.13). The house sits amid smooth lawns graced with scattered trees and clumps in the approved landscape style, conforming to the Brown formula of a house show-cased by lawns, relieved by trees, and set off like a jewel by the water. The use of a weir produced contrasting effects for the viewer: still, calm water above the weir, and the excited, noisy rushing water as it passed over the weir and beyond. It is unlikely, however, that this effect was the primary reason for constructing a

FIGURE 4.13 Chatsworth, Derbyshire, engraving by W. Watts, 1779. *Courtesy of Robin Simon.*

river-lake. The most cogent reason was probably topography (a gentle fall of the river bed) plus the proximity of the house to the river. An irregular lake created simply by putting a dam south of the house would have brought the water close to the house, flooding Queen Mary's Bower. Furthermore, the River Derwent absorbs spates from the adjacent moors,[64] and weirs are more suitable for coping with erratic flow than dams, as sluices do not have to be opened and closed to control the flow. The river-lake was therefore a stylistic choice governed by the topography and the site of the house.

Few river-lakes were made before *c.* 1760, and Brown's lake at Belhus (mid-1750s, Figure 5.11) is apparently one of the earliest.[65] The Chatsworth lake was in place by 1763 so it is possible that the idea had occurred to Brown when he was working at Chatsworth in the late 1750s to early '60s. However, it is difficult to assess the 'lead time' of projects such as these, and the basic idea for Chatsworth may have been discussed some time earlier. On river courses with a greater fall, weirs would not work to pond back water as they would have to be higher or very numerous, and weirs are not robust enough to be high and stable.[66] If the fall of the land is significant, or a larger body of water is required, a dam becomes necessary.

Weirs

A number of weirs may be necessary to make a river-lake, and at least two were usually required. The lake at Kedleston, Derbyshire, shows this clearly (Figure 4.14). The weirs are labelled on the OS map, and the bridge was constructed on a further weir (Figure 4.15).

Though weirs are less complex to construct than dams, they still require some expertise. Figure 4.16 shows the general principle: a low barrier is made to pond back water, thus raising the water level and widening the river. The vital difference between a dam and a weir is that a weir is designed so that the water overtops it, whereas a dam must not on any account be overtopped, as this might lead to it being breached – the current would begin to erode the dam itself, causing it to start crumbling. Because dams usually retain much larger bodies of water, a failure would be catastrophic.[67] Their construction takes account of this, as we have seen, and their shape is different: sloping on both sides to withstand water pressure, with sluices to prevent overtopping. Weirs do not need sluices, and special attention is paid to the construction of the crest to prevent water breaching the weir through constant friction, as Figure 4.17 shows. The weir is largely made of brick and stone, in contrast to an eighteenth-century dam, which was largely made of earth.

A weir has to be somewhat higher than the natural river level. The water is ponded back, and the river spreads out upstream, whereas a dam is significantly higher than the natural river level and significantly wider than its natural course. In practice though, there may be little difference between a lake produced by a high weir or a low dam on a fairly flat site, as at Kimberley (retained by a weir, Figure 4.18). The

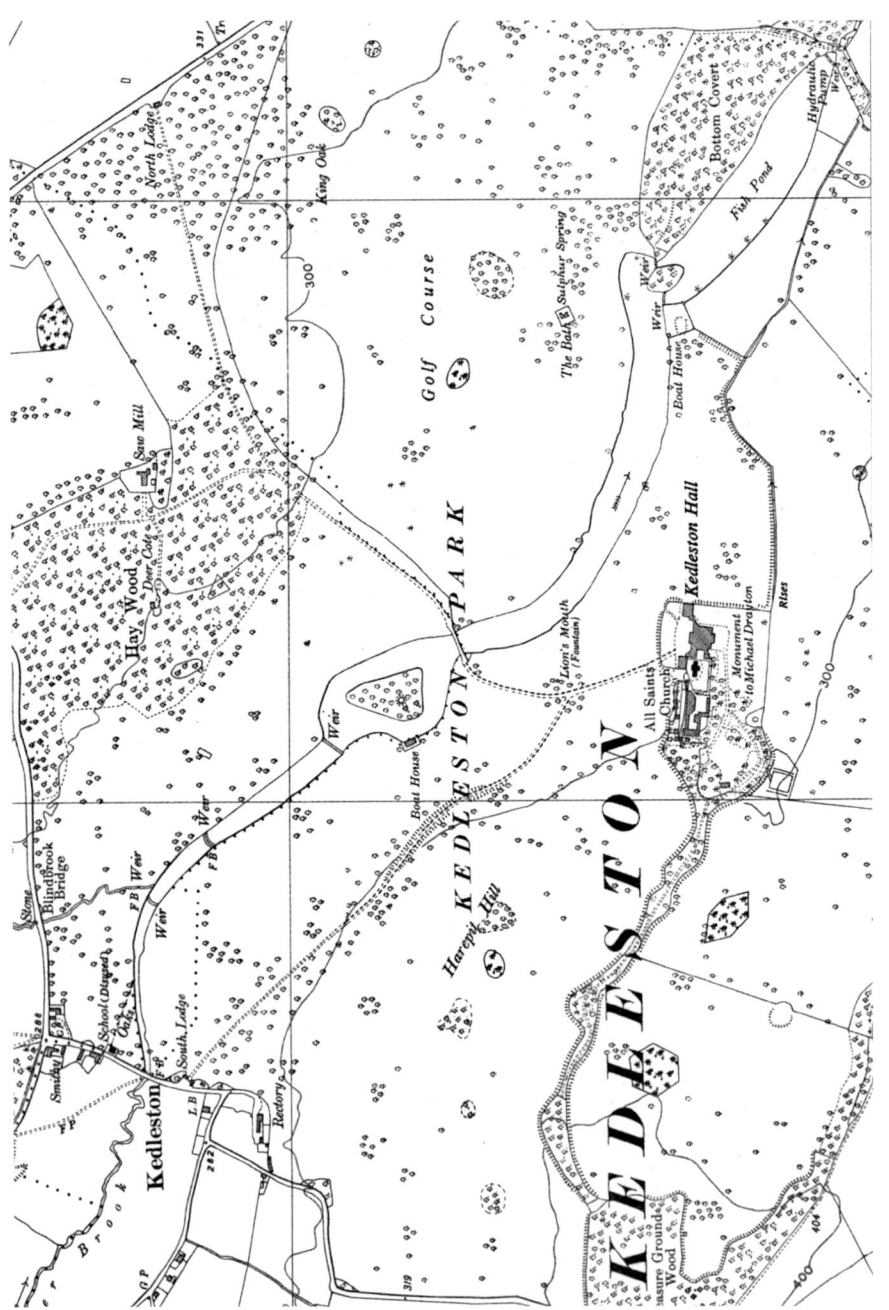

FIGURE 4.14. Kedleston Park, Derbyshire, OS 6" map, pre-1930–'55. By permission of the National Library of Scotland.

FIGURE 4.15 The bridge–weir at Kedleston, designed by Robert Adam in 1764 but not built until *c.* 1771.

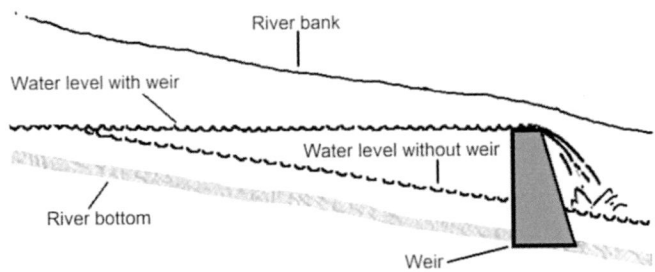

FIGURE 4.16 A diagram of a basic weir construction.

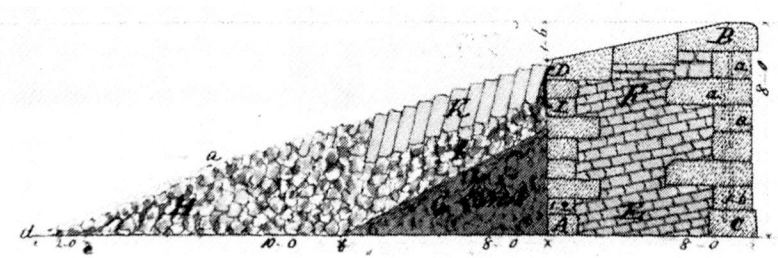

FIGURE 4.17 John Smeaton's 1776 design for a weir on the River Coquet, Northumberland. Note the stone crest 'B' and the stone facing 'a'. The water is flowing from left to right.

FIGURE 4.18 The weir, probably modern, retaining the lake at Kimberley Park, Norfolk.

difference becomes significant as the topography becomes hillier, and the mode of construction becomes more critical.

Not a great deal is known about eighteenth-century weirs. However, the Environment Agency's *River Weirs – Good Practice Guide* throws some light on weir construction in the past.[68] A previous method was to drive wooden piles upright into the river bed, as at Northenden, Manchester, where there has been a weir on the site since 1607. Repairs in the early twenty-first century revealed sawn timbers probably dating from the eighteenth century, and ashlar blocks forming the weir's stone crest of *c.* 1800. However, underlying these is a gravel and silt mound with stakes and posts which is older.[69]

Sites for lakes

Where is the best place to put a lake? This is a very important question as putting a lake in the wrong place could easily result in failure – the water supply might be insufficient, or the bedrock might not support the dam securely, with the risk of it starting to move. It is a complex subject, and the two main factors which govern it are the geology and topography. Sociological factors such as fashions in garden and landscape design, or the relationship of the lake to the house also have a bearing and these are discussed in Chapter 6. Although it is common for garden history features to be analysed in economic and sociological terms, looking at them in relation to geology is a fresh approach, but one which is particularly pertinent to lakes. Examining the geological and topographical factors which operate in the siting and construction of lakes helps us to understand where and why lakes were made successfully.

Geology

Geology has a direct impact on the making of lakes. It is particularly significant given that lakes do not usually have a clay lining, and require constant replenishment. It is not possible to make a lake where there is no water supply, as North hints at, and it is the geology of an area – the superficial deposits as well as the bedrock geology – which largely determine whether surface water is available. Ideally, there should be a good supply of surface water which can be captured, and for this to be the case, the superficial deposits need to be impermeable at or fairly near the surface, or water will not be available. To take an extreme example: an upland limestone area will have little or no surface water, and little chance of gathering it naturally, as the limestone is very porous. An ideal situation for a fishpond of *c.* 2 h/5 acres (or an ornamental lake), as North and Switzer suggest, would be where the surface deposits are clayey, or where there is an impermeable lens (often clay) not far beneath the surface, acting as a water-proof layer, and causing water to appear on or near the surface. Areas with very sandy soils, with no layer of clay near the surface, or lacking a robust water source such as a river, would also prove difficult, and therefore expensive, for making lakes, as North indicates:

> but yet the two great Distinctions, are Clay and Sand, or standing Water and Springs [and] My concerns are in a Clay Country.[70]

Bramshill Park, Hampshire, gives a good idea of the geological complexities underlying the making of lakes. The lake was made by 1699 and is a typical hybrid lake of *c.* 6.4 h, having three straight sides, forming right angles, the remaining 'sides' being irregular in shape (Figure 4.19). It was created by

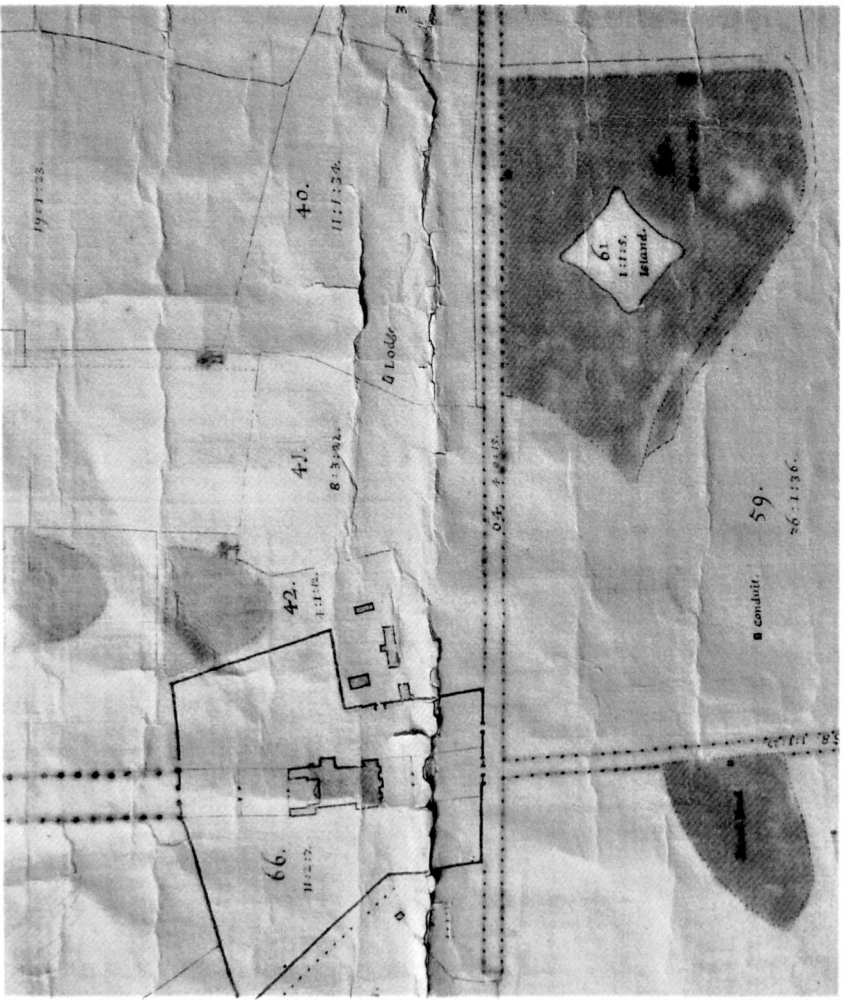

FIGURE 4.19 Detail of the 1699 map of Bramshill Park, Hampshire, by Isaac Justis, showing the lake and three ponds. *Private Collection.*

building large dams, 3–4 m high, forming the straight sides, and is fed by a stream coming in on the eastern side. It lies predominantly on the Barton, Bracklesham and Bagshot Beds of north east Hampshire, with outcrops of London Clay and the Harwich Formations on the peripheries.[71] These Barton, Bracklesham and Bagshot Beds consist mainly of fine to medium grained sand with thin layers of clay and clayey silt. It is these lenses of clay that contain the layers of sand which act as aquifers, and springs occur at the junction with underlying clays, so there is plenty of surface water.[72,73] The predominant geology of the area was ideal for making a large lake: the sandy nature of the soil would have made it relatively easy to move in large quantities, and the thin layers of clay in the Barton, Bracklesham and Bagshot Beds possibly provided clay for water-proofing the dams. These are large at Bramshill, as Figure 4.20 shows. They would have been earth dams constructed according to the historical methods described by North *et al*, unless they have been rebuilt at some point. A new spillway in the north east dam suggests modifications have been made in recent times. Hybrid lakes of this kind are spreading in plan view, rather than elongated, as a result of the flattish areas where alluvial deposits, or

FIGURE 4.20 The north-west dam at Bramshill House, Hampshire.

lenses of clay, occur. The lake at Kimberley (made by 1739, Figure 1.3) is a case in point, being in a basin-like area inclined to marshiness. Hybrid lakes are not usually found in hilly areas.

The geology of an area – the bedrock and superficial deposits – is the principal determinant of the topography. The superficial deposits generally consist of sands, gravels, clays or alluvium, which have been formed through glacial or river action, for example. They do not occur everywhere. In Norfolk, for instance, it is noticeable that most lakes seem to occur in areas of chalk bedrock, which is surprising.[74] However, when the superficial geology is examined, a different picture emerges. Almost without exception, lakes do not occur on the diamicton (glacial deposits of sands, gravels, clays), but they do occur where other deposits border it, and these are often alluvial.[75] This is partly because the diamicton represents the higher ground, so streams and rivers are less likely to occur there. The low-lying area of The Broads, east of Norwich, has very few lakes, probably because this area is not suitable for making parks.

In Wiltshire there are almost no lakes on the chalk deposits, and this roughly corresponds to Salisbury Plain, where there is no surface water other than winter-bournes.[76] The exceptions are a few lakes on alluvial deposits at the edges of the Plain. Unlike Norfolk, there are no superficial deposits on the Plain, which almost certainly explains why there are lakes over the chalk bedrock in Norfolk, but not in Wiltshire.[77]

The eighteenth-century lakes around the edges of Salisbury Plain tend to be river-lakes, or river-like lakes, as at Longleat and Wilton. Very few villages existed on the Plain, and it has been an army training ground since at least 1757.[78] This points to the geology determining settlement patterns which, in turn, meant that no lakes were made in the area. Where alluvial deposits occur, river-lakes would have been a feasible option, though some parks may have been too small, particularly some new parks made in the nineteenth century, such as Ferne Park, Wiltshire.

At Tottenham Park, Wiltshire, the lack of a lake is puzzling, as there appears to be a good scattering of ponds, suggesting that a reasonable amount of surface water could be retained. Alternatively, the presence of *only* ponds may point to the difficulty of retaining water other than by making ponds lined with puddled clay. The limits of this kind of map-based survey, even when linked to geology maps, are illustrated by the 'lake' (0.3 h) at Newark, Gloucestershire. The OS map (1:25,000, 2016) suggested that the irregular pond would have been difficult to make as it is on a steep limestone scarp slope below the house. However, on the ground, it was seen to be on a natural 'terrace' on the scarp, with multiple springs feeding it, which did not appear on the OS map. The size of the water may indicate that puddling was necessary to retain it, and that a larger piece of water (a lake) was not feasible because of this.

Several points emerge from this discussion of geology. Firstly, where there are superficial deposits overlying bedrocks such as sandstones, chalks and limestones, lakes have been made, as in Norfolk. Where these are absent, as in the chalk upland area of Salisbury Plain in Wiltshire lakes, other than river-lakes on alluvial deposits, are generally not feasible. Secondly, geology has a bearing on where lakes can be made because it determines the topography and the resulting drainage systems, which are discussed later. It is likely, however, that specific local factors have the strongest bearing, as Grundy's reports for a lake at Grimsthorpe highlight. These emphasise that lakes were only made where they were feasible, where the ground would literally hold water. Best practice, followed by men such as Grundy and Brown, was to inspect the site personally, and glean all available information.

Topography

The drainage systems produced by factors such as rivers and glaciers operating on the underlying geology are of primary importance in the siting of lakes. Because lakes are dependent on rivers and streams to replenish them, they have to be made where these occur. Often, house sites were adjacent to rivers, so creating a lake in the vicinity was feasible, as we can see at Bowood, Wiltshire. The original eighteenth-century house was *c*. 200 m from the original stream, which was dammed in 1766 to produce a lake adjacent to the house (Figure 5.19). The *type* of lake which can be made is dependent upon the general topography, although more than one lake type may be feasible in a given area. In fact, as might be expected, there is a changing relationship between topography and lake type.[79]

Irregular lakes are the most flexible in terms of possible sites, because they can be adapted to most land forms. Though often they are constructed to be reasonably deep and not particularly wide, they can also be relatively shallow and spreading. River-lakes, on the other hand, require a river with a gentle gradient, otherwise dams are required to retain the water. Geometric lakes are the most constrained by topography: a nearly flat site is ideal, because of the expense of making sides straight, whereas semi-geometric lakes are more flexible because they are not symmetrical and can be adapted according to the topography, as can hybrid lakes.

Highclere Castle, Hampshire, (also known as Downton Abbey!) is a good illustration of the interlinked effect of geology and topography on the siting of lakes (Figure 4.21). The house, on a site dating from medieval times, is on a chalk ridge, with no surface water within *c*. 500 m, and was probably originally supplied by wells and dewponds. Approximately 500 m north of the house, the chalk gives way to London Clay deposits and the land dips generally towards the north. Being softer, the clay deposits have been eroded by water action, producing a drainage basin, with streams and rivers. The result of these factors was that a lake could not be made near the house. The feasible sites were some distance away, in the river valleys, and this is where the lakes have been made: Duns Mere and Milford Lake.

FIGURE 4.21 Highclere Castle and Park, Hampshire, OS 6" map, 1909. *Reproduced by permission of the National Library of Scotland.*

Conclusion

In summary, the fundamental difference between ponds and lakes is that lakes are constantly replenished by a water source, but ponds do not have to be, and usually are not. Thus, ponds are often lined with clay, but lakes are not. Lakes are made by damming a water source, and the different types of lake are constructed in slightly different ways, as we have just seen. Irregular lakes, which are basically the same as large fishponds (*vivaria*) in construction, can be made in almost any type of terrain, providing there is a water source. River-lakes are constructed with weirs, in gently falling river valleys, and are useful where a spreading lake will not work, or funds are limited, as they are usually the least expensive type of lake to construct.

Geology and topography are the two primary factors in determining where and how lakes can be made. At the macro level, geology determines the porosity of rocks and where drainage basins occur. At the local level, topography governs where suitable water sources are, and the types of lakes which can be made. A comparison of the geology in Norfolk and Wiltshire shows that lakes are unlikely to be made in areas where the bedrock is chalk or limestone without any superficial deposits. A further point is that the usual geology maps are too small scale to show enough detail for gauging whether an area would support a lake. On-site surveys or inspections would be necessary, and these were carried out by Grundy, Brown and his contemporaries. Given the parameters at hand, lakes could be made in most areas, but realistically, expense restricted certain types of lake to certain types of topography.

Notes

1 Bishop, op. cit.
2 C. Currie, op. cit., p 30. *Vivaria* could be very large: the Bishop of Winchester's largest pond, Frensham Great Pond, is still *c.* 29h today. Given in E. Roberts, 'The Bishop of Winchester's Fishponds in Hampshire, 1150–1400: Their Development, Function and Management', *Proceedings of the Hampshire Field Club Archaeological Society*, Vol. 42 (1986), pp 125–138, p 125.
3 Steane, op. cit., p 40.
4 I am indebted to Roger Cortis for this information about James Briggs.
5 This phrase, 'surveying and levelling' also occurs in the accounts for Thoresby Park, and I have taken it to mean conducting a survey on the ground as to where the best site for a dam is, and then working out the heights required for various items such as the top of the dam, the subsequent level of the water, the sites for sluices etc. £9. 1s possibly equates to approximately £927 (2017). In this case, two men travelled for two days and, as well as the surveying, produced a design for the dam and plans for it.
6 Letter from Henry Merewether to Lord Shelburne, 13th March, 1769, in Bowood Muniments.
7 North, op. cit., p 14.
8 Bishop, op. cit.
9 Ibid., pp 4–6.
10 North, op. cit., p 8.
11 Ibid., p 12.
12 Binnie, op. cit., p 39.
13 T. Hinde, *Capability Brown: The Story of a Master Gardener*, p 62. Brown did this at Burghley.
14 J. Taverner, *Certaine Experiments Concerning Fish and Fruite Practised by Iohn Taverner, Gentleman* (London: William Ponsonby, 1600) p 3.
15 Ibid., pp 3–6.
16 J. Lawrence, *The Modern Land Steward in Which the Duties and Functions of Stewardship Are Considered and Explained . . .* (London: Various, 2nd ed. 1806) p 312.
17 Switzer, *Hydrostaticks*, op. cit., p 10.
18 Ibid., p 130.
19 Ibid., p 128.

20 S. Switzer, *A Universal System of Water & Water-works* (London: Thomas Cox, 1734) p 131. The practice Switzer mentions is that of applying clay in two distinct layers, with a layer of slaked lime between to deter earthworms.

21 Diary of John Kyrle-Ernly, WSHC 1720/742.

22 M. Symes & S. Haynes, *Enville, Hagley and the Leasowes: Three Great Eighteenth-Century Gardens* (Bristol: Redcliffe Press, 2010) p 96.

23 Switzer alludes to the extra expense of puddling the whole construction. See earlier.

24 North, op. cit., p 16.

25 Switzer, *Ichnographia*, op. cit., Vol. I, p 303.

26 Ibid., pp 306–307.

27 Ibid., p 128.

28 Richard Woods, quoted in F. Cowell, *Richard Woods (1715–1793): Master of the Pleasure Garden* (Woodbridge, Suffolk: Boydell Press, 2009) p 113.

29 Binnie, op. cit., p 39.

30 A. W. Skempton, *A Biographical Dictionary of Civil Engineers in Great Britain and Ireland . . . 1500–1830* (London: ICE Publishing, 2002), p 278.

31 John Grundy, *Slopes Ponds and Reservoirs and Engine and Piping to supply the House and Offices and other Works done at Grimsthorpe in Lincolnshire 1745 to 1748*, p 143, in Report Books, Vol. 2, ref. 1740 GRUSLR, held by the Institution of Civil Engineers, London.

32 Ibid., p 189.

33 Ibid., p 190.

34 Currency conversion given at www.nationalarchives.gov.uk/currency/default2.asp and Binnie, op. cit., p 70 and Grundy, op. cit. However, using Roderick Floud's conversion method, this figure would be over £2.86 million: R. Floud, *An Economic History of the English Garden* (London: Allen Lane, 2019).

35 North, op. cit., p 5.

36 Bishop, op. cit., p 19.

37 Contract of June, 1755 between L. Brown and Lord Egremont regarding work at Petworth; West Sussex Record Office, ref. PHA 6623.

38 Treswell's 1610 Map of Petworth Park, West Sussex Record Office, ref. PHA 3574.

39 Contract of May, 1756 between L. Brown and Lord Egremont regarding work at Petworth; West Sussex Record Office, ref. PHA 6623.

40 This is written on the Crow survey of 1779, and the lake perimeter is the same today.

41 At *c.* 0.7 h, this piece of water hardly qualifies as a lake, according to the criterion adopted in this book. Bentonite is a form of clay and can be sandwiched between membranes to form a water-proofing layer.

42 Barry Smith, Garden and Estates Manager. Stowe, personal communication, 6.4.2016.

43 G. B. Clarke, *The History of Stowe*, Part 14, p 21 quoted in Hinde, op. cit., p 27.

44 Brown's letter to Cobham, 24th February, 1746, Soane Museum, Stroud papers. Quoted in S. Shields, *Moving Heaven and Earth: Capability Brown's Gift of Landscape* (London: Unicorn Publishing, 2016) p 40.

45 William Smellie, ed., Encyclopaedia Britannica (Edinburgh: C. Macfarquhar and A. Bell, 1768–71).

46 Diamicton (also 'till') is an unsorted sediment ranging from clay particles to boulders in a matrix of muds or sands, resulting from dry erosion of the landscape.

47 H. Woodward, *The Geology of Water-Supply* (London: Edward Arnold, 1910) p 117.

48 Ibid., p 117.

49 Binnie, op. cit., p 73.

50 N. Smith, *A History of Dams* (London: Peter Davies, 1971) pp 179–180.

51 Andy Hughes, Chairman of the British Dam Society, personal communication, 9.12.2015. The British Dam Society is an association of the Institution of Civil Engineers.

52 Henry Hewlett, ed., *Improvements in Reservoir Construction, Operation and Maintenance* (London: Thomas Telford Publishing, 2006) and Binnie, op. cit.

53 It is also possible that some lakes were partially lined; this may apply to the Copper Bottom Lake at Stowe.

54 The proviso is that there was a constant water supply.

55 Lists of all types of lakes can be found in the Landscape Database.
56 BGS Geology of Britain Viewer, at http://mapapps.bgs.ac.uk/geologyofbritain/home. html (2014).
57 Clearly, the wishes of the owner, and local conditions, might make a lake unusually expensive. Because of the relative paucity of data, comparing costs of lake-making is fraught with difficulty.
58 RCHME, 'Archaeological Sites in Central Northamptonshire', *An Inventory of the Historical Monuments in the County of Northamptonshire*, Vol. 2 (London: HMSO, 1979) pp 61–71.
59 This would not be the case, of course, if the owner wanted lots of alterations to the river banks.
60 Historic England listing: Wallington Park and Garden 2014.
61 T. Williamson & J. Barnatt, *Chatsworth: A Landscape History* (Macclesfield: Windgather Press, 2005) p 108.
62 Ibid., p 116.
63 Quoted in Williamson & Barnatt, *Chatsworth*, p 116.
64 T. Williamson, personal communication, 2016.
65 See Appendix II.
66 Modern dams are sometimes a combination of weir and dam. The main construction is a dam, with a section of the crest being lower and acting as a weir. At Clumber, a modern spillway/weir has been inserted into the dam (Figure 1.8).
67 News footage of the near failure of the Toddbrook Reservoir dam in 2019 graphically illustrates this.
68 C. Rickard, R. Day & J. Purseglove, *River Weirs: Good Practice Guide R&D Publication W5B-023/HQP* (Bristol: Environment Agency, 2003) p 55.
69 University of Manchester, *Northenden Mill: Manchester An Archaeological Desk-Based Assessment* (Manchester: University of Manchester, 2004) p 13.
70 North, op. cit., in the Introduction.
71 BGS Geology of Britain Viewer op. cit. These are largely deposits of laminated clays, silts and sands.
72 C. S. Cheney et al., *The Physical Properties of Minor Aquifers in England and Wales* in Hydrogeology Group Technical Report WD/00/04 Environment Agency R&D Publication 68 (Nottingham: British Geological Survey, 2000) p 40.
73 The London Clay Formation and Harwich Formation are described in the BGS Lexicon on line at www.bgs.ac.uk/lexicon/lexicon.cfm?pub=LC These formations mainly consist of blue-grey or grey-brown, slightly calcareous, silty clay and clayey silt, with some layers of sandy clay.
74 Bishop, op. cit., p 280.
75 Ibid., p 281.
76 Ibid., p 282.
77 Ibid., p 283.
78 Pete Cox (AC Archaeology), *Anyone Lost a Hillfort? – Archaeology Results from the Wessex Water Salisbury Transfer Pipeline Scheme 2014–2016*, lecture at the WANHS annual Archaeology in Wiltshire Conference, 2017.
79 Bishop, op. cit., p 289.

5

THE LAKE MAKERS

Lancelot Brown

The man responsible for making the greatest number of lakes was Lancelot 'Capability' Brown. The basic facts relating to Brown are well known: born in Northumberland in 1716, he died Lord of the Manor of Fenstanton in 1783, having designed or part-designed some 250 landscapes, mainly for the aristocracy or the well-to-do. His commissions ranged throughout England, with a handful in Wales, and were characterised by lakes, often crossed by approach drives, and also perimeter belts and circuit drives, tree clumps, and smooth, uncluttered lawns surrounding the house.

In spite of recent work highlighting the importance of his contemporaries, Brown merits a section to himself because of what he achieved in making lakes in the eighteenth century.[1] When he began working, the concept of a lake – a large, ornamental piece of water – barely existed. By the mid-1740s though, when he was working for Lord Cobham, Brown was also making irregular lakes for Cobham's friends. These were men such as Lord Denbigh of Newnham Paddox (lake, 1748), the Duke of Grafton at Wakefield Lodge (lake, c. 1745), Lord Brooke, later 1st Earl of Warwick (Warwick Castle, lake c. 1751), Lord Coventry of Croome Court (lake c. 1753).[2] These lakes were of various sizes, from 1 h to 8.6 h, with Wakefield Lodge being the biggest, so Brown was engaging in substantial lake building even in this early stage of his career. This raises the question of how he acquired his engineering expertise, to which there is no clear answer, although it may have been at Grimsthorpe, Lincolnshire.[3] We know that at Bowood he engaged a dam-building expert but how often he did that is not known. It could have been his usual practice, or it may have been exceptional. As it occurred when he was at the height of his career, it seems feasible that it was a common practice of his, possibly owing to pressure of work.

Once Brown had set up his own practice in 1751, after Cobham's death, his workload increased and he had made about a dozen lakes by the end of the decade. In all, Brown had made over 70 lakes by the time he died in 1783, which was approximately one third of the lakes made in his lifetime.[4] It may be claimed that in some instances, he merely altered an existing piece of water. That was certainly true, but in many instances he increased the size of the water and significantly remodelled it, effectively producing something new.

Brown may not have been particularly innovative in the sense that there were other people (Richard Woods, Nathaniel Richmond, William Emes) making landscape parks, and given that parks were not new features – deer parks had existed for centuries.[5] However, in other ways, he was an innovator. He extended the concept of 'garden' into the whole park, and he designed landscapes on a much larger scale than most other people, because his clients were the foremost landowners of the time.[6] He also popularised a new way of experiencing the landscape – moving rapidly around it by carriage or horse – meaning that large areas could be seen at once, not just narrow vistas framed by plantations. He was particularly original in the way he manipulated the landscape, especially the approach drives, to make the house the focus of attention, and he 'manicured' the land surrounding major features, such as the house or lake to set them off.[7] However, it was his use of water that was the most innovative.

A context for lakes

Capability Brown worked for Lord Cobham at Stowe from 1741–50, where he encountered two vital features: the 'ha-ha' introduced by Charles Bridgeman, according to Walpole, and the deformalized gardens designed by William Kent.[8]

FIGURE 5.1 Part of Brown's second plan for Kimberley, dated 1778. North is at the bottom. *Courtesy of Tom Williamson.*

Brown absorbed these aspects into his own designs, which led to his new approach of blending the gardens with the park. He treated them as a whole: there was no *obvious* demarcation between the gardens around the house and the park. The informal ornamental water was the element which linked the two and unified them into a holistic design. The lake tended to be large and it was viewed *in* the park, especially on the approaches to the house, as well as *from* the house, making the link between house and park. The ha-ha was the key to this unification (Figure 7.22). As well as designing the garden area within the ha-ha, Brown also planted alongside carriage drives in the park, as well as planting clumps and individual trees in the park. He also broke up hard edges of existing tree planting with thinning and new planting. Along rides and drives Brown's selection included Lombardy poplars, weeping willows, and cedars of Lebanon. Cedars are long-lived trees, and these are the ones which have survived until today.[9] Brown's plan for Kimberley illustrates this clearly (Figure 5.1). As well as providing a pleasure ground (bottom left), the exotic planting is evident along the carriage drives or rides and paths crossing and circling the lake, and winding through the woodland (Figure 5.2). Kimberley is a late example of Brown's work, but this concern with tree planting is evident in his landscape planning from the beginning, as in his first sketch for alterations at Packington, Warwickshire, of *c.* 1750, for Lord Guernsey (Figure 5.3). We can see that Brown has

FIGURE 5.2 A detail of Brown's second plan for Kimberley, dated 1778. North is at the bottom. *Courtesy of Tom Williamson.*

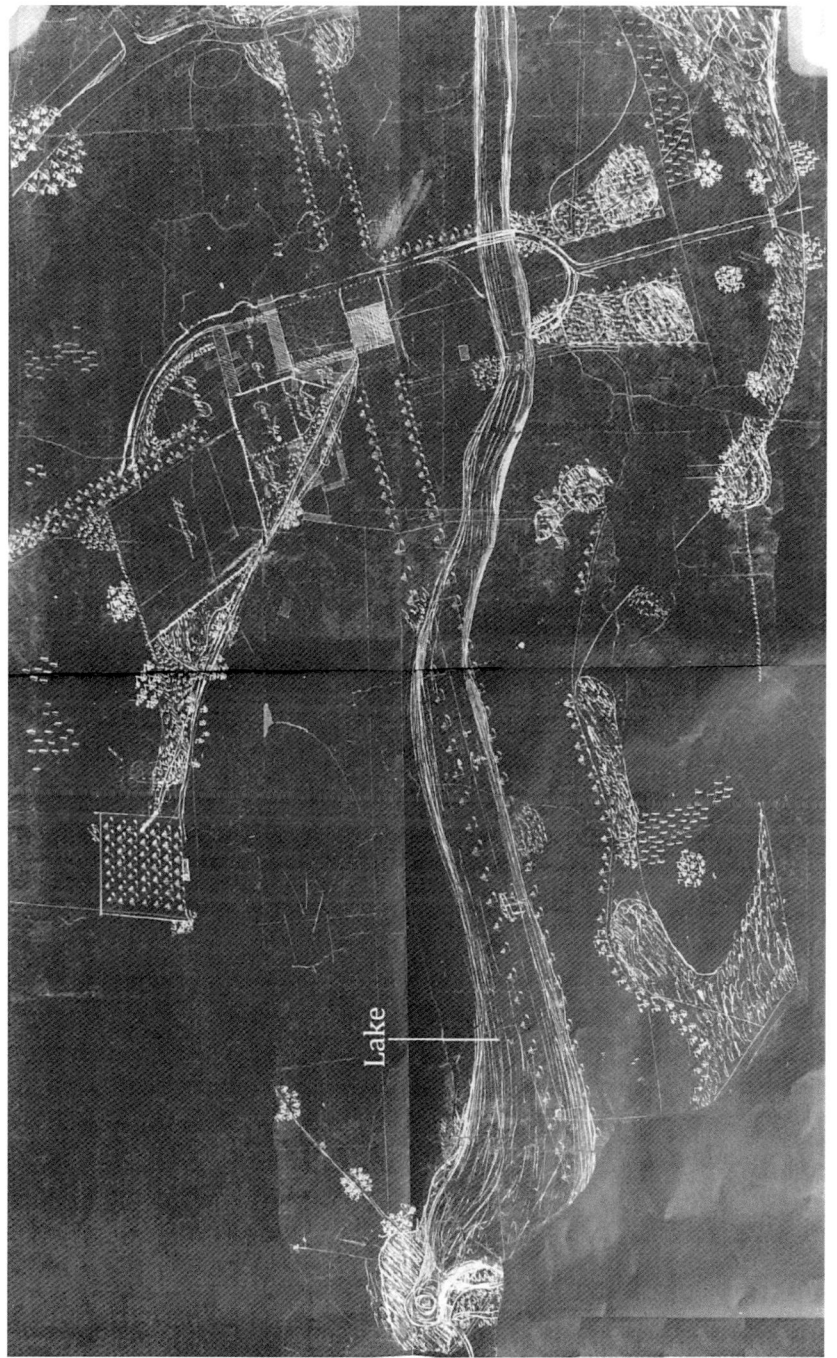

FIGURE 5.3 A detail of Brown's sketch plan for Packington, Warwickshire, of *c.* 1750, showing the planned lake. North is at the top. *Image courtesy of Packington Estate.*

FIGURE 5.4 Brown's design for a cascade at Packington, 1751. *Image courtesy of Packington Estate.*

clearly differentiated between deciduous and evergreen trees, and that the carriage drives are largely lined with deciduous trees. There is a group of conifers in the lower right quadrant of Figure 5.3, whilst deciduous clumps line the drive in the upper left. Brown is planning to put in a lake in place of the existing avenue, making the water carry the eye to and from the house instead of the avenue doing so. (Note the boat on the planned lake – upside down in this image.) On his 1751 plan for Packington, Brown indicates the planting around the cascade – Cedars of Lebanon – in this elevation (Figure 5.4). It shows Brown using exotic species in the middle of the park, as well as in the pleasure grounds and near the house. This suggests that Brown, and his clients, are beginning to regard the whole landscape, centred around the lake, as an extended area for leisure pursuits such as carriage driving, riding, walking perhaps, with destinations for those activities – a view from a bridge or seat, a cascade to admire, a lake to boat on – a concept which Tim Mowl has likened to a Disneyland of the eighteenth century, and leisure was certainly a factor in the evolution of lakes.[10]

Brown mainly worked for the richest and most prestigious clients whose estates were large, which gave him the scope for making large pieces of water, and creating approach drives and circuit drives. Because he 'cornered the market' of these clients, other improvers such as Richard Woods had less scope for their designs, as they worked with fewer and generally smaller landscapes.[11] Brown's designs enabled people to experience the park three-dimensionally, and to follow circuits round it, with opportunities for varying prospects, including different views of the house. In many cases, other designers were addressing only one part of a landscape, as was the case with Richard Woods' work at Cannon Hall, Yorkshire. Perhaps only Nathaniel Richmond designed on a similar scale: Beeston St. Lawrence, Norfolk (1773–7) was *c.* 172 h, and Stoke Park, Buckinghamshire was *c.* 153 h. Erlestoke Park, Wiltshire by William Emes was *c.* 259 h. Neither designer routinely dealt with such extensive landscapes though. Working on the large scale meant that Brown was able to make lakes of significant size, and to place them prominently in the landscape. Why did he have the opportunity to do this from the beginning of his career? The answer can only be surmised. He had a very influential patron in Lord Cobham, who clearly had confidence in Brown as he allowed him to work for other influential men, as we have seen. However, that alone would not have been sufficient if Brown had not been an able man in many spheres: surveying, landscape and architectural design, horticulture, drainage, water engineering. Had any of his lakes failed in the 1740s, it is unlikely that many other large commissions would have followed. Also, he was undoubtedly a good business man. He had the acumen to set up a 'design and build' business, plus good personnel skills, which enabled him to capitalise on the excellent client contacts he had. With few exceptions, Dickens of Branches, Suffolk, being one, we do not hear of him falling out with rich clients, unlike Vanbrugh, or of problems with his foremen, though that may be due to lack of information.[12] He also seems to have possessed a good appreciation of just what the 'capabilities' of the land were and the ability to match them to his clients' expectations and pockets. No mean skill.

A new focus: house and lake

One of the hallmarks of Brown's landscapes was the way he focussed on the house from the park, often with the lake as a foil. This reversed the previous rationale of the house at the centre of the landscape, with axes radiating from it, as Marc Girouard described at Badminton.[13] He carefully orchestrated the approach through the park, and ensured that there were views of the house across the lakes he made. With his curving approaches and circuit drives, Brown then displayed the whole landscape, with a particular focus on the house and the water, carefully choreographing glimpses of the lake or house between trees. Trees would be planted, hills flattened (as at Bowood), water created in order to give tantalising glimpses of the house as the visitor or owner approached, with final impressive views, often over the lake, of the house itself. We can see Brown aiming to do exactly this in his Packington plan of 1751 (Figure 5.6). He wanted to route the main approach to the house northeastwards over the cascade, then north-westwards to the house. This would have given glimpses of the lake and house as the visitor entered the park, then hidden that view. The visitor would then pass on through denser (possibly evergreen) woodland to the drama of the cascade, before emerging from these trees to a full view of the house. Clearly, this was not favoured by Lord Guernsey as, in the key to the plan, Brown has written "The great Road, which would in my opinion be much better turned". It seems from this that Lord Guernsey favoured a direct approach from the park entrance, over the lake to the house. Brown's plan indicates a bridge at that point, and it appears that owner and designer were tussling over this, as OS maps today and in 1886 show a track of some kind (Figure 5.5) on that direct route.

FIGURE 5.5 Packington Park, Warwickshire, OS 6" map, 1886. *Reproduced by permission of the National Library of Scotland.*

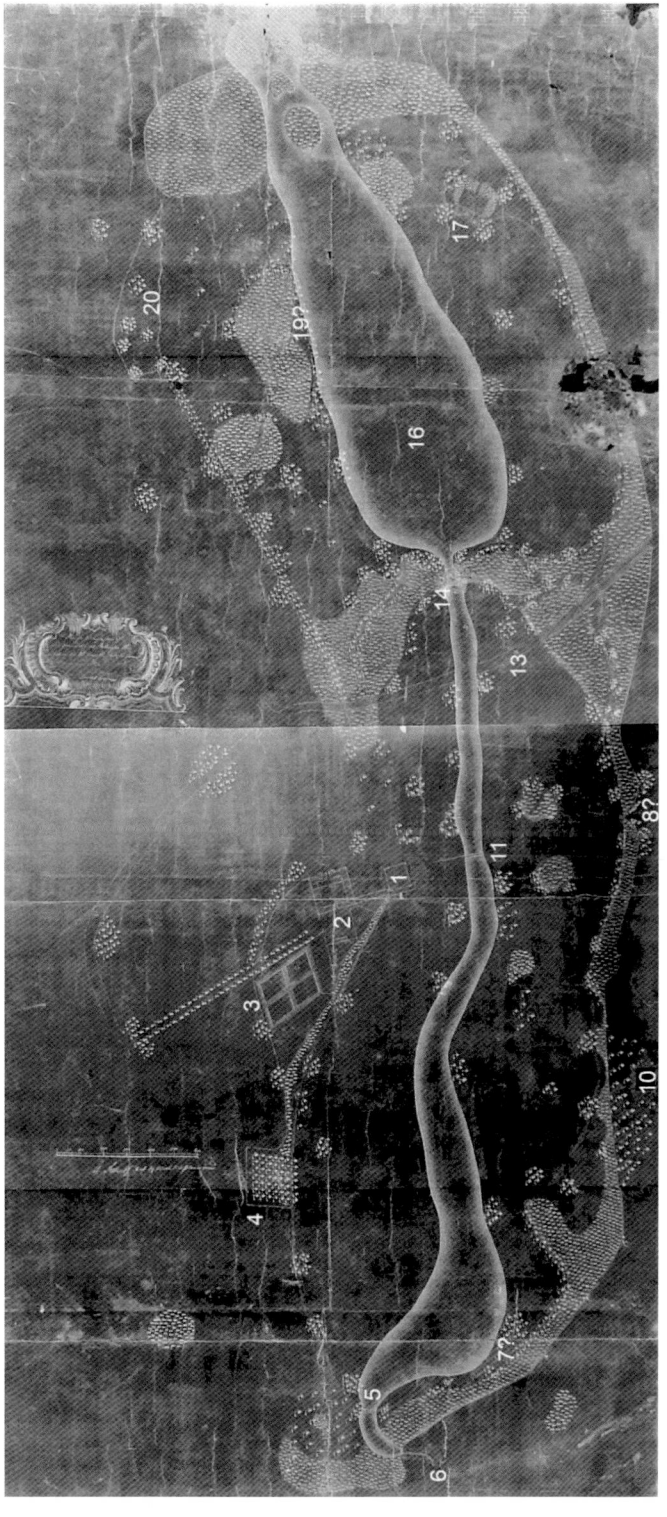

FIGURE 5.6 Brown's 1751 plan of Packington, Warwickshire. 1. The House 2. Courts 3. Kitchen Garden 4. Seats Firs Chasse? 5. Wooden Bridge 6. A Mill 7. A seat 8. Gatehouse? 10. Oler Plantation 11. Bridge 13. The Great Road 14. Cascade 16. The Lake 17. My Lady's Lodge 19. A Seat? 20. The Church. *Image courtesy of Packington Estate.*

However, Brown seems to have prevailed, as there is still an approach evident today on the line of his planned route. It is noteworthy that he could take issue with the owner in this way, and hope to prevail, especially as he did not have the great weight of fame on his side at this early stage of his career. He did have a remit for designing a large part of the landscape, which was not often the case for his contemporaries. When Woods did have a remit for more of the landscape as at Cannon Hall (1760s) and Wardour Castle (1760s), his approach drives were much more direct (Cannon Hall) and not accompanied by the subtle manipulation of tree planting to choreograph the views.[14]

Brown's inspiration of putting in perimeter carriage drives enabled people to use the new light carriages (phaetons) which were becoming popular, to see the landscape. This meant that they could move more easily through it, and also that larger areas of the park could be seen in one go, rather than offering constricted views along narrow vistas. Importantly, this also meant that varied views of the lake were possible. There was an opening out of the visual effect of the park, and the house, and it was the irregularity and informality of Brown's designs which led to those views being unpredictable and surprising, unlike the geometric vistas of old. At Packington (Figure 5.6), the carriage drive follows the edge of the perimeter belt (bottom left of plan), then winds through a denser area of woodland, with the choice of going north and across the cascade, or continuing east around the park, with views down to the lake, and on to 'My Lady's Lodge'. The centrality of the view back to the house was new, and a stable ingredient in Brown's designs from the 1750s.[15] Wherever possible, Brown would route the main approach to the house over water, as at Packington, Bowood, Chatsworth, Burghley, Shortgrove, or alongside it if that was not possible, as at Croome and Blenheim. This was very much akin to the medieval phenomenon of approaching elite residences with water on either side and, in subsequent centuries, looking down over fishponds. The underlying rationale remained the same: to impress the visitor and confer status. The superficial rationale had changed though: lakes were made, and positioned prominently, for their visual qualities, although they were probably also stocked with fish in most instances.[16]

Manicuring the setting

It is tempting to say that it is the water which marks out Brown's landscapes from others. In fact, something much subtler was of equal importance: the key to his landscapes is the levelling work he did and this is what makes them instantly recognisable. It had a great impact on the house, and also the lake, as the smooth, uncluttered banks emphasised the visual qualities of the water. Contracts at Bowood and Longleat show that Brown did a great deal of levelling of land sloping down to the lakes he created, as well as in other parts of the park (Figure 5.7). This had a significant impact: it changed the landscape near the house from being bitty and compartmentalised into a homogeneous area, integrally linked with the park, and usually a lake. This simplification of the landscape had the effect of making both

FIGURE 5.7 Longleat, Wiltshire: High Wood is *c.* 50m behind the viewer; the hall door is just up the steps.

the house and the lake very prominent. He moved all the extraneous 'offices', kitchen gardens and stables away from the house, to leave it uncluttered, and did extensive earth-moving to make the land around it look as though it was completely natural and undisturbed. It was a vast change. We can see exactly how he did this in his contracts of 1757 and 1758 with Viscount Weymouth at Longleat. As well as altering the sharp turns of the 'serpentine Water' (the serpentine canal adjacent to the east front of the house, Figure 3.20), Brown was

> To lay the two Canals next the serpentine Water into one, and to lower the Surface of them so much as to make the Surface of them have a natural, correspondent, Level, with the Ground on each Side of it

and:

> To begin at the Hall Door and to give a proper Levell to the Ground from it to High Wood, through where the Mill stood and from High Wood down to the Water which was altered last year.[17]

He achieved almost exactly the same gentle gradient in the landscaping around the chain of fishponds which he created adjacent to his new kitchen garden, 900 m from the house (Figure 5.8).[18] This work is also mentioned in the contract. A

FIGURE 5.8 One of Brown's fishponds at Longleat. The angle of the slope indicated is typical of Brown's landscaping.

similar gradient can be seen at Bowood, where the angle of the slope is very similar. Figure 5.9 shows the 'lawn' dropping down from the house to the lake, and the marks of the extensive drains put in by Brown. His contract with the Earl of Shelburne, dated 10th August, 1762, makes clear how carefully graded these slopes were:

> To Level, Drain, alter, Plant, and sow with Grass seeds all the Ground on the South Front, down to the Water.[19]

Again, the contract is quite specific:

> To level all the Ground between the Kitchen Garden [immediately north of the house] and the Water, and also to Drain, plant, and sow with Grass seeds all such Parts as shall be thought Necessary to be in Grass making the Whole compleat.[20]

This is exactly what can be seen in Figure 5.10: the gently graded grass sloping down from behind the walled kitchen garden to the water. The 6th Article in the contract deals with the opposite side of the lake, from the dam southwards, which was also to be levelled. Planting was to be done (grass?),

FIGURE 5.9 Bowood House, Wiltshire, from the east side of the lake.

FIGURE 5.10 An adjacent view of Bowood from the east, showing the dam.

trees and bushes were to be removed "as shall be thought proper" and a sandy walk made along the side to connect across the dam with the one coming from the house, "in the best Direction for Shade and Prospect". This walk was significant as it provided views back to the house across the water, as Figure 5.9 illustrates.

This 'uncluttering' of the landscape and extensive levelling[21] work was an aspect of Brown's design present in his early work, as his contract at Petworth with the Earl of Egremont in 1753 explains:

> To reduce the Terraces & shorten them: so as to give the Ground on which they stand a natural Form, making it correspond properly with the Park & Level on Front of the House.[22]

This is one of the clearest, direct references to Brown removing residual formal elements around the house. Although this was not always the case in the 1750s, he did often succeed in persuading owners to accept his minimalist alternative, as at Trentham, Staffordshire.[23,24] Why he was so successful at doing so is an interesting question. Probably, it was a reflection of how influential his patron, Lord Cobham, was on the formation of 'taste' in landscape making. Stowe had been a landscape about moral improvement, which the Patriots such as Cobham and General Dormer at Rousham, Oxfordshire, favoured, almost incidentally shaped in an 'arcadian' form. Brown's landscapes, however, were about personal responses. His designs took the concept of a sylvan landscape and developed it in a completely new direction. The aim was to create something of beauty for its own sake, not for its message, to show off the house and lake.

Brown's 'lawns' were innovatory in another way: he used Dutch Clover (*Trifolium repens*) as well as grass seed in the mixture. This had two important advantages. The clover remained green throughout the winter,[25] tolerating frost well, making the lawns attractive even in winter, and they considerably enhanced the nutritious value of the lawns if the owner wanted to graze them. This would have applied to the area on either side of the pools at Longleat that Brown made for example (Figure 5.8). Other advantages of this white clover, with its smaller leaves than forage varieties, were its ability to spread quickly and its low habit of growth, making it a good choice for a lawn.[26] As far as is known, Brown was the first person to do this in a *designed* landscape.[27] This three-pronged approach – levelling, draining, grass planting – formed the bones of a Brown landscape from the beginning. These items appear in contract after contract, which is why Brown's landscapes are so recognisable, and may well be why many have been allowed to endure for so long.

Manipulation of water

It was Brown's design of ornamental water, however, which had the greatest impact, and there are several aspects to this. He extensively manipulated the water, which was often on several different levels, yet succeeded in making it look completely 'natural' and like one lake. He also used, and probably invented, the river-lake to enable him to provide a lake in difficult topography, or where money was limited. His lakes were often directly in view of the house, and they often appeared larger and more interesting than they actually were because he concealed the ends. These elements were present in his designs early in his career, as we have seen with his plan for Packington.

Because Brown's lakes appear to be so natural, it is tempting to think that he decided where to build the dam and roughly how big the lake would be, and commenced building, allowing the water to fill the valley and form an irregular lake. Instead, the evidence suggests that Brown was very careful to ascertain just how high the water would be, and where it would extend to. His staking out of the landscape is nearly as well-known as his legendary remark about its 'capabilities'. In fact, Brown was meticulous about measurements.[28] At Blenheim, he had to work out how high the dam (in the form of a cascade) needed to be in order to flood Vanbrugh's bridge to the desired height, even though the bridge and dam were out of sight of each other.[29] Clearly, the process was to talk to the client whilst walking over the landscape, and to put stakes in to signify the extent of the water and the various levels, with ground to be made up or excavated. Brown's contract with Lord Weymouth (Longleat) refers to this:

> to repair the Head [dam] and to plant and alter it [the water] according to the Idea talked on with Lord Weymouth and also to lengthen it across the Fosse and the Road according to the Stakes put in for that Purpose and to make the Cascade proposed there and to plant the same.[30]

One possibility is that at a critical point in the Blenheim landscape, for example, such as Vanbrugh's bridge, Brown and the owner would discuss how high the owner wanted the water to be. A stake would be planted to record that on one side of the valley, and another one mirroring it on the other. Then they would walk, following that level (or contour) along the valley, planting stakes, until a reasonable site for a dam was reached, depending on factors such as the width of the valley and the feasibility of building a dam there, plus whatever was known about the underlying rocks, soils and springs.[31] It is likely that the owner would have some idea of these latter factors. It is also likely that the owner would already have some idea of how much land he was prepared to take out of production, and that Brown had that information from the outset. This was a basic process which others would have used, not just Brown. Having sketched an outline, as it were, Brown would then have had to make many decisions, based on the water sources and soil types that he found, about what kind of dam construction to use, whether subsidiary dams were necessary to create sufficient extents of water in various places, whether digging out banks was necessary. We have seen Brown doing exactly that with the chain of ponds at Longleat, raising the dam of one and lowering the end of the next pond in order to make one larger pool out of the two. The fact that Brown adapted or extended existing pools does not belittle his achievements. Working with an existing water source was sound practice: at least it was known that the ground would hold water there, and how reliable the water supply was. As the lake was being filled, adjustments might be needed and were probably commonplace, especially in connection with any bridges being built, and especially in the context of Brown's detailed attention to the angle of slopes.

It was this ability as a water engineer that was one of the keys to Brown's success. He had to be able to dam rivers and channel water successfully to make lakes.[32] Possibly, he spent some weeks in Lincolnshire, perhaps in 1738–9, with the Duke of Ancaster at Grimsthorpe.[33] If so, it is also possible that he encountered John Grundy (1719–83) of Spalding who was working there. Known for his engineering work in the Fens, and as a founder member of the Smeatonian Society,[34] Grundy was at home with designing and building dams, using the clay core method with a cut-off trench (Figure 4.2) at Grimsthorpe and it is clear from this, and Richard Woods's instructions for Cusworth, as well as a diagram for the dam at Bowood, that this was the widely accepted method used at the time.

As we have seen, the dam is usually the most important part of a lake, and the most difficult to construct. Figure 1.7 is an interpretation of the type of dam Brown used, based on the plan in the Bowood archives and what is known about the dam for the main lake at Petworth (1755), through field investigations and the contract with Lord Egremont.[35] It indicates that Brown was well abreast of contemporary construction techniques. Despite this, he still had problems with various lakes leaking, as at Bowood in 1768. At Harewood, Yorkshire, in 1777, Brown experienced another problem: the water ran out as fast as it ran in when the 'plug' [sluice] in the dam was closed. That dam had to be completely opened up and the clay core was found to be leaking, which, as at Bowood, would have entailed draining the lake.[36] This problem of dams leaking when they were first made – we have seen how concerned Vanbrugh was about his dam at Castle Howard – was clearly not uncommon, and Brown does not seem to have been unduly concerned when it happened, though evidence is slender.[37] It is possible that there was a much greater acceptance in the eighteenth century that water features might not be immediately successful, which would not be countenanced today. Brown completed contracts for lakes for many clients, which illustrates that he was not just a fashionable landscape designer but also a successful water engineer. Had his lakes not held water in the long term, he would not have been employed by so many people.

For reasons of natural topography, Brown did not always manage to position his lakes in direct view of the house – Melton Constable and Wimpole are places where he did not – but in many instances he did. At Wardour Castle, he planned to increase the size of the lake to bring it directly into view from the house, although this was never implemented. Another significant way in which Brown was able to provide a 'lake' directly in view of the house was by creating a river-lake, although this ingenious solution depended on the house being situated near a river. A river-lake was a reasonably straight-forward way of making a lake, with weirs being used to pond water back, making the river wider and giving the appearance of a lake. Unlike serpentine canals, which were geometric and imposed on the topography, river-lakes were irregular and, on suitable sites, involved less earth moving. It appears that Brown was the first person to make a lake in this way, possibly at Latimer, Buckinghamshire, in the late 1750s, but definitely for Lord Dacre at Belhus (Figure 5.11). The intention of making a river appear to be a lake is clearly shown by Lord Dacre, writing to Sanderson Miller about Belhus, in 1761:

> I have a number of expenses on me this year and yet I doubt whether I shall have prudence enough to abstain from meddling with my water in the lower part of the park; . . . I know that that coarse meadow and moory sided canal might be converted into a very pleasing scene: And Brown is of the same opinion: we now have another scheme; it is to make it in the river stile instead of the lake.[38]

Several interesting points emerge from this letter. It sheds light on the process of making lakes: Dacre is clearly in discussion with Brown about how to make a lake on the site. There is also a suggestion that the land in question was not of great value (financially or aesthetically) as it was 'moory', and that a lake would improve it. More importantly, it pin-points the concept that a river could be made to fulfil the role of a lake – it could be made to look like a lake. Lastly, it implies that a river-lake was cheaper to make, and Lord Dacre's letter is useful evidence that this was the way

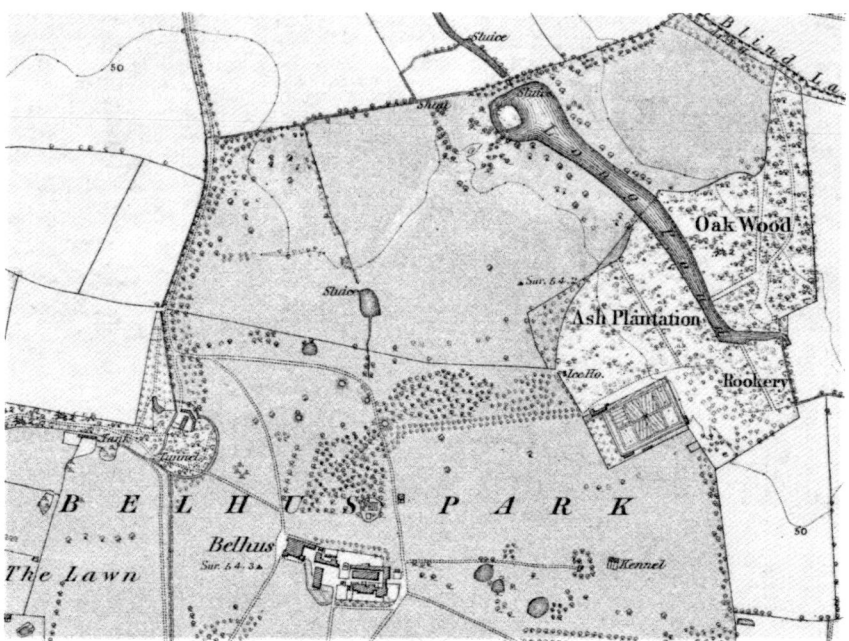

FIGURE 5.11 Belhus Park, Essex, OS 6" map, 1862. Richard Woods modified the ends of the lake in 1770. *Reproduced by permission of the National Library of Scotland.*

in which contemporaries were thinking. A river-lake followed an established water course, so obviated the necessity for a great amount of expensive earth moving; it did not involve dams which might give rise to problems, and it took a smaller area out of production.[39] Brown made about a dozen river-lakes out of perhaps 23 (Appendix II) which were made in the second half of the eighteenth century. Two are well-known: Chatsworth and Audley End. Chatsworth is discussed in Chapter 4, but it is worth pointing out that the bridge designed by James Paine over the river-lake provides the perfect view of the house (Figure 6.13). Brown created the lake over which to put the bridge, and angled the approach drive to create the 'big shot' view of the house across the water.[40] At Audley End, and also a little further south at Shortgrove, Brown widened the River Cam to produce river-lakes for Baron Braybrooke and the Earl of Thomond respectively.[41] At Shortgrove, one weir appears to have been sufficient to pond back the Cam and produce a similar lake to Audley End. In both places, minimal work was needed to produce the river-lake because of the gentle gradient of the river (it drops *c.* 8 m in 4 km), so that any ponding back, and possibly some digging out of the banks, would produce a significant widening of the river.

A significant part of Brown's skill with water was the way in which he created large and impressive looking lakes almost regardless of the terrain. There were two key elements to this: he disguised the ends of the lakes, with planting and by making the lake end just out of sight, as we have seen, and he created 'split-level' lakes which appeared as one lake, by disguising 'the joins'. The extent to which Brown did this is not easy to evaluate in a map-based survey, as the changes in level are frequently not significant enough to be indicated on maps. However, an examination of factors such as topography, the shape of lakes, the use of weirs and 'bridge-dams',[42] plus known sites, makes it possible to make a provisional assessment. Where a bridge crossing a lake coincides with a contour line and a 'pinched' shape to the lake, for example, it suggests a change in level, with the bridge placed at that point to disguise it, as well as to cross the lake. Brown appears to have been the first person to make lakes in this way, though others soon followed suit. Of course, fishponds had been made in series dropping down the course of a valley for centuries, but no attempt to link them aesthetically had been made. Sometimes geometric ponds and lakes were also made in a series, as at Dyrham Park, Gloucestershire, and Londesborough, Yorkshire, but there was no intention of disguising the ends in order to create an illusion of one large piece of water. This only occurred when lakes became informal. This strategy of masking the 'join' between two lakes was important because it allowed a much greater degree of flexibility in the construction of lakes, which Brown took full advantage of. He was able to create what appeared to be a lake in places which were not suitable for one, or to make two pieces of water appear continuous and therefore more impressive for a client. Prior Park, just outside Bath, where Brown produced a plan for Ralph Allen, illustrates the way this worked (Figure 5.12). In a narrow, deep valley, what were probably originally medieval fishponds have been made into two ponds of *c.* 0.7 h each, or a split-level lake.[43] As the map shows, the dam dividing them has been shaped to mirror the shape of the lower pond, and the illusion of a larger

FIGURE 5.12 The 'lake' at Prior Park, Somerset, with an extract from the OS 6" map, 1902. *Author and the National Library of Scotland.*

piece of water was very successfully created, as the photograph illustrates. It should be noted that, despite the plan, there is no evidence that Brown actually worked at Prior Park, though perhaps his plan was influential.

Brown's use of bridge-dams to disguise different levels of water may have been learnt at Stowe. A tour round the Stowe landscape reveals calm, uncluttered, gently sloping banks down to lakes. The main lakes were still geometric when Brown was there, but the 'lakes' (irregular ponds) of the Elysian Fields had been made. Bridgeman's plan of Stowe, *c.* 1735, indicates that the area of the Temple of Ancient Virtue was largely a clear lawn then.[44] The Shell Bridge in the Elysian Fields (Figure 5.13) clearly appears to be a bridge over a lake, whereas it is actually disguising a dam and a change in level.[45] Brown used a scaled up version of this bridge at Wotton, Buckinghamshire, to disguise a change in level between The Warrells lake and the informal canal which he constructed in the late 1750s (Figure 5.14), connecting The Warrells and The Lake. During significantly wet weather, water does flow through the central upper arches from the informal canal beyond.[46] Remains of a sluice gate to control water levels still exist in the central lower arch; this has been replaced, probably in the

FIGURE 5.13 A 'lake' in the Elysian Fields, Stowe, Buckinghamshire.

FIGURE 5.14 The Five Arch Bridge, Wotton Underwood, Buckinghamshire.

mid-nineteenth century, by an open spillway. The Five Arch Bridge does not attempt to deceive the viewer about the change in level but it is so dramatic that it does just that, by distracting the person walking round it, who finds that the water apparently resumes its course. Whateley, in his *Observations*, came to the same conclusion:

> out of it [The Warrells lake] issue two small streams, winding towards a large river, which they are seen to approach, and supposed to join. A real junction is however impossible, from the difference of levels; but the terminations are so artfully concealed, that the deception is never suspected; and when known, is not easily explained.[47]

Brown's work with water at Wotton is highly complex and, at a technical level, is extremely competent. The landscape is low lying and undulating at best, formed from underlying mudstones, with the house on a low ridge, and a few small streams. Brown created *c.* 22 h of water out of this unpromising area, including an informal canal which linked the two large lakes (Figure 5.15). The system basically works

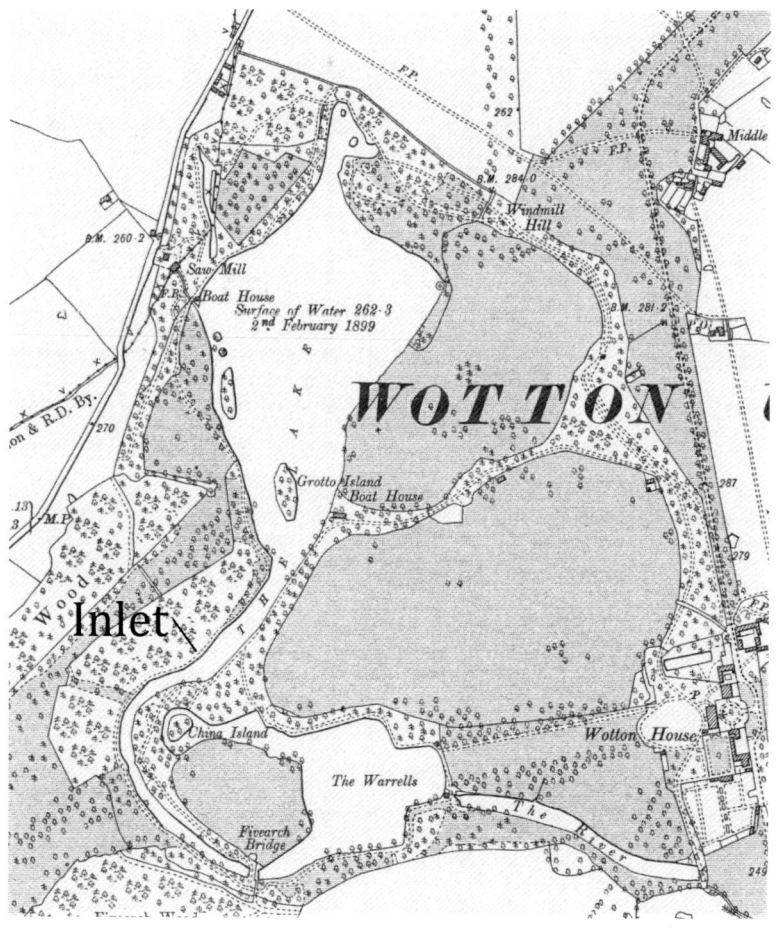

FIGURE 5.15 Supply inlet for the lakes, Wotton Underwood, Buckinghamshire, OS 6" map, 1898. *By permission of the National Library of Scotland.*

like an enormous bath, with the water supply coming in about halfway along the canal. It fills up the 'bath', and overflows in both directions, north and south. The water 'flows' in either direction, depending on the wind.[48] Currently, the over-flows/control points are open spillways in the centres of the dams at either end of the system, but these are almost certainly nineteenth or twentieth century, having replaced sluices within the dams (Figure 5.16).[49]

As Thomas Whateley says in his description, the circuit followed by the visi-tor works powerfully on the senses, with views back to the house across water at several points, glimpses or views of buildings in the landscape, or bridges across the water. His analysis of the landscape is very insightful. His comments on the island – Grotto Island – which marks the transition between the canal and The Lake are particularly relevant:

> an island near the conflux, dividing the breadth, and concealing the end of the lake, moderates for some way the space; and permitting it to expand but by, raises an idea of greatness, from uncertainty accompanied with increase. The reality does not disappoint the expectation; and the island, which is the point of view, is itself equal to the scene; it is large, and high above the lake; the ground is irregularly broken; thickets hang on the sides; and towards the top is placed an Ionic portico.[50]

FIGURE 5.16 An open spillway in the Warrells dam at Wotton Underwood, Buckinghamshire.

Brown used two other islands in his design to mask the ends of the lakes: one in the western arm of The Warrells, and one in the north of The Lake.

Many of the things Whateley says in *Observations* 1770 regarding the successful treatment of water could have been taken from Brown's landscapes: turning the end of a lake out of sight to make it appear to continue, as at Bowood, or planting around the end of it for the same reason, using an island to mask a junction, or make a lake look bigger, planting naturally alongside a river-like piece of water in "a just imitation of cultivated nature".[51] As *Observations* was published in 1770, it is reasonable to conclude that Brown's plans largely pre-date Whateley's ideas, and that Whateley's conclusions were based on, or at least influenced by, Brown's landscapes. His publication of them may also have promoted Brown's work among clients. In one respect, Whateley may have had a particular influence: the use of islands in lakes increased towards the end of the eighteenth century, and this may be ascribed to Whateley's analysis of how they increased the apparent size of lakes, if used circumspectly. Holkham is a case in point. An island was made in the southern part of the lake by the mid-eighteenth century. Towards the end of the century, two more islands were made at the northern end, probably in 1784–5 when William Emes modified it. Similarly, the group of three islands in The Lake at Wotton was made in the late eighteenth century (after 1789), one of their functions being to act as a duck decoy.[52]

Brown's treatment of the water at Wotton is notable because of the level of difficulty involved. As at Croome, he created a lake on an unpromising site.[53] In places which were less challenging – less flat, with a better water supply – Brown made use of a 'bridge-dam' to conceal a drop in level and make two lakes appear to be one. Figures 5.17 and 5.18 illustrate how this works. Figure 5.17 shows the filter pond which Brown constructed at Bowood, which has prevented the main lake from silting up since it was made.[54] This necessitated a significant change in level between the two lakes, and Brown provided decorative side channels to cope with any flood water, similar in principle to the one at Wotton. It is marked on his plan of 1763 (Figure 5.19). As we have seen, Brown probably absorbed this idea from the Shell Bridge at Stowe, but he then developed it into a structure which carried a drive. This appears to be the primary reason for the bridge in Figure 5.18, whereas actually, the main reason for the bridge-dam being in that place was to create this subsidiary lake.[55] This gave Brown much more flexibility in designing his lakes: as well as creating the illusion of one large lake, it also took less land out of production, and Brown could juggle the requirements of topography, water retention and approach drives (or other carriageways) to give the best result. This is one of the answers to the question of how Brown so frequently managed to achieve spectacular views of the house across water as the visitor approached. Instead of making one enormous lake, he linked smaller pieces of water together, giving him more flexibility with siting the water, and creating the illusion of one lake. The bridge-dam was a very successful new technique in the mid-eighteenth century, possibly invented by

FIGURE 5.17 Bridge–dam at Bowood, Wiltshire, upper side.

FIGURE 5.18 Bridge–dam at Bowood, Wiltshire, lower side.

Brown, and he took full advantage of it. The viewer is either walking across on the bridge, and sees water on either side, or is walking alongside the lake and becomes distracted by the cascading water on the downstream side. In either case, the fact of the difference in water level does not obtrude on the consciousness. If Brown had not used these interim dams at Bowood, he would have had to make a much bigger dam at the north end of the valley (it is *c.* 4 m high) to create a lake which stretched so far south (Figures 5.19 and 5.20) and, as well as being a much more challenging dam to construct, the lake itself would have filled the valley and reached towards the walls of the house. Clark's Hill, which

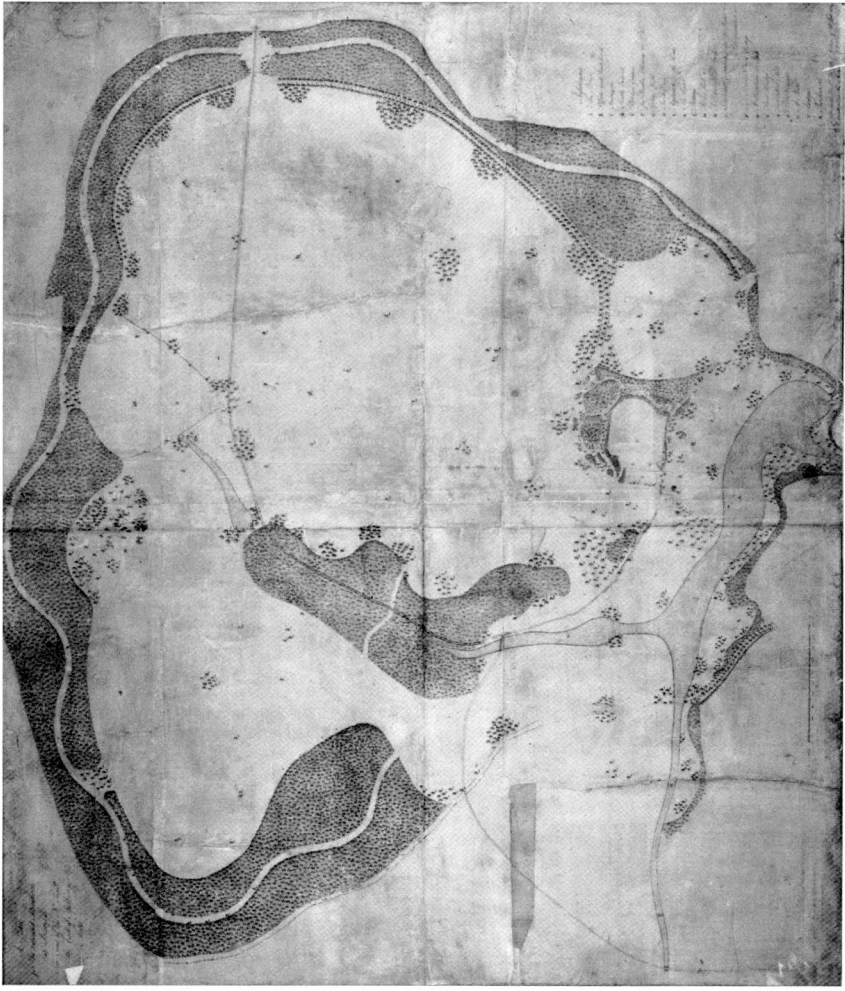

FIGURE 5.19 Brown's 1763 plan for Bowood Park, Wiltshire. *Courtesy of the Trustees of the Bowood Collection.*

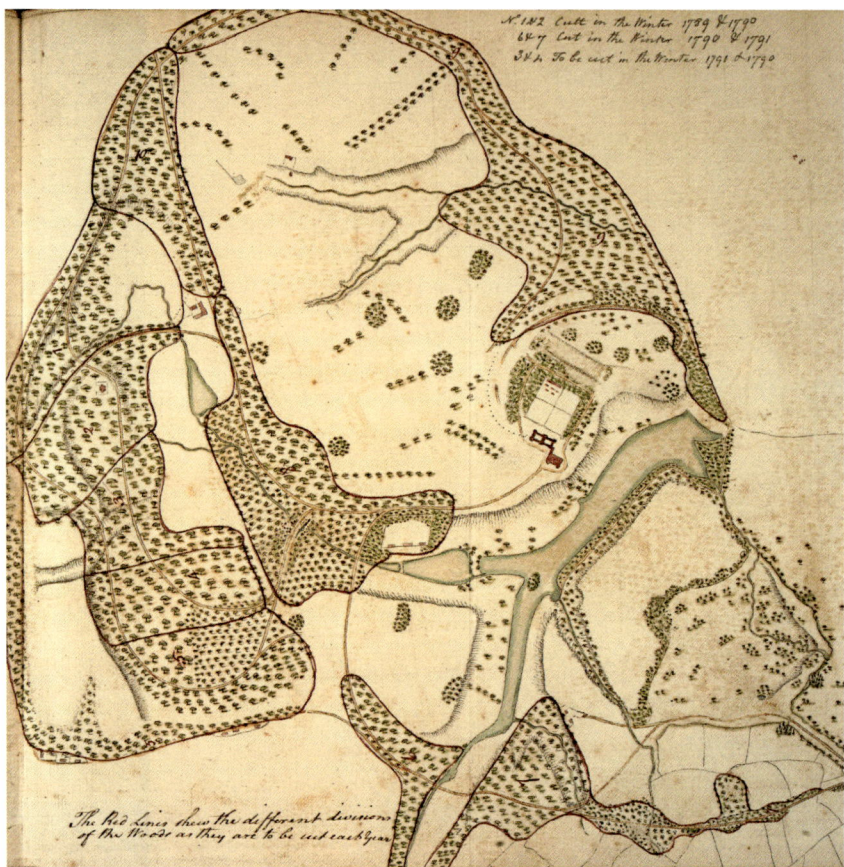

FIGURE 5.20 A 1789 plantation management plan for Bowood Estate, Wiltshire. *Courtesy of the Trustees of the Bowood Collection.*

rises behind the trees in Figure 5.17, was lowered by Brown; the main approach is behind it at this point.[56] Why the hill was lowered is a mystery, unless it was to create a suitably flat area for *al fresco* entertainments whilst looking up the lake and back to the house (Figure 5.21). It did not create views from other points, such as the approach drive.

Two major lakes: Trentham and Blenheim

Brown carried out many prominent commissions and a number have already been referred to, but the lakes at Trentham and Blenheim should also be mentioned, both of them being very large. At Trentham, Brown signed a contract with the 2nd Earl Gower,[57] and constructed a 27-h lake as part of the works (Figure 5.22).

FIGURE 5.21 View from Clark's Hill back to the house and lake at Bowood. The star marks the site of the original house, demolished in 1955. *Courtesy of Catherine Bishop.*

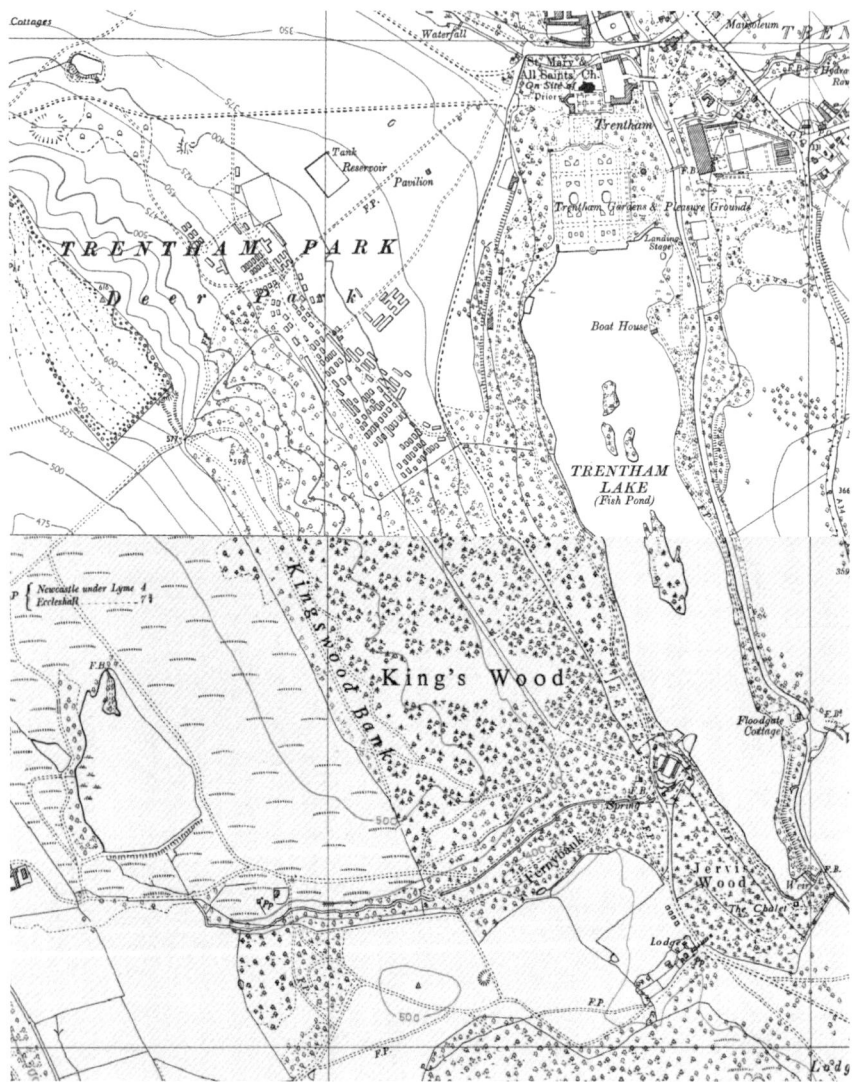

FIGURE 5.22 The lake at Trentham Park, much as Brown planned it, OS 6" map, pre-1932–55. *Reproduced by permission of the National Library of Scotland.*

Originally, a fishpond dating back at least to the sixteenth century occupied the area just north-west of the present formal gardens.[58] By the early 1700s, the area south of the house had been drained and two 'canals' constructed, one of which was quite lopsided (Figure 5.23), though intended to be geometric, according to an earlier 1700s plan (Figure 4.12). In this low-lying, marshy area, Brown incorporated the two existing pieces of water into a vast irregular lake extending southwards, following the western edge of the higher land, marked largely by the

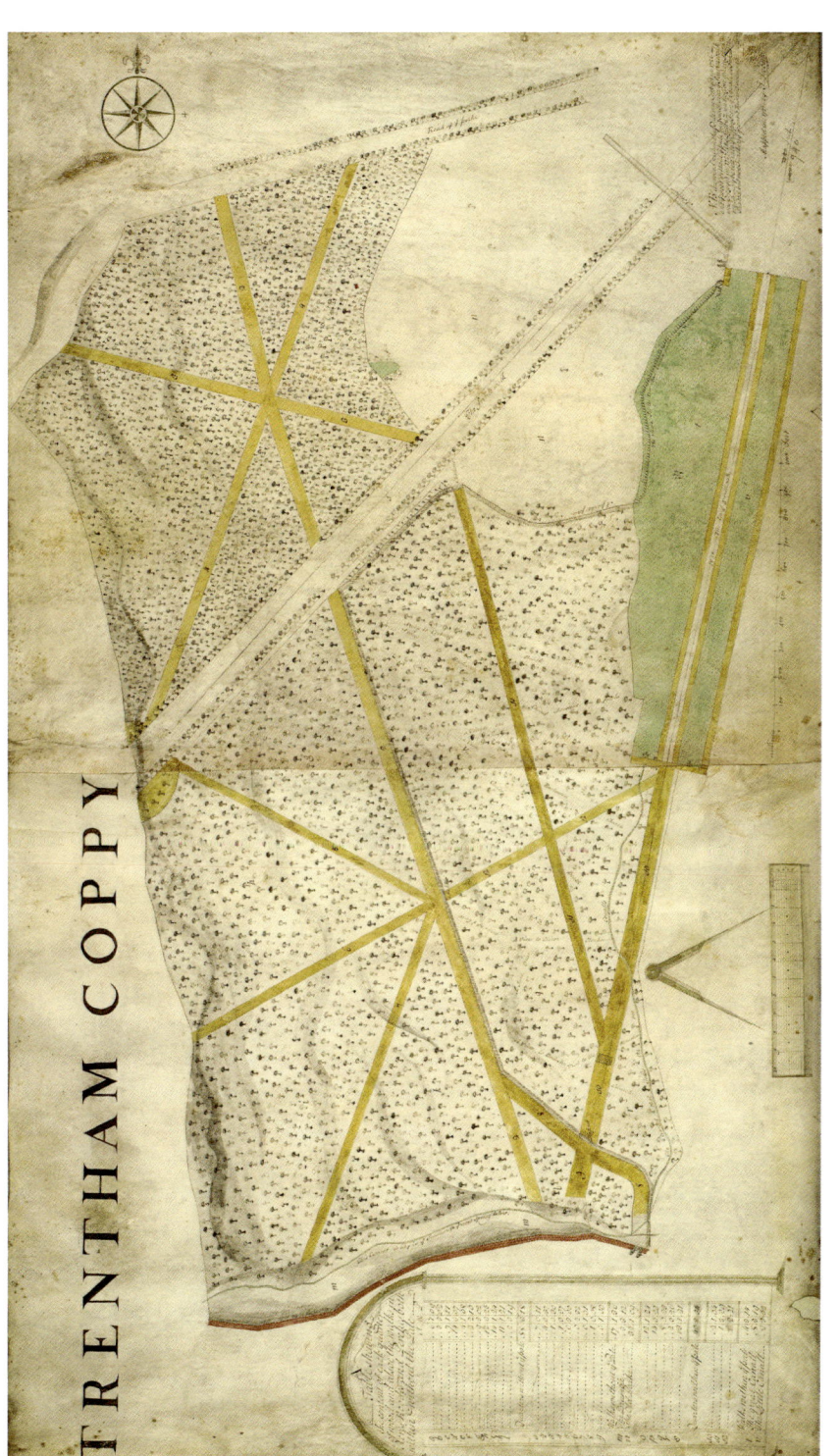

FIGURE 5.23 The 1727 'Coppy Map' of Trentham, Staffordshire. The house is just off the top right corner. *By permission of Staffordshire Record Office. Ref. D593/H/13/37.*

woodland, and he also re-routed the river.[59] The north-west shore of his lake follows the shoreline of the 1727 lake exactly.[60] He also created an island on the site of the previous causeway.

As mentioned earlier, working with a proven water source was good practice, and many of Brown's lakes were re-workings of existing ponds or canals. This applied to all of his 'top ten' lakes:

TABLE 4 Brown's 'top ten' lakes

Date circa	Place	County	Lake size (h)
1753	Croome	Warwickshire	7
1757	Petworth	Sussex	6
1757	Longleat	Wiltshire	8
1759	Wotton Underwood	Buckinghamshire	15
1761	Trentham	Staffordshire	27
1761	Castle Ashby	Northamptonshire	3
1766	Bowood	Wiltshire	14
1767	Blenheim	Oxfordshire	47
1770	Compton Verney	Warwickshire	13
1780	Burghley	Lincolnshire	4.5

In some cases, the existing pond was relatively small, as at Bowood, and at others, relatively large, as at Compton Verney and Blenheim.

From the point of view of the lake, two things stand out about Brown's work at Blenheim: his ability to persuade Marlborough to flood Vanbrugh's bridge, and the Lince dam. Boydell's views of Armstrong's water-works hint at why he was successful, as the palace appears stranded above the water (Figures 5.24 and 5.25). The Lince dam was 900 m long, and Brown built it to canalise the River Glyme and create an imposing river-lake in the park, to replace the stream. Building a dam of that length, which time has proved to be successful, was a considerable technological achievement.

Whilst discussion may continue about how pioneering Brown's landscapes were, the nature of his work with water is outstanding and impressive. The way in which he consistently produced significant lakes at places like Croome and Wotton, and managed to position lakes in full view of existing house sites, as at Burghley and Petworth, or to create the impression of larger areas of water using split-level ponds or lakes, as at Castle Ashby, marks him out from his contemporaries, and commands admiration. He had the skills and confidence to undertake and successfully complete some truly pioneering projects, like Blenheim, and his use of river-lakes was significantly innovative, enabling him to create satisfactory lakes where there were limiting factors such as topography (Audley End) or money (Belhus). The irregular lake was the most significant feature of the English Landscape, and Brown was a master in the creation of lakes.

FIGURE 5.24 Boydell's 1752 view of Blenheim, Oxfordshire, from the north-east. *Courtesy of Blenheim Palace; Jeri Bapasola.*

FIGURE 5.25 Boydell's 1752 view of Blenheim, Oxfordshire, from the south-west. *Courtesy of Blenheim Palace; Jeri Bapasola.*

Other improvers

Brown made about a third of the lakes created in the eighteenth century, and this accounts for his pre-eminent position as a lake maker.[61] However, this means that about two-thirds of lakes in the eighteenth century were made by other people. A small number (about 10 percent) were made by other known designers and the rest were made by unknown improvers – possibly the owners themselves.[62] The other improvers or lake makers were men such as Henry Flitcroft, 1697–1769, John Grundy, 1719–83, Francis Richardson, active 1748–60, Nathaniel Richmond, 1732–84, Richard Woods, 1715–93, William Emes, 1729/30–1803 and Thomas White, 1736–1811. (Humphry Repton is included in Chapter 7.) The lakes attributed to these people are shown in the tables that follow (in Tables 5–8, brackets indicate lakes planned but not made). Of these men, only Richardson, Richmond, Woods and Emes are known to have designed more than one or two lakes, the focus being on 'new' lakes, rather than tweaking existing ones. Richardson appears to have made a lake of *c.* 4 h at Welbeck for the 2nd Duke of Portland in *c.* 1750, though when the other lakes there were made is not known. We have seen that Vanbrugh planned a substantial one (16 h) there in 1703. Stephen Wright

TABLE 5 Lakes by Francis Richardson

Garden Name	County	Approx. date	Date	Feature	Lake Size (h)
Blagdon Hall	Northumberland	1755?	1755	Irregular lake	1.5
Welbeck Abbey	Nottinghamshire	1748–56	1756	Irregular lake	4
Atherton Hall	Lancashire	1759–	1765	Irregular lake	*c.* 10

Source: Data extracted from the Landscape Database.

TABLE 6 Lakes by Nathaniel Richmond

Garden Name	County	Approx. date	Date	Feature	Lake Size (h)
Shardeloes	Buckinghamshire	1757	1757	Irregular lake	13
Stoke Park, Stoke Poges	Buckinghamshire		1762	Irregular lake	2, 1.7
Danson Park	Kent	1763	1763	Irregular lake	8
Himley Hall	Staffordshire		1770	Irregular lake	1.5, 1, 0.7
Beeston St. Lawrence	Norfolk	1773–7	1777	Irregular lake	3.7

Source: Data extracted from the Landscape Database.

TABLE 7 Lakes by Richard Woods

Garden Name	County	Approx. date	Date	Feature	Lake Size (h)
Cannon Hall	Yorkshire	1760–5	1763	Irregular lake	2.8
(Goldsborough)	Yorkshire		1764	Ponds	
Cusworth	Yorkshire		1764	Irregular lake	1, 0.7, 0.3
Belhus Park	Essex		1770	Lake altered	(1.9)
(Boreham House)	Essex		1771	Pond	
Brocket Hall	Hertfordshire	1772–4	1773	Lake altered	8.5
Wivenhoe	Essex	1777	1777	Irregular lake	1.1
Wardour Castle	Wiltshire	1770s?	1778	Irregular lake	2.4, 2.4

Source: Data extracted from the Landscape Database.

TABLE 8 Lakes by William Emes

Garden Name	County	Approx. date	Date	Feature	Lake Size (h)
Keele Hall	Staffordshire	1768–70	1769	Ponds	
(Dudmaston	Shropshire		1777	Hybrid lake)	
Badger Hall	Shropshire	1780	1780	River-lake	1.5, 0.8
Hawkstone Park	Shropshire	1783–7	1787	Irregular lake	19.5
Radbourne Hall	Derbyshire		1790	Irregular lake	1
Erlestoke	Wiltshire	By 1803	1790	Irregular lake	4.1
Locko Park	Derbyshire		1792	Hybrid lake	4
Chippenham Park	Cambridgeshire	1792–1805	1800	Irregular lake	2.5

Source: Data extracted from the Landscape Database.

is connected with the large lake (currently *c.* 42 h) at Clumber Park, designing the bridge for the Duke of Newcastle in *c.* 1774.[63] Both these estates are adjacent to Thoresby, and lie in valleys suitable for making large irregular lakes.

Virtually every self-respecting 'improver' was confident enough to make a lake, and this is illustrated by the one constructed for the Duke of Cumberland and extended for George IV: Virginia Water (*c.* 56 h today). Henry Flitcroft senior made the original lake for the Duke in the early 1750s. It was the biggest ornamental lake of the eighteenth century, even bigger than Blenheim (47 h).[64] In addition, "A grotto was included in the rockwork of the 1754 dam".[65] Perhaps this was over-ambitious as the dam was overtopped in 1768. Eventually (1797), a new dam was built further east, by Thomas Sandby, increasing the size of the lake, and he incorporated a new cascade into the dam, made from massive boulders.[66]

Richard Woods 1715–93

Amongst the lesser known improvers, Richard Woods appears to have been relatively prolific in lake-making (six lakes), though his lakes were generally small, being of 1–3 h, and often involved enlarging ponds or modifying existing lakes, as at Brocket Park.[67] There, in the 1770s, he re-shaped and extended the lake, creating a wide river-lake using a series of weirs. One of these was a bridge-weir, designed by James Paine. This is remarkably similar to the bridge-weir designed by Robert Adam for Kedleston in 1764 but built in *c.* 1771 (Figure 4.15)[68] which is reminiscent of, though more elegant than, Brown's at Bowood. This use of weirs by Woods, and Adam or Emes at Kedleston, illustrates how relatively easy it was to create a satisfactory lake – one that was big enough to impress – in a valley which was fairly flat. In all three places, the main approach was carried over the lake by a bridge-weir, providing views and the sound of falling water, sometimes with a glimpse or view of the house. Where the land was more undulating or hilly, dams would be required rather than weirs, and these could also be disguised, as bridges. By the later 1760s and 1770s, this format had become popular and reflects the common approaches to landscape design by the various improvers.

In general, Woods would not be regarded as a significant maker of lakes or a designer of large landscapes as he is mainly associated with improvements or modifications to existing landscapes and pleasure grounds.[69] He was involved with only about six landscapes of over 100 acres (40 h) and his reputation was for designing pleasure grounds, rather than whole landscapes.[70] This was the case at Wardour Castle, Wiltshire, where, although Woods submitted a comprehensive plan, only parts were implemented (Figure 5.26). (Brown's plans suffered a similar fate.) At

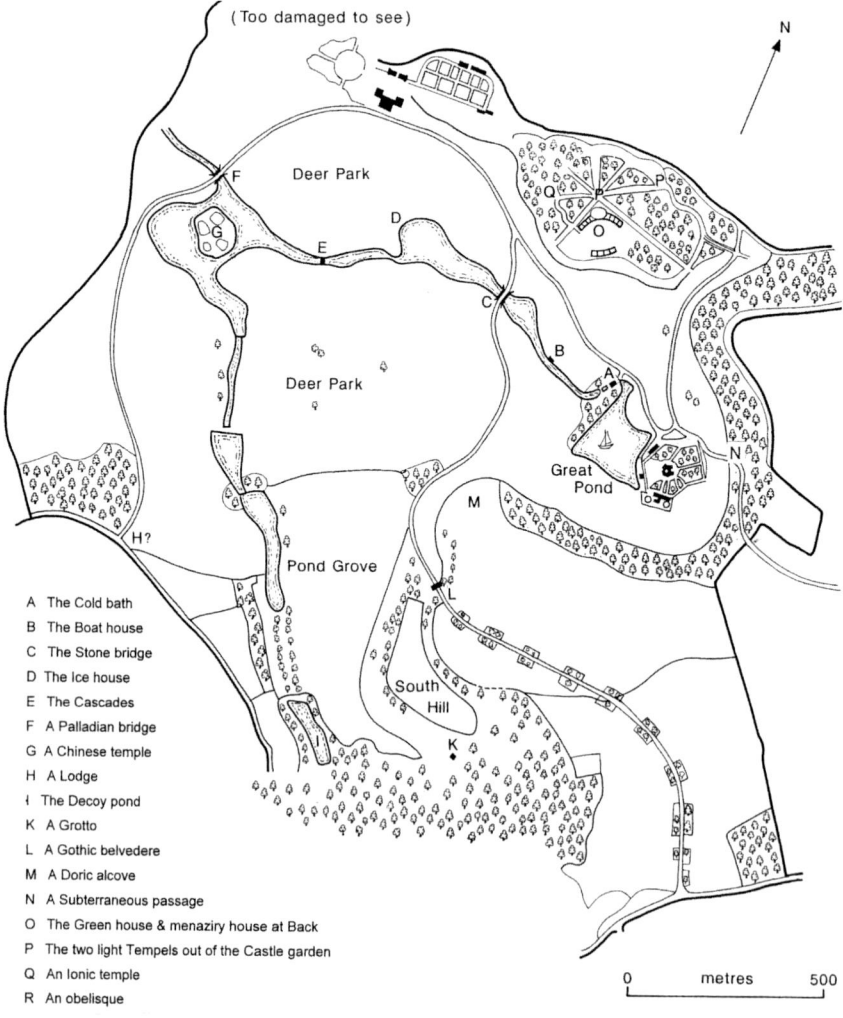

A   The Cold bath
B   The Boat house
C   The Stone bridge
D   The Ice house
E   The Cascades
F   A Palladian bridge
G   A Chinese temple
H   A Lodge
I   The Decoy pond
K   A Grotto
L   A Gothic belvedere
M   A Doric alcove
N   A Subterraneous passage
O   The Green house & menaziry house at Back
P   The two light Tempels out of the Castle garden
Q   An Ionic temple
R   An obelisque

FIGURE 5.26 Richard Woods' plan for Wardour Castle, Wiltshire, 1764. *Courtesy of Fiona Cowell; Philip Judge.*

Cusworth, Woods created a new lake, of *c.* 1 h, with two subsidiary pieces of water (*c.* 1762), amounting to *c.* 2 h overall.[71] His diagram for field drains makes it clear that he was aware of the importance of draining the adjacent boggy ground, and of sculpting the land into pleasing 'soft swells' where stakes were set for that purpose. He said:

> in forming the mount you'l use all your engineowitty [ingenuity] to give the ground as much variety and life as possible by rowling and waveing it about in the manner I described to you.[72]

Though not working on the same scale at Cusworth as Brown was at Bowood, Woods was clearly aware of the prevailing fashion for smooth, undulating turf. Whether he would have achieved a similar result at Wardour with his large planned lake in full view of the house as Brown did at Bowood we will never know. Unlike Brown though, he separated the house at Cusworth from the lake with a Hanging Lawn, which was something Brown would have avoided. However, it is clear that Woods was particularly concerned that the lakes would also be visible from the house, and from various walks and seats:

> having leavel'd and formed the water line next the park, you must then . . . cut down four other pateron lines [lime trees?] and let them be so shaped as to let you see the edge of the water from the house.[73]

Similar levelling of the lawn was specified by him at Wivenhoe in order to ensure the water would be seen from the house (Figure 5.27). Woods's work illustrates that there was a uniformity about what clients wanted their landscapes to be like in the

FIGURE 5.27 Wivenhoe Park, Essex, by Constable, 1816. *National Gallery of Art, Washington.*

mid-eighteenth century: a (large) piece of water visible from the house, with smooth lawns going down to it, and further opportunities to view it from walks, drives around the park and, most importantly, when approaching the house along the main drive.

Nathaniel Richmond 1732–1784

Nathaniel Richmond, another 'improver' working at much the same time as Brown and Woods, made five lakes, on current evidence, but one of these, Danson Park (8 h), was substantial. It is likely that he also made the 13 h lake at Shardeloes (begun in 1757) for William Drake, to replace a canal and ponds.[74] He had some experience of working with Brown, at Warwick Castle and Moor Park, Hertfordshire, in the early 1760s.[75] There is a slight question mark over Richmond's lake-making. His designs, as David Brown points out, often feature a 'curlicue' shape to the dam (Figure 5.28),[76] but this made the dam unnecessarily complicated. Earth dams were very basic structures, and not highly stable; it would certainly have been unwise to over-complicate the business end of a lake in this way. In fact, it seems likely that these dams were not constructed as shown on the plan as at both Danson and Shardeloes subsequent maps show much straighter, conventional dams (Figures 5.29 and 5.30). This indicates that either Richmond

FIGURE 5.28 Nathaniel Richmond's 1763 plan (attributed) for Danson Park, Kent. *Courtesy of David Brown and Tom Williamson.*

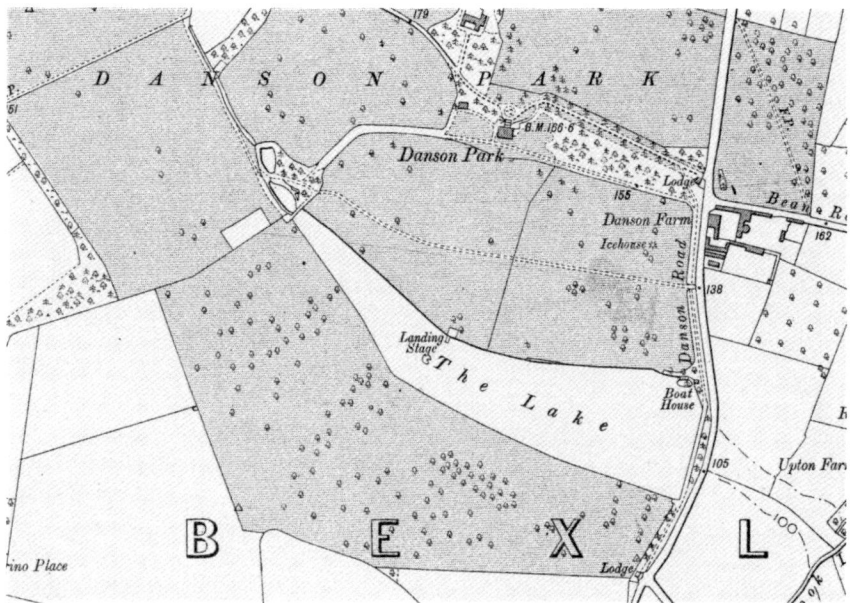

FIGURE 5.29 Danson Park, Kent, OS 6" map, 1908. The lake is basically the same shape as shown on the 1799 OS drawing. *Reproduced by permission of the National Library of Scotland.*

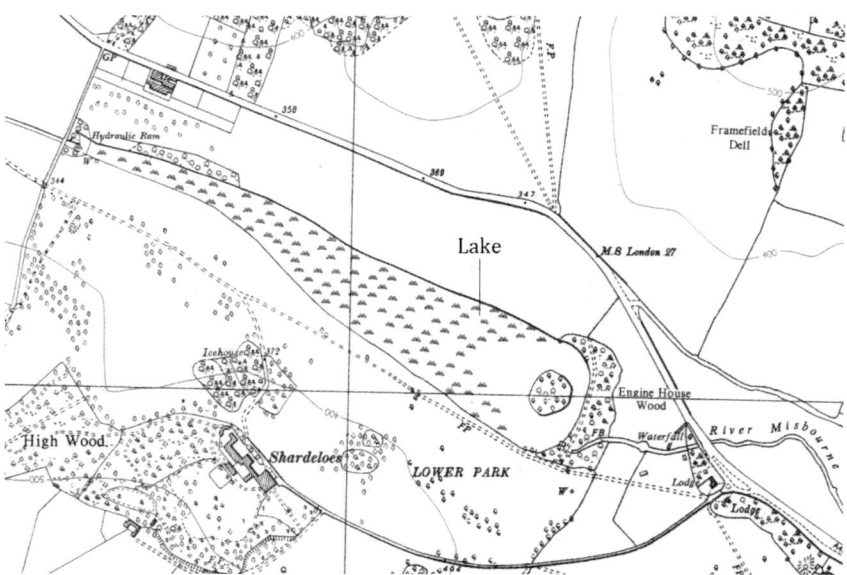

FIGURE 5.30 Shardeloes Park, Buckinghamshire, OS 1:10,000 map, 1930–59. *Reproduced by permission of the National Library of Scotland.*

produced more pared down and rational plans for construction, or that someone else was in charge of making the lakes. The 'curlicues' at the ends of his dams appear to be merely a kind of 'signature' element in his plans, perhaps to impress clients.

William Emes 1729/30–1803

In contrast to the men discussed earlier, William Emes was a more notable lake-maker. He first came to notice as head gardener at Kedleston in 1758, where he was attributed with designing the lake (Figure 5.31), which was then implemented by Robert Adam, who subsequently ousted Emes completely.[77]

FIGURE 5.31 Estate plan of Kedleston Park, Derbyshire, by George Ingman, 1764. *Copyright National Trust Images/John Hammond.*

The style certainly belongs to Emes, as a comparison with his 1786 plan for Erlestoke shows (Figure 5.33). Drawn by George Ingman, the Kedleston plan may well be a composite effort as the northern part of the long circuit walk south of the house is very much like Adam's plan for a shorter one drawn in 1759.[78] Whether by Emes or Adam, the plan has the requisite ingredients of lake, lawns, hills, clumps, perimeter planting and an approach drive crossing the water. A glance at today's OS map of the 13 h river-lake shows that it was made almost exactly as planned – using four weirs to pond the water back, and Adam's elegant bridge-weir (Figure 4.15).

Emes, like his contemporaries, modified existing water as well as creating new pieces. Woods added the fashionable serpentine ends to Brown's river-lake for Lord Dacre at Belhus, and Emes did likewise for Coke at Holkham in the 1780s.[79] Unlike most of his contemporaries, except Brown, Emes was responsible for some sizable lakes: Erlestoke (4 h), Hawkstone Park (Hawk Lake, 19 h and Menagerie Lake, 2.5 h), Oulton Park (14 h) and Locko Park (4 h). Emes was working for Sir Richard Hill at Hawkstone in the mid-1780s and both lakes he made there were slim and sinuous.[80] The larger, Hawk Lake, was very much like a contour canal in construction, with a dam 1.3 miles long. At Oulton Park, Emes worked in conjunction with his pupil, John Webb, and produced an unusual design for a lake. It was a roughly rectangular area of water with two very slim arms extending from either end, suggesting that an existing large area of water was being adapted.[81] However, as previously mentioned, lakes in Cheshire have to be approached with caution, as the effects of salt extraction can lead to subsidence and the appearance of new lakes.

The design for Locko Park, which Emes produced in 1792, appears to con-form to the Brownian blue print of house looking down on the lake and approach drive crossing the water, but with Emes's rather square perimeter planting, which can also be seen in his plan of 1786 for Erlestoke.[82] What is different about Locko is the squarish, hybrid lake. The Historic England entry mentions, "Vari-ous drawings of stretches of water amongst estate papers appear to be alternative designs for the shape and outline of The Lake", and it may be that Emes was not particularly happy with this uncharacteristic shape.[83] Possibly there was an exist-ing piece of water which he was trying to adapt. Usually, Emes's lakes tapered elegantly away and curved into woodland.[84] Perhaps he was constrained by his client's wishes: a river-lake of the Kedleston variety would not have reflected the house satisfactorily (and does not do so at Kedleston), but Emes's lake at Locko does (Figure 5.32).

The most complete picture of Emes's design style emerges from his work at Erlestoke. The series of maps which has survived – before, during and after – makes it particularly valuable because not only can we see what Emes planned, but we can also see what was implemented and this tells us a lot about what was fashionable in lake design towards the end of the eighteenth century, *after* Brown. The *c.* 1782 parish map shows a small park with a house on the low land, and a village street extending south of today's main road, with two village ponds. Joshua Smith built a

FIGURE 5.32 Locko Park, Derbyshire, 2020. *Copyright Alan Murray-Rust (via Geograph Project www.geograph.org.uk).*

new house on the high ground to the south-east and commissioned Emes to rede-sign the landscape.[85] Emes's plan (Figure 5.33) shows that he intended to convert a stream and the ponds into a long, sinuous lake of *c.* 4 h. Tree planting, adapted from a previous formal plantation, would mask the dam at the head of the lake, and woods wrap around the tail end. Emes depicted this lake as one continuous piece of water, but the ground falling from south to north dictated that this should be constructed in several pieces, with 'seven cascades', as John Britton remarked.[86] Wooded walks bordered the southern half of the lake on the plan, with at least two crossing points. The northern part of the lake was made in flat, marshy ground, necessitating minimal work in forming embankments and the low dam. A 'Small Banqueting Room for Fishing', which Emes had tucked away near the farmyards, appeared in a much more prominent (and useful) position alongside the lake, near the lowest cascade or weir, by 1825 (Figure 5.34).

Almost all the wooded areas on the parish map were incorporated by Emes into his plan, and modified or extended by him. Of two main approach drives planned by him, one from the east simply entered the park and went straight to the house, though possibly giving a view from this higher ground over the park and lake. The westerly approach crossed the lake on a three arched bridge and went through the park before ascending to the house. Perhaps this was the reason for moving the church in the 1880s: it may have obscured the view of the house whilst

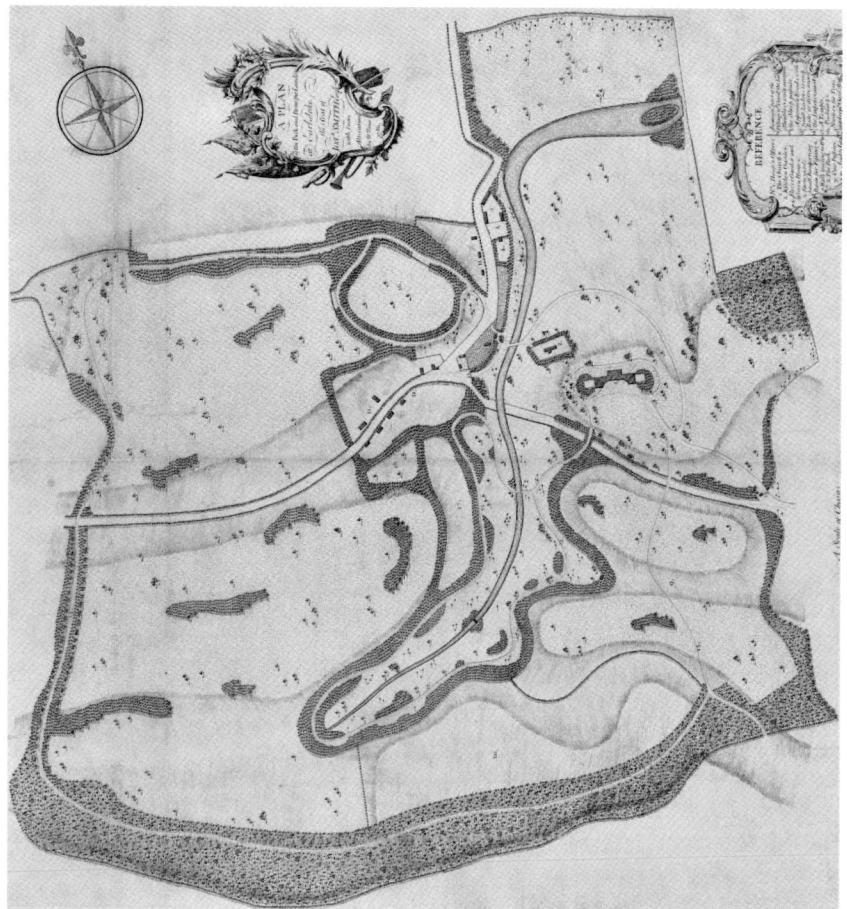

FIGURE 5.33 A plan of Erlestoke Park, Wiltshire, by William Emes, 1786. *RIBA Collections. Ref. RIBA96297.*

crossing the lake. "A Ride or drive around the Improvements", going southwards from the house providing a complete circuit of the park on the high ground, was implemented as planned. Emes differed from Brown here: there was no apparent thinning of trees to facilitate views over the estate.

Joshua Smith (died 1819), the owner of Erlestoke, was a local MP and a rich man, though not a member of the aristocracy, and he clearly regarded William Emes as a suitable improver to design his new estate.[87] He obviously approved of the design whole-heartedly as he implemented it very faithfully, presumably feeling that it had every modern feature – circuit drives, lake, cascades, fishing pavilion, garden 'temple', greenhouse, detached kitchen garden, not to mention the park itself. This 'three dimensional' view of Erlestoke tells us a lot about prevailing fashions in *c.* 1790. The 'Brownian' formula, which Emes implemented, of park

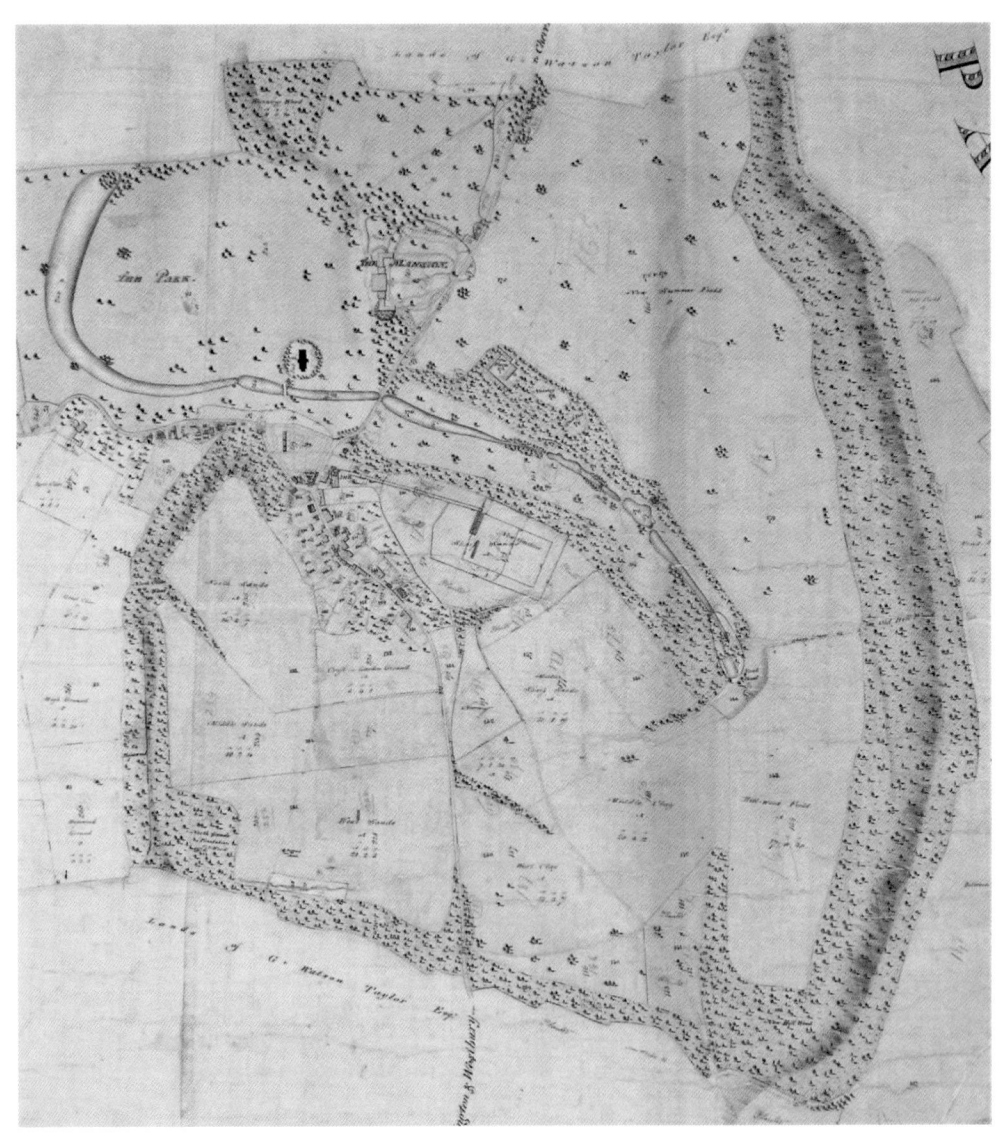

FIGURE 5.34 An 1825 estate map of Erlestoke Park, Wiltshire. *Courtesy of Wiltshire and Swindon Archives. Ref. 1553/92.*

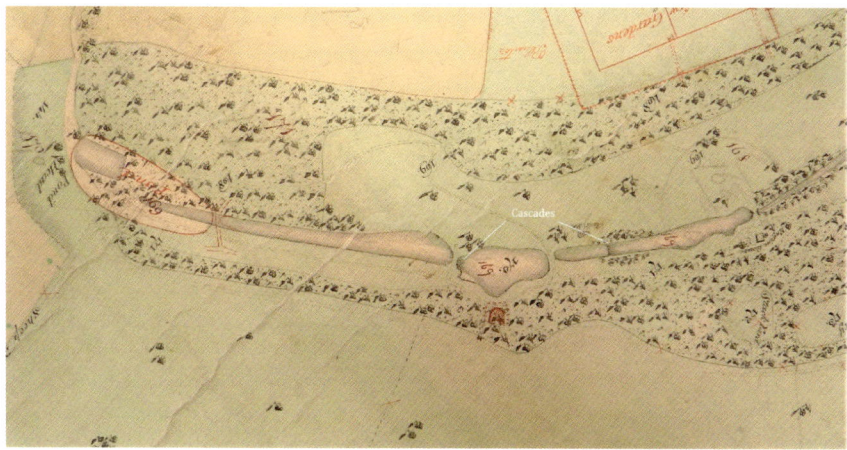

FIGURE 5.35 A detail of an 1825 map of Erlestoke Park, indicating the cascades. *Courtesy of Wiltshire and Swindon Archives. Ref. 1553/92.*

with perimeter planting, clumps, lake, and approach drive over water was still desired and the elongated, irregular lake was still fashionable. The southern part of the lake had parallel sheltering belts of mixed deciduous trees and conifers, with some denser planting at the tails of ponds, and by a possible cascade (Figure 5.35). However, the planting was not as subtle or varied as Brown's, as illustrated in the Kimberley plan (Figure 5.2) and, apart from the conifers, there is no indication of any specimen trees. What this adds up to is a formulaic landscape design, albeit a high quality one. What is original though is the sinuous design of the lake, which is the epitome of Emes's style.

Conclusion

In the first half of the eighteenth century, it was the owners of great estates who drove the evolution and creation of lakes, aided by Vanbrugh, who also designed some very significant buildings. His ideas about ornamental water and early plans for irregular lakes pre-dated Brown's by several decades. By *c.* 1750, 'improvers' were emerging, such as Brown, Richmond, Woods and Emes, to implement the 'landscape style' for a wider circle of clients, some of whom were of lower social rank. Brown was pre-eminent, however, not least because of the numbers of land-scapes he designed. His approach to creating lakes was innovative, and he stamped his style on house and surroundings, creating a Brown 'marque' of house, smooth lawns and lake. His successful commercial operation ensured that his design style became widespread, and widely desired. The other designers were also successful in much the same style, but they did not make many lakes compared with Brown, although they had the technical competence to do so. What also emerges is that many lakes were made by unknown people, probably owners working with their head gardeners and estate labour. It is possible there was a significant body of skilled men, probably at local level, who could make dams for large fishponds and lakes but there is very little information about them. Similarly, at national level there are very few indications of men who were dam making experts for ornamental lakes. Up to c. 1790, designers, including Brown, were largely following the consensus of what constituted a desirable landscape: 'natural' looking contours, 'natural' looking water, clumps and perimeter tree cover to control views, and to provide seclusion and a sense of ownership. In the 1790s, this was beginning to seem *passé* and the ideas of Sir Uvedale Price and Richard Payne Knight were starting to influence opinions, as was Humphry Repton's practice, the two being largely opposed, as we shall see.

Notes

1 Such as Fiona Cowell's book on Richard Woods.
2 The Historic England listing gives varying dates for Kiddington: *c.* 1740 or 1760s for Brown's lake there, so this has not been included here.
3 S. Shields, *Moving Heaven and Earth: Capability Brown's Gift of Landscape* (London: Unicorn Publishing, 2016) pp 29–30.
4 The Landscape Database can be queried in order to show which lakes Brown (or anyone else) made, with details of them.
5 T. Williamson, personal communication, September 2015.
6 D. Brown & T. Williamson, *Lancelot Brown and the Capability Men: Landscape Revolution in the Eighteenth Century* (London: Reaktion Books, 2016) p 101.
7 Ibid., p 95.
8 Hinde, op. cit., p 20.
9 T. Williamson, personal communication, September 2015.
10 Mowl & Earnshaw, op. cit., p 49.
11 T. Williamson, personal communication, October 2015.
12 Hinde, op. cit., p 97.
13 Girouard, *Life in the English Country House*, op. cit., p 145.
14 Bishop, op. cit., p 151.
15 Brown & Williamson, op. cit., p 116.
16 Lawrence, op. cit., p 314.
17 L. Brown's contracts with Viscount Weymouth, 10th October, 1757; November, 1758, ref. Thynne Papers, Box XXX, vol. LXXVII, Longleat House.
18 Bishop, op. cit., p 156.
19 L. Brown's contract for work at Bowood with 2nd Earl of Shelburne of 10th August, 1763, ref. Box VI (a), Bowood Muniments, by kind permission of the Marquis of Lansdowne.
20 Ibid.
21 In the context of landscaping works in the eighteenth century, 'levelling' appears to mean both grading a slope or surveying an area.
22 Brown's contract for Lord Egremont at Petworth, dated 1st May, 1753, ref. PHA 6623, West Sussex Record Office.
23 Brown & Williamson, op. cit., p 78.
24 T. Mowl & D. Barre, *The Historic Gardens of England: Staffordshire* (Bristol: Redcliffe Press, 2007) p 172, Figure 76.
25 William Blake, farmer, personal communication, October 2015.
26 As J. Phibbs points out, clover for forage had been used by great estates in crop rotations since the late seventeenth century: J. Phibbs, 'The Use of Plants in Eighteenth Century Gardens', *Garden History*, Vol. 38, No. 1 (2010), fn. 68.

27 There is no mention of this practice by Switzer or Batty Langley, or George Mason in *An Essay on Design in Gardening* 1768.

28 Savernake Archives, ref. Ailesbury, 1300/1910–3266, WSHC, given in Hinde, op. cit., p 112 and fn. 20.

29 Hinde, op. cit., p 120.

30 L. Brown's contract of 10th October, 1757, with Lord Weymouth of Longleat, ref. Thynne Papers, Box XXX, Vol. LXXVII.

31 In fact, Lawrence alludes to a similar process in his book, op. cit., pp 315–316.

32 The construction of lakes is discussed in detail in Chapter 4.

33 J. Brown, *Lancelot 'Capability' Brown: The Omnipotent Magician* 1716–1783, p 36.

34 Binnie, op. cit., p 67.

35 Hinde, op. cit., p 51.

36 Binnie, op. cit., p 65.

37 1768, 28th May, letter from Henry Merewether to Lord Shelburne about the dam, Bowood Muniments, ref. Box 5 (x) (c).

38 L. Dickins & M. Stanton, eds., *An Eighteenth Century Correspondence* 1910, p 416, quoted in Hinde, op. cit., p 59.

39 Considerable earth-moving might be done when making a river-lake to achieve a particular, desired result. This may have been the case at Panshanger. See Chapter 7.

40 T. Williamson, lecture to the Norfolk Gardens Trust, 2015.

41 Bishop, op. cit., p 164.

42 A 'bridge-dam' is a dam built of stone (usually) which appears to be a bridge rather than a dam. A 'bridge-weir' is a modification of this, for example Adam's at Kedleston.

43 Historic England listing: Prior Park and Garden, Somerset.

44 Jackson-Stops, *An English Arcadia*, p 65.

45 This is an 1879 reconstruction: G. Clarke et al., *Stowe Landscape Gardens: A Comprehensive Guide* (Swindon: National Trust Enterprises, 1997, rev. 2005) p 31.

46 Michael Harrison, estate manager, personal communication, 14.10.15.

47 T. Whateley, *Observations on Modern Gardening* (London: T. Payne, 2nd ed. 1770) p 85.

48 Michael Harrison, estate manager at Wotton, personal communication, 14.10.15.

49 'By-lead' or 'by-wash' (or 'spillway', the American term) is given a date of 1885 in the OED online, "by-lead *n c.* (a) (ii)", accessed May 2017. In addition, the brickwork seen in Figure 3.60 looks modern.

50 Whateley, op. cit., p 85.

51 Ibid., p 74.

52 Michael Harrison, estate manager at Wotton, personal communication, 14.10.15: letter from the 1st Duke of Buckingham and Chandos regarding the state of the net tunnels for the duck decoy.

53 Tim Mowl has pointed out that Phipps made a 'New River' at Croome before Brown worked there: *Historic Gardens of Worcestershire* (Stroud: Tempus Publishing, 2006) p 46. This appears to be a canal-like piece of water, according to the Doherty map of Croome of *c.* 1751.

54 The filter pond was dredged for the first time in 2014, according to the 9th Marquis of Lansdowne.

55 Before I started researching lakes, I spent several years puzzling over why a mere footpath had such an elaborate bridge!

56 Letter from Lancelot Brown to Lord Shelburne, 13th December, 1768, ref. Box VI item d, Bowood Muniments, by kind permission of the Marquis of Lansdowne.

57 His mother was Evelyn Pierrepont, daughter of the 1st Duke of Kingston, and his sister became Duchess of Bedford.

58 A large 'poole' is present on a 1599 estate map in Staffordshire Record Office, ref. D593-H-3–339.

59 Mowl & Barre, op. cit., p 61.

60 Bishop, op. cit., p 175.

61 Ibid., p 144.

62 Statistics extracted from the Landscape Database.

63 Historic England listing: Clumber Park and Garden.
64 Lady Shelburne's diary, 13th May, 1766, ref. Vol. III, p 57, Bowood Muniments, by kind permission of the Marquis of Lansdowne.
65 Historic England listing: Virginia Water Park and Garden.
66 J. Jacques, *Georgian Gardens: The Reign of Nature* (London: B. T. Batsford, 1983) p 119.
67 Cowell, op. cit., pp 121–122.
68 Historic England listing: Kedleston Park and Garden.
69 Cowell, op. cit., p 40.
70 Ibid., p 183.
71 Figure 4.4 shows his plan for the dam.
72 Richard Woods's *Memorandum 5* quoted in Cowell, op. cit., p 113.
73 Richard Woods's *Memorandum 3* quoted in Cowell, op. cit., p 112.
74 A letter from Charles Lyttleton to George Lyttleton, 3rd August, 1757, refers to 'a very extensive lake being form'd', Bucks. R. O., D/DR/2, and John Britton & Edward Wedlake Brayley, *The Beauties of England and Wales*, Vol. 1 (London, 1801) p 361 quoted by D. Brown in *Shardeloes*, PhD thesis, UEA.
75 Brown & Williamson, op. cit., p 117.
76 Ibid., p 144.
77 O. Garnett, *Kedleston Hall* (Swindon: National Trust Enterprises, 1999, rev. 2009) p 30.
78 Sketch plan by R. Adam in Jackson-Stops, op. cit., p 97.
79 This may have been John Webb's work, although he and Emes often worked together at this time, according to David Jacques, op. cit., p 143.
80 Historic England listing: Hawkstone Park and Garden, and Jacques, op. cit., p 116.
81 A plan of this design is given in Mowl, *Historic Gardens of Cheshire* (Bristol: Redcliffe Press, 2008).
82 Historic England listing: Locko Park and Garden.
83 Historic England, ibid.
84 Jacques, op. cit., p 116.
85 Ibid.
86 John Britton, *Beauties of Wiltshire*, Vol. 2 (1801) p 203.
87 VCH, *A History of the County of Wiltshire: Volume 8, Warminster, Westbury and Whorwellsdown Hundreds* (London, 1965) p 84. Smith was brother to Drummond Smith, of Tring Park, given in D. Le Faye, ed., *Jane Austen's Letters* (Oxford: Oxford University Press, 4th ed. 2011) pp 572–573.

6

WHY LAKES EMERGED

Lakes as we normally think of them are extensive bodies of water which are irregular in form, and the question of why they emerged as a wholly new landscape feature in the first half of the eighteenth century is an interesting one, to which there is no straight-forward answer. Several strands are relevant, though. One of the key elements was the increasing scale of landscapes. Another was the influence of Italy and the Grand Tour. Linked to this was the role played by paintings, which had often been painted in the seventeenth century. A fourth strand was the development of leisure activities in parks, and the part played by women in those activities.

'Unbalancing' of landscapes

Wildernesses and plantations were elements which were in the process of changing in the early eighteenth century. Although these elements were becoming less symmetrical, the first truly irregular element to appear was the irregular lake, made possible to a large extent by the 'unbalancing' of landscapes. This is contrary to the widespread assumption that landscapes became informal in the mid-eighteenth century, and that ornamental water became irregular to fit in with that change of style. While landscapes were intrinsically linear, with symmetry as the underlying ethos, it was very difficult to fit lakes – even geometric lakes – into them without disturbing those things. Ornamental water basically had to be in the form of canals to fit into linear, geometric landscapes. However, water changed out of all recognition, in size and shape, in the 1720s–'30s. Figures 6.1 and 6.2 illustrate this change well. The ornamental canals at Belton in the early eighteenth century fitted easily into the geometric design, whereas the lakes, made by *c.* 1750, would not have fitted into that at all, and were made further out in the landscape. Exceptional circumstances, such as wealth and topography, did enable some geometric lakes to be made, as we have seen at Boughton and Welford, but they were usually small because of the

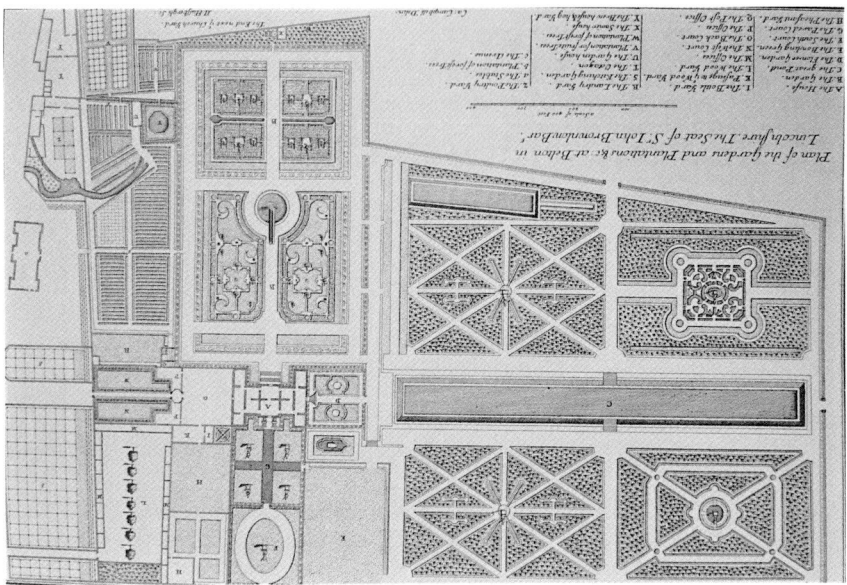

FIGURE 6.1 Belton House, Lincolnshire, in *Vitruvius Britannicus*, 1725. *Historic England.*

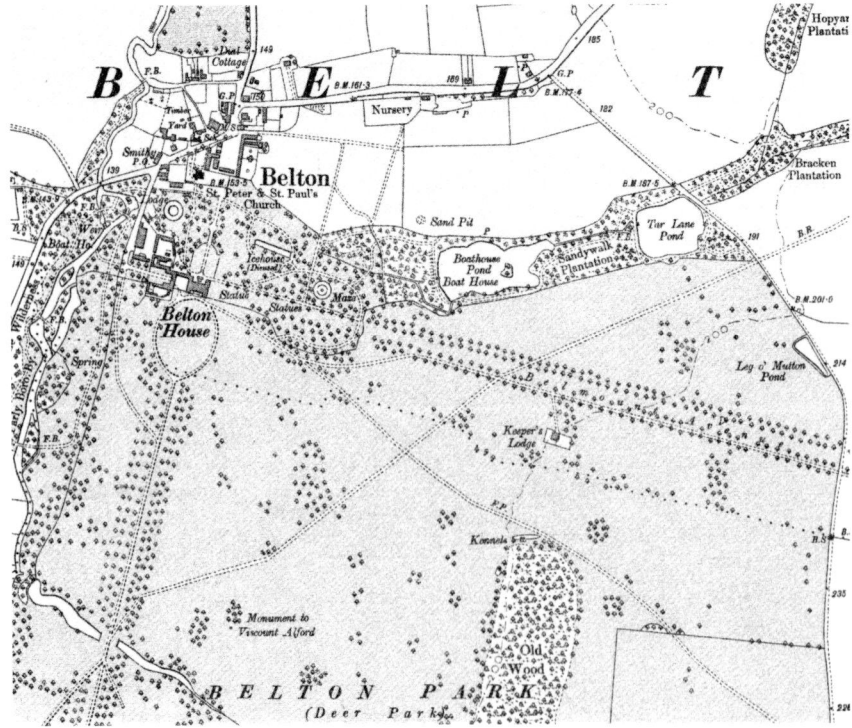

FIGURE 6.2 Belton Park, Lincolnshire, OS 1:10,000 map, pre-1930–55. *Reproduced by permission of the National Library of Scotland.*

expense of making them, and because replacing at least one parterre was usually the only way to fit them in, so this was not a common occurrence. Once the straitjacket began to loosen, however, it became easier, psychologically as well as physically, to fit lakes into the scheme of a design. By *c.* 1720, the degree of asymmetry and unbalancing in landscapes was increasing, as we have seen at places like Thoresby and Castle Howard. These designs were still geometric in character and axial, but strict symmetry was being relinquished and the central axis was less dominant, with marginal elements developing. One such marginal element for example, was the garden at Moor Park with serpentine paths, which appears to have Rococo elements, being detached from the house and a sub-set, as it were, of the main gardens (Figure 3.13).

Plantations especially tended to become much larger. Though still geometric, they were no longer symmetrical in shape. As size increased, it became impossible to maintain symmetry or even balance in the design, and landscapes became 'unbalanced'. This process had begun with Le Nôtre's work at Versailles, and reached England in the early years of the eighteenth century, as images in *Britannia Illustrata* of places such as Grimsthorpe, Cashiobury, New Park (Surrey) and Longleat show. Balance in the design might still be aimed at, but even this became difficult to achieve in very large landscapes. This 'unbalancing' was increased by the advent of sinuous paths in wildernesses and irregular outlines of plantations. These sinuosities began to appear as early as the 1710s, and by the mid-1720s were more prominent features (Figures 3.5 and 3.9).[1] Wildernesses were also increasing in size, as at Badminton and Chatsworth, with the possibility that this may have been linked to maintenance, as they were perhaps less onerous to maintain than parterres.[2]

The value of the timber made large plantations attractive, which was a significant factor for many owners. Cirencester Park illustrates this process well. It was perhaps one of the most well-known of Alexander Pope's projects, with Lord Bathurst (1684–1775, inherited in 1704). As Mowl points out, Bathurst planted the timber as a valuable crop.[3] Pope corresponded with and stayed with Bathurst from 1718 onwards, the two minds sparking off each other to produce a vast forest landscape interspersed with buildings, including Alfred's Hall, the "earliest recorded Gothick garden building in the country" (1722–32).[4] Bathurst's natural inclination for straight avenues was modified by Pope's inclination for winding paths and informality. These were introduced in various parts of the park, notably in the 1730s, when a Rococo style layout was introduced near the house and along the then northern boundary of the park, as shown in Samuel Rudder's map of 1779.[5] Rococo gardens, often at the margins of landscapes, were typical of these transitional geometric landscapes, and contributed to further 'unbalancing'. (See Figure 3.8, north-west of Thoresby House.) By the mid-1730s, Cirencester was just such a landscape, with an irregular lake (*c.* 3.2 h) south west of the house by 1736. However, unlike Thoresby and Holkham, it was not planned as such, but had evolved from a largely geometric landscape as the owner responded to changing fashions.[6]

Cirencester, like Thoresby, was a large landscape: the main axis, from the house to the western edge of the park, was 4.5 miles long. It is clear how the scale affected the symmetry or balance of the landscape, areas being added piecemeal over Bathurst's long lifetime. The 'unbalancing' made it all the easier to develop different areas in different styles and 'allowed' the incorporation of the irregular lake in the 1730s, as these lakes became increasingly *de rigueur* for fashionable landowners. It is important to recognise how revolutionary this development was. Up until then, ornamental water had been equated with geometry, something obviously fashioned by man.

When pieces of ornamental water became larger, they tended to be irregular in form, as we have seen. Whilst it was possible to maintain an idea of geometry in irregularly shaped plantations by cutting linear rides and vistas through them, this could not be done with water; there was no way of disguising its irregularity. Also, the larger a landscape element was, the more difficult it was to fit it into a design, and there were far fewer choices about where to put it. In addition, a lake would have to be made where the water source was (see Chapter 4). Also, to some extent, in the 1720s, irregular water features *had* to be big to justify their irregularity: a small, irregular pond would simply have looked unfinished or out of place (a fish-pond) in the context of a largely geometric landscape. As such, irregular lakes were the drivers of change once the scale of landscapes started to increase. It became too difficult to maintain geometry or even symmetry with large pieces of water, and attitudes to gardens and landscapes were also beginning to change.

This theory is borne out by the chronology of irregular lakes. As we have seen, there was only a handful of lakes of any kind in the 1720s, but by the 1750s, numbers had increased significantly. By then, fashion had changed and new landscapes were being laid out in an irregular and informal style, as at Stourhead and Pain-shill, whilst many older landscapes were being updated, for example Longleat and Petworth. Once landscape design had caught up with the evolution in ornamental water, the number of lakes increased rapidly, peaking in the 1760s and '70s.

Italy and the Grand Tour

It was men like Kingston, Carlisle and Coke who made the first lakes in these large, 'unbalanced' landscapes, with Vanbrugh exerting a significant influence, but these men alone were not responsible for the sea-change in ornamental water. The influence of the Grand Tour and Italy has to be taken into account. With the ending of the War of the Spanish Succession in 1713, Continental travel became attractive again, and so did the Grand Tour. As is well-known, it was regarded as an important part of a young gentleman's education and, whilst many undoubtedly went through the motions of acquiring a cultural education, others like Coke and Burlington took it seriously. There were several routes to Italy, and a popular one was via Mt. Cenis, with its glacial lake, which was taken by Coke (1713–18), as well as men like Joseph Addison and Thomas Nugent.[7] The views, the buildings, the paintings, all had an impact on the landscapes they laid out when they returned, and of specific importance was the combination of villas adjacent to lakes which they saw, a combination which men such as Coke were keen to imitate.

Coke travelled extensively in Italy, atypically applying himself studiously to Ital-ian, Latin and Greek, diligently studying architecture, collecting books to establish a good library on his return, and buying pictures, probably from Kent in 1714, who formed a strong connection with him.[8] He had a deep interest in all cultural aspects of the classical world, spending a month in Vicenza in 1714, the birthplace of Palladio. On his return to England, Coke began laying out a new landscape at Holkham, as discussed in Chapter 3. It seems to have been a mixture of elements which he saw on his Grand Tour: the natural lakes with forested mountain slopes of the Alps, and the Palladian buildings of Italy. These impressions were reinforced by depictions of classical scenes in 'natural' landscapes by Claude Lorrain (Figure 6.3).

Like Burlington, Coke apparently wanted to re-live a Palladian dream, putting his knowledge of architecture to work in the house, probably designed by himself, Kent and Burlington. That Coke had a deep interest in the design of his landscape is demonstrated by this extract from a poem written by him:

> Here Kent and I are planting Clumps
> Not minding when our Monarch Rumps
> Or what Sir Robert's doing . . .
> Contented I enjoy my home,
> Design a Temple, Build a Dome,
> Or raise an Obelisk.[9]

Coke's comments, in a letter to Lord Burlington in November, 1736, on the dullness of geometric gardens, echo Pope's criticism of their predictability:

> But to think of those damned dull walks at Jo. Windhams, those cold and insipid straight walks which would make the Signor sick, which even Mr. Pope himself could not by description enliven,[10]

C. W. James, who quotes this letter, identifies the 'Signor' as Kent. These two quotations tell us several important things about Coke: he obviously wanted to move away from what he saw as the stultifying geometry of 'straight walks', and presumably felt that he had achieved this in his landscape, despite the straightness of the vistas within the plantations and the geometry of the 'hippodrome' to the south of the house (Figure 3.10). Secondly, they reinforce the idea that Kent was a guiding hand in the matter of landscape design, presumably promoting clumps, but deferring to Coke's desire for vistas. Possibly, he also created an island in the lake, and a 'serpentine water' (1743) to link the lake and the basin.[11] Thirdly, the poem makes clear that for Coke, architectural structures in the landscape were important, perhaps mirroring in his mind what he had seen in Italy: a 'natural' landscape interspersed with classical buildings.

Coke's conception of the landscape at Holkham seems to have been a mixture of elements which he saw on his Grand Tour. Whilst, by no stretch of the imagination can Holkham be considered mountainous, the plantation along the north-west side of the lake, with vistas aligned on the lake, may have represented those forested slopes. Meanwhile, the opposing bank shelves sharply enough down to the water to appear steep (Figure 8.1). The plantation finishes halfway along the western bank, allowing an imposing view from rising ground back to the house, echoing the houses and villas which Coke saw bordering the Swiss and Italian lakes.

One of the 'spin-offs' of the Grand Tour was the art work which men brought back with them, either originals or copies, both of which spawned many engravings.

The artists commonly mentioned in this connection are Claude Lorrain, Nicholas Poussin, Salvator Rosa and Gaspar Dughet, who all painted in the seventeenth century, mainly in Italy, and the gentlemen and aristocrats who made the Tour would have been well aware of them. Coke, despite his youth – he was 15 when he embarked in 1713 – bought many art works, including Claude Lorrain paintings.[12] Holkham was one of the first three non-geometric lakes to be made – the other two were Thoresby and Londesborough – and although the landscape of Norfolk is nothing like Italy, it seems that in Coke's mind, water of an irregular nature was linked to the Palladian concept. It is no coincidence that in Lorrain's paintings, 'natural' water with adjacent classical structures is a common theme (Figure 6.3) and that Coke was an enthusiastic collector of Lorrain's paintings.

The significance of the landscape paintings these men brought back from Italy lies in what they depicted: often quite rural scenery, with water of some sort – a river, the sea, a port – often in the middle distance, plus classical looking buildings, and there does appear to be a link between this formula in the paintings and the landscapes which these men made on their return. Although we do not know if Kingston had any classical garden buildings, and Walpole did not have a lake at Houghton, Norfolk, many of the notable landscapes of the 1720s–'40s did have a classical house or classical garden buildings adjacent to a lake. Initially, the water might be somewhat hybrid or geometric, as at Castle Howard, or Studley Royal, or Claremont, or Stowe, but by the 1750s most of these places had irregular lakes to complement their classical architecture and, as with the paintings, the earlier lakes were often not immediately in the foreground.[13]

Henry Hoare 'the Magnificent' was another on whom the Grand Tour had a significant impact, and one of the most celebrated examples of the influence of paintings on landscapes is Stourhead. Kenneth Woodbridge ties the landscape development to Hoare's possession of Italian pictures, especially Claude, Poussin and Dughet. Hoare, who owned two large landscapes by Dughet, said himself, "the View of the Bridge, Village & Church altogether will be a Charm[in]g Gasp[ar]d picture at the end of that Water"[14] (Figure 6.4). The influence of Italian paintings should not be overstated, though it was undoubtedly significant at Stourhead. However, in conjunction with the Grand Tour, and experiences of the Italian lakes, perhaps they served to embed those experiences, and acted as visual reference points for the returned tourist when laying out or improving a landscape. Perhaps more importantly, once the lake had been made, and the classical buildings erected, whether house or temple, they served as evidence to visitors of the owner's cultural knowledge and awareness of fashion, at the same time authenticating both. Whilst by no means all men returning from the Tour gleaned as many benefits from it as Coke and Hoare, it did produce a common knowledge of 'classical' landscapes, which many gentlemen could 'read' in the muted interpretations back home. The diaries and accounts of Tours also contributed to this.

Joseph Addison was a pivotal figure in this respect. He travelled in both Italy and Switzerland (1700–2) and, whilst Italian travels are frequently commented on today, the impact of Switzerland has been noted much less in relation to

FIGURE 6.3 Claude Lorrain *Landscape with the Rest on the Flight into Egypt* 1654. *The State Hermitage Museum, St. Petersburg; photograph by Svetlana Suetova, copyright The State Hermitage Museum.*

FIGURE 6.4 The Temple of Flora and the Palladian Bridge, Stourhead, Wiltshire.

eighteenth-century travellers. His descriptions of his travels in *Remarks on the Several Parts of Italy Etc. in the Years 1701, 1702, 1703* may hold the key to developments in the landscapes of these men in the early decades of the eighteenth century.[15] He spent five days sailing around Lake Geneva, noting the prospects of woods, vineyards, meadows and corn-fields which bordered it, and at the Carthusian convent at Ripaille on the lake shore, he made this observation:

> They have a large forest cut out into walks, that are extremely thick and gloomy, and very suitable to the genius of the inhabitants. There are vistas in it of great length, that terminate upon the lake.[16]

This is very similar in concept to the vistas in the plantations alongside the lakes at Thoresby, Londesborough and Holkham (Figure 3.6), as these terminate on the water. Addison commented a number of times on the Swiss and Italian lakes:

> There is nothing in the natural face of Italy that is more delightful to the traveller, than the several lakes which are dispersed up and down the many breaks and hollows of the Alps and Apennines.[17]

He also sailed on Lake Constance, as well as visiting Lake Albano, a natural lake near Rome, linking his travels with those of Horace:

> In our return from Jensano [Gensano] to Albano, we passed through La Ricca, the Aricia of the ancients, Horace's first stage from Rome to Brundisi. There is nothing at Albano so remarkable as the prospect from the Capuchin's garden, which for the extent and variety of pleasing incidents is, I

think, the most delightful one that I ever saw. It takes in the whole Campania and terminates in a full view of the Mediterranean. You have a sight at the same time of the Alban lake, which lies just by in an oval figure of about seven miles round, and, by reason of the continued circuit of high mountains that encompass it, looks like the area of some vast amphitheatre. This, together with the several green hills and naked rocks within the neighbourhood, makes the most agreeable confusion imaginable.[18]

Significantly Addison is, unconsciously or otherwise, linking gardens with views over large bodies of water.

It is difficult to avoid the conclusion that Carlisle, Coke, Burlington (and probably Kingston), having travelled in Italy as Addison did, and seen very much what he had seen, were spurred on by his *Remarks* to try to recreate something of what they had seen and experienced. This would also go some way towards explaining the increase in the size of ornamental water which was taking place, and perhaps the increase in the popularity of boating as a leisure activity (discussed later). These men copied the classical buildings they saw, re-inventing them as garden buildings or Palladian mansions, and it seems reasonable to suggest that they did likewise with elements of the landscapes they travelled through, specifically the natural lakes.

Addison was particularly influential because, as well as having a voice for his ideas in *The Spectator* he also, like Vanbrugh, moved among these men as a near equal, being a member of the Kit Cat Club, and later an MP. Writing in *The Tatler* and *The Spectator* in 1710–11, he admired the beauties of wild landscapes such as he had seen in the Alps on his travels (1700–2).[19] In addition to the opinions implied in his *Remarks*, Addison's ideology encompassed the concept that gardens should be freed from constraints.[20] This *laissez faire* attitude complemented the relatively untamed scenes he had encountered in the Alps.

In his piece in *The Spectator* of 12th April 1711, cast in the form of a dream, Addison describes an idyllic garden in Leonora's country seat. It had grottoes, woods, bowers, and murmuring springs "collected into a beautiful Lake, that is inhabited by a Couple of swans, and empties itself by a little Rivulet which runs through a green Meadow".[21] Whilst it is unlikely that Addison was directly advocating the making of irregular lakes here, he promulgated the beauty of wild, irregular landscapes:

> There is something more bold and masterly in the rough, careless Strokes of Nature, than in the nice Touches and Embellishments of Art. The Beauties of the most stately Garden or Palace lie in a narrow Compass, the Imagination immediately runs them over, and requires something else to gratifie her; but, in the wide Fields of Nature, the Sight wanders up and down without Confinement, and is fed with an infinite variety of Images, without any Stint or Number.[22]

It is also in this letter that he introduces the concept of treating the whole estate as a garden: "But why may not a whole Estate be thrown into a kind of Garden by frequent Plantations, that may turn as much to the Profit, as to the Pleasure of the Owner?"[23] Switzer subsequently took this up and termed it 'rural and extensive gardening'.[24] In his writings, Addison mentions Chinese ideas of garden-making – *sharawadgi* – and was much against topiary. These were nascent concepts in the 1710s and early '20s, but men such as Vanbrugh, Carlisle, Marlborough, Newcastle, Burlington, Manchester, Coke, Walpole and Kingston would have been aware of them; many would have visited Italy and seen the landscapes Addison referred to at first hand.[25]

Another influential work was Castell's *Villas of the Ancients Illustrated* 1728. Sponsored by Burlington, it was an attempt to reconstruct the gardens and landscapes which the Younger Pliny wrote about; again, Italy was the inspiration. As Dixon Hunt and Willis point out:

> Castell's reconstructed plans enforce his written commentary in suggesting the happy juxtaposition of two sorts of garden styles that characterised both Roman villas and such English estates as Stowe at the time Castell was writing.[26]

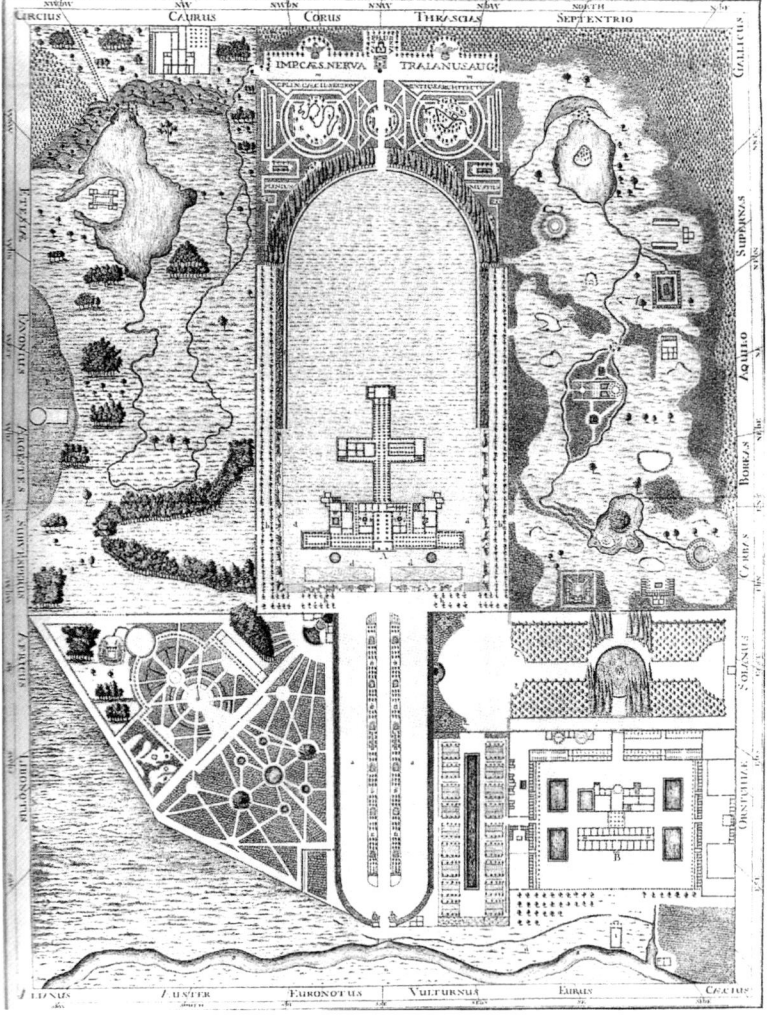

FIGURE 6.5 A reconstruction of Pliny's Tuscum Villa by Castell, 1728. *Courtesy of Robin Simon.*

What is noticeable in Castell's Tuscum Villa plan (Figure 6.5) is the juxtaposition of the house, adjacent formal gardens and service area with the informality and irregularity of the rest of the estate. It is very much in accordance with what we have seen occurring in the transitional geometric landscapes of the 1720s; it is the water which is noticeably irregular, although there are several formal water features which are smaller. Both Thoresby and Holkham conform to this 'formula'. There are also several detached gardens in Castell's plan, both formal and informal, within the informal part of the landscape. Both appear to be surrounded by water, and are suggestive of Rococo gardens, with their irregular elements, separation from the main house, and suggestion of intimacy. Also of note is the informal nature of the woodland which consists of sinuous outlines and clumps, and brings Kent and Brown to mind. As previously mentioned, Rocque's 1737 plan of Claremont bears a strong similarity to the upper right quadrant of Castell's, especially in the relative positions of the 'rococo'

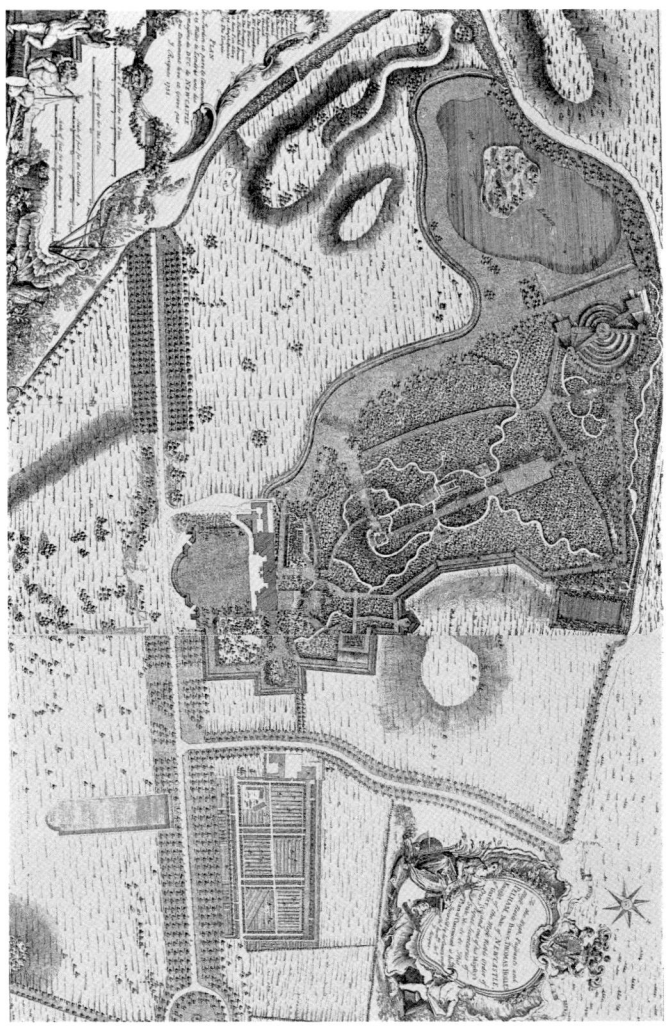

FIGURE 6.6 J. Rocque's plan of Claremont, Surrey, 1738. *Private Collection.*

garden, the water, the house, and also the kitchen garden and adjacent avenue. The water in Roque's plan (Figure 6.6) is not as irregular as in Castell's though.

Whilst Burlington was trying to ensure that his buildings and landscapes were conforming to the Palladianism which was his preferred version of what he had seen in Italy, Castell, abetted by Kent, was feeding him irregularity (water and woodland) in the guise of the villa concept: the house and gardens embedded in the surrounding (productive) landscape. Whilst Chiswick certainly was not productive, its context being urban, Londesborough, with its chain of fishponds and kitchen garden, was much more so, and its lake (c. 1729) was noticeably irregular (Figure 3.17). By linking this interpretation of landscape with Pliny, Castell undoubtedly popularised the developing Palladianism favoured by Burlington (his patron) and formed perceptions of what Classical landscapes were like. In doing this, Burlington and Castell were effectively validating their view of what landscapes in England should be like.

The writings of men such as Pope, Switzer and Batty Langley, who fostered the concept of irregularity in their works, may also have affected attitudes to landscape style more generally between 1700 and 1730. Langley's texts made much of the beauty of irregularity but this was not generally reflected in the designs he produced in *New Principles of Gardening* in 1728. These were still geometric in character, although cut through with many wiggly paths. Though they may have had some influence, as did Switzer's, they do not account for the changes which were happening in the 1720s and '30s. As early as 1713, Pope (1688–1744) was writing about

> the amiable Simplicity of unadorned Nature, that spreads over the Mind a more noble sort of Tranquillity, and a loftier sensation of Pleasure, than can be raised from the nicer Scenes of Art.[27]

He favoured landscapes which he *perceived* as not having been shaped by man in any way: 'unadorned Nature'. In his letter to Martha Blount of c. 1724, his description of a visit to Sherborne Castle makes his concern with irregularity plain:

> The Gardens are so Irregular, that 'tis very hard to give an exact idea of 'em but by a Plan. Their beauty rises from this Irregularity, for not only the Several parts of the Garden itself make the better Contraste by these sudden Rises, Falls, and Turns of Ground; but the Views about it are let in, & hang over the Walls, in very different figures and aspects.[28]

Here, Pope is admiring the accidents of topography, as the gardens themselves were still formal at that time. His *Epistle to Lord Burlington* in 1731 was an exposition of what was desirable in garden design, and cogently expressed concepts which had emerged in his own writings, as well as those of others, in the preceding two decades (Addison, Switzer, Langley, Castell). Irregularity and the beauty of Nature (unspoiled by man), plus the sterility of topiary, were the key ideas. Pope exercised an appreciable influence over landscape style through his writings and also by example, because his gardens at Twickenham became famous during his lifetime.[29] However, this may be overstating the case, as his own garden evolved over 20 years and may have been as much a response to new trends as the maker of them. In comparison with other landscapes discussed, it was small, and it did not have any ornamental water (apart from in the grotto, eventually), although the Thames formed the prospect on the other side of the house. What may have been of greater influence was the visit of Father Matteo Ripa to London early in 1724, and the presentation to George II of 36 plates of the Chinese palace gardens at Jehol.[30] It is known that Burlington owned a copy, and the preoccupation with 'Chinese' features – boats, buildings and winding paths – which emerged in the 1720s and '30s may have been a stronger catalyst than Pope's writings.

The Grand Tour continued to be popular, although many travellers, including Richard Pococke, who travelled in Europe in the 1730s and '40s, either just listed the lakes they saw, without comment, or merely included their dimensions. However, the antiquary Thomas Nugent (c. 1700–1772), who published his *The Grand Tour* in 1749 (so presumably travelled in the middle to late 1740s) was more expansive. Whilst his description of Lake Albano possibly relied heavily on Addison's, his comments on Lake Geneva are more original. After mentioning the particularly large trout, he goes on to say:

> The city has three or four small frigates with sails and oars, in which they often entertain princes upon the lake. . . . Travellers generally divert themselves here with fishing.[31]

He also commented on the glacial Mt. Cenis Lake at the top of the pass which Grand Tourists used *en route* between Lyons and Turin:

> there is a pretty large lake near the road, formed by the melted snow; in the middle of this lake the king of *Sardinia* has a handsome house for his diversion.[32]

Clearly, the lake as a setting for the house was noteworthy, and he responds in a similarly positive way to the 'charming' position of the town of Annecy alongside its lake.[33] Likewise, he notes the gardens of Isola Bella and Isola Madre in Lake Maggiore.

It is widely accepted, from the evidence of the paintings they brought back, the houses they built and the classical garden buildings they erected, that men were significantly influenced by their experiences of the Grand Tour. On this basis, it can also be argued that the natural lakes they saw in the Alps and in Italy influenced their attitudes to ornamental water from the second decade of the eighteenth century, especially as travel on the Continent became easier again with the end of the War of the Spanish Succession in 1713.

Landscapes and gender

Whilst such stylistic interpretations are important, the way in which landscapes were used also had an impact on the development of lakes. As we have seen, irregular lakes appeared in parks in the 1720s, but they were not isolated in the park, as fishponds had been. They were relatively close to the house, forming a link between the gardens and the park, and were often placed obliquely in relation to the house. This linking of park and gardens with a lake was reinforced by Brown, and his contemporaries, as he positioned the lake prominently in relation to the house and the approaches, and used the ha-ha to obscure the demarcation between park and gardens. This is relevant to the use of the park by women.

Traditionally, men had used parks for riding, hunting, field sports and angling, but it may well be that, in the early eighteenth century, women were beginning to use parks more extensively than they had before, though information is fragmentary before the middle of the century. In *A Discourse*, North talks of angling as a suitable entertainment for women, which would also encourage them to go out of doors, and he mentions the pleasures of boating for young people, and as a family entertainment.[34] If women were beginning to participate in activities in parks, a closer relationship between gardens and parks would have been an advantage.

The family at Thoresby provides some of the clearest evidence of how women, and men, used parks in the early eighteenth century. The painting by Peter Tillemans (1726), shows the Duke shooting gamebirds (Figure 6.7).[35] People are also enjoying the lake in three boats and a closer examination shows that some of them are women (Figure 6.8). Two of the boats are craft of the type described as shallops by Felus, with a small rowing boat in attendance.[36] The boat on the left has several men in it, whilst the one on the right has four women, seated under a red awning, and two men, being rowed by four more men, with a fifth in attendance, so a reasonably large, elegant craft. This is a pleasure party of some size, suggesting the entertainment of visitors as well as family. The painting is apparently the earliest evidence of women boating for pleasure, and could almost have been composed in answer to North's recommendations:

> Or if the Female Part are so grave, to decline that Course of Life [going out visiting], must they always be within? Or if they stir out, have nothing but mere Air to invite them?[37]

It appears from this painting that a significant part of the reason for making the large lake at Thoresby was the aim of providing sport and entertainment for visitors and the family.

FIGURE 6.7 Pieter Tillemans's painting of Thoresby Park, Nottinghamshire, *c.* 1725. *Courtesy of Gregor Pierrepont.*

FIGURE 6.8 A detail showing boats in Pieter Tillemans's painting of Thoresby Park, Nottinghamshire, *c.* 1725. *Courtesy of Gregor Pierrepont.*

Very little is known about the use of the park by women in the early eighteenth century, and this picture of women boating is one of the few pieces of direct evidence of their activities. Lady Mary Wortley Montagu (Kingston's daughter) provides some of the earliest evidence (to date) of women hunting on horseback: in a letter of 1711, she says:

> I've had a general Hunting Day last Tuesday, where we had 20 Ladys well dressed and mounted, and more Men. The Day was concluded with a Ball. I rid and danc'd with a view to Exercise,.[38]

This is clearly a normal pursuit for her and her friends, and presumably took place in the parks of her father and his friends, and possibly the open countryside.[39] In 1725, she mentions that she is stag hunting in Richmond Park with the Prince of Wales.[40] Queen Anne was known to hunt, in a specially built chaise (she was too fat to ride); she had wide tracks cut through Windsor Forest to enable her to do so, and Pope deplored, in *The Guardian* of 1713, that the knife was handed to 'ladies of quality' to cut the cornered deer's throat.[41] Likewise, Sophia Western hunts in *Tom Jones* (1749).[42] Women who were less adventurous, or less able, also followed the hunt in carriages, as a detail in the Thoresby painting also reveals as well as in a number of other eighteenth-century paintings.[43]

As discussed earlier, angling was popular with men as a recreation from the beginning of the seventeenth century at least, but there is some evidence that women were beginning to become involved: in a picture of Chiswick by Rysbrack (*c.* 1730), a woman is possibly being taught the art, and several mid-century paintings by Thomas Robins and Edward Haytley show women and children angling (Figure 6.9). This indicates a developing interest of women in outdoor pursuits, which increased later in the eighteenth century. This is based on the premise that the format and content of paintings (both landscapes and portraits) follows fashion; it does not dictate fashion.[44] For example, paintings did not usually depict women in the wider landscape unless it was actually acceptable for them to be there, so they did not depict women boating unless (until) it was acceptable for them to do so.

An analysis of images, mainly paintings and engravings, of landscapes from *c.* 1650–1800 suggests that there are three main trends in the ways women used landscapes.[45] Firstly women were not routinely portrayed outside gardens until the 1730s, other than an occasional woman riding. From the 1740s, women began to have their portraits painted outside the formal garden, in the wider landscape, sometimes 'relaxing', for example talking, using a telescope, drawing and angling. Secondly, women begin to appear in boats relatively often from the 1730s onwards. This continued to be a popular activity, judging by the accounts of it at Wrest in the 1760s and 1790s (see later), but in the paintings it is often not possible to determine the gender of people in distant boats. Thirdly, a small number of pictures show women driving across country in light carriages, often drawn by six horses, in the 1730s and '40s. By the 1750s, this activity appears to have ceased. This suggests that women wanted more access to the wider landscape and its activities, and that carriage drives may have developed in response to this. One woman in a carriage at Dogmersfield appears to be part of a hunt, and there also appears to be women watching the fowling from a carriage in the Thoresby painting (Figure 6.7).

From the evidence of the image analysis, we can see that women were beginning to engage with the landscape beyond the formal gardens, and specifically lakes, by the third decade of the eighteenth century.[46] At Holkham, the gravel walk or carriage drive was not laid down until 1801–3 so until then the further reaches of the park were basically accessible only to more intrepid women.[47] However, the lake provided easy access by boat to the further shore, enabling women to disembark and walk up the slope to appreciate the 'reverse' view of the house (Figure 8.1).

By the mid-eighteenth century, walking or driving around parks, especially lakes, had become popular and landscapes began to be laid out with circuit walks and drives, including the lake if there was one, such as at Stourhead and Kedleston in the 1750s. As improvements in carriage technology occurred (better suspension), carriage drives through the landscape became *de rigueur*.[48] The landscape, and lakes in particular, began to be experienced in quite new ways. No longer were they looked down upon and appreciated from a distance, or from selected viewpoints, such as from a roof terrace or along a vista. They were experienced directly: the size and shape of lakes, as well as their sound and texture, were experienced at close quarters, especially if on foot, and the view of the water was ever changing, elements which can be seen in Brown's plan for Packington (Figure 5.6). This contributed significantly to the increasing popularity of lakes. Planting around them, as well as the shaping of the lakes themselves, became very important, as these factors enabled views to be constructed to entertain and surprise. As Kate Felus suggests, there was an almost cinematic quality to these drives, as views – of the water, or a temple on a nearby rise – were revealed and then concealed by careful planting alongside lakes.[49] The placing of buildings and monuments or structures of various kinds enhanced views and also provided further interest in the form of destinations. These landscapes developed in the 1730s and '40s, and were subsequently described by historians as Arcadian because of their classically inspired buildings. Stourhead is an archetypal example, having an irregular lake (1754), circuit walks, a carriage drive and numerous buildings and structures in the Classical style. One aspect of garden buildings which Felus draws attention to is their function as a retreat from visitors and servants.[50] Stowe and Stourhead were both well provided with them. This was not a new concept: William Cecil had built a lodge near his new ponds at Theobalds in the late sixteenth century so that he could spend time undisturbed.[51]

This phenomenon of classical buildings in a landscape, probably reached by carriage, is well illustrated by Beachborough Hall. Figures 6.9 and 6.10 show the pond, as painted by Edward Haytley in 1745. Although the water there was only *c*. 0.17 h (0.4 a), it seems to have pretensions to being a lake, as we have seen, and the painting shows an interesting array of activities. The ladies are painting or sketching the scene as well as angling, and the scene includes a small classical 'temple', which almost certainly appears on the OS map of 1872 (Figure 3.22).

FIGURE 6.9 View towards the temple: *The Brockman family at Beachborough* Kent, by Edward Haytley, 1745. *National Gallery of Victoria, Melbourne.*

FIGURE 6.10 Men fishing: *The Brockman family at Beachborough* Kent, by Edward Haytley, 1745. *National Gallery of Victoria, Melbourne.*

Haytley's second picture (Figure 6.10) shows a boat in the foreground, and two men netting the pond for fish, which confirms what Switzer says about fish being kept in ornamental lakes and pools.[52] The picture could almost have been painted to illustrate what North said about the benefits of healthy, outdoor exercise: he pointed out that angling kept young men away from card playing and similar pursuits. Looking at the evidence as a whole, it must be concluded that providing entertainment was an important sociological factor in the development of lakes.

One point of interest is that up to *c.* 1720, if people appeared in landscape paintings, both men and women were largely portrayed walking in the gardens.[53] By the mid-eighteenth century, this had been replaced by walking in the park, and being portrayed there, either as families or as individuals. By the 1740s, activities in parks were becoming more diverse, including drawing or painting, and playing cricket.[54] It is likely that men continued to ride and hunt just as much, but were less often depicted doing so. However, it should be noted that the number of images of landscapes began to decline towards the end of the eighteenth century, and 'portraits' of gardens and plants had become popular by the mid-nineteenth century, so it is less easy to pinpoint activities in parks from then on. This may itself suggest that those activities had declined, at least for women.

Boating

Clearly, boating would be a more amenable activity for women if the lake was relatively near the house, rather than the further reaches of the park. Size was also a factor. Felus suggests that the desire for boating led to an increase in the size of lakes.[55] Her focus was on naumachia, which became particularly popular in the 1740s, but although they are undoubtedly relevant, Kingston's painting (*c.* 1725) and the gradually increasing numbers of boats on lakes in paintings from the 1730s onwards point to this factor beginning to operate before the 1740s. John Whitney's comments, in 1700, indicate that it may have begun even earlier. Talking about a pond at 'Sundridg' (Kent) of 300 by 100 feet, he says,

> then in the middle of the Pond a most delightful *Summer* House to go to by Boat, twelve foot long and ten foot broad, with a Fountain in the middle, where the Water plays in sundry Figures.[56]

Felus also draws attention to the importance of boating as a social activity to entertain family and guests, and this echoes Nugent's comments about the way in which princes visiting Geneva were entertained in boats on the Lake in 1749.[57] We have seen that two yachts were bought for the Royal Family once the Serpentine was completed in 1731, and that Viscount Weymouth ordered one in 1736.[58] Boats of considerable size were commissioned for use on lakes at Stowe, Wrest Park and Newstead Abbey. They were used for the pleasure of sailing on the lake, and there is some evidence that the canals at Wrest increased in size, and that larger lakes were created in order to facilitate boating. As well as in Haytley's painting, boats appear in numerous other pictures of the eighteenth century, one of the earliest being a topographical painting of Wingfield Manor in *c.* 1700, where the water was *c.* 0.4 h, or nearly an acre.[59] It shows a large house looking down on a geometric shaped pond with a boat on it, possibly with a mast. The pond was almost certainly a fishpond, but its central position in the painting, and the women in the boat, suggest it is in the process of evolving into a piece of water which was as much ornamental as functional. Up to *c.* 1720, very few landscape paintings showed ornamental water large enough to be a lake.[60] This changed in the 1720s, with many paintings showing significantly large ornamental water (a canal for example, or a geometric pond). Initially, a few pictures showed boats on these pieces of water, but from the 1750s approximately half with water did depict boats. This strongly suggests that boating was not a significant leisure activity in the first half of the century, although it was beginning to become popular, and there is a strong correlation between the increasing size of lakes and the depiction of boats.[61] In the second half of the eighteenth century, many more pictures showed boats, the ornamental water having become larger, as well as being irregular in shape. Clearly, an irregular lake is likely to be better for boating (or even sailing) than a narrow geometric canal. These facts suggest that boating was one of the driving factors in the emergence and subsequent popularity of lakes.

Boating as a high-status activity was noted at Thoresby, where Kingston was an early innovator. His example was followed by other aristocrats from the 1720s onwards, notably at places like Studley Royal, Wrest, Chiswick and Stowe. As Felus points out, the boats at Stowe were elaborate and numerous.[62] Figure 6.11 shows a

View from the Head of the Lake.   View prise sur le Bord, ou à la Teste du Lac.

FIGURE 6.11 *View from the head of the lake at Stowe* by Rigaud & Baron, 1733. *Courtesy of Stowe School/SHPT.*

galley of a similar design to the barge which Kent designed for Frederick, Prince of Wales, in 1732, but with a sail. This boat was of significant size, requiring a crew, and capable of carrying a number of people. It was spacious enough to carry provisions – picnics – and possibly musicians as well, the covered poop providing shelter from sun or rain. This activity occurred on the Eleven Acre Lake (4.5 h) and Rigaud's view suggests that the water was large enough to pick up some speed and enjoy 'sailing'. The Octagon Lake was not big enough, though perhaps rowing in smaller boats, kept in vaulted boat-houses on either side of the cascade, may have been possible.[63]

A 'Kanooe' is mentioned in the accounts at Stowe in the 1730s, but the 'ship' does not appear until 1750. This may well be the 'man of war' illustrated by George Bickham (Figure 6.12).[64] From the 1750s, the majority of boats depicted in landscape paintings have sails, although many pictures showing water did not have any boats.[65] Also, the large majority of the places with boats were elite residences, demonstrating that boating on your own water was a high-status activity. One of the later pictures of a boat on a lake, with an adjacent house, is an engraving published in 1787 of a picture by Paul Sandby of Brocket Hall.[66] This was a Chinese boat built for Sir Benjamin Truman at a cost of £300, which emphasises the importance attached to boating, as well as illustrating the popularity of the 'Chinese' style.

That boating as a recreation encouraged the development of lakes, and led to an increase in the size of ornamental water, is vividly illustrated by the activities of Jemima, Marchioness Grey, at Wrest. She inherited a landscape of formal canals from her grandfather in 1740, and by 1748 had commissioned Thomas Wright to 'serpentine the end of the peripheral canal, bringing it into the ground to join the Lady's Canal'.[67] This, together with her commissioning Brown to make the water at Wrest less formal in *c.* 1760, and her purchase of a galley type boat, strongly support this theory. The inspiration for this 'Chinese' boat may well have come to Jemima from her brother-in-law, Admiral George Anson, at Shugborough, who advised her on the best type of boat to have – a galley because it could be rowed and also have sails, and it could be lavishly decorated.[68] Felus also makes the point that a number of Chinese boats complemented Chinoiserie summerhouses or temples, and this was the case at Shugborough, and the Duke of Cumberland's lake at Virginia Water.[69] Jemima provides further evidence of the pivotal role of lakes in recreation in the later part of the eighteenth century: her daughter, Amabel, describes the launching and maiden voyage of the Chinese boat in 1766, and a grandson, Thomas, Lord Grantham, does likewise in 1790, though little is known about that boat. Both parties obviously enjoyed themselves considerably:

> Having passed every straight & doubled every cape without the least accident, & being arrived at the open sea behind the pavilion, we landed under a clump (which I should have called a Wood) & left the vessel to proceed to its moorings,[70]

and

Grandmama has got a great boat which we saw launched on Monday; it was afterwards brought to the bank we all have been aboard of her and fished in her. The men that got her out of the house were a great while a doing it.[71]

The humour and light-hearted joy which these accounts show is delightful.

The boats at Stowe were also used in the evening entertainment, as part of the stage-set which was centred on Kent's Grotto, often having musicians in them. Presumably, these were smaller craft, whereas the 'ship,' decorated with lights and containing the musicians, was stationed at the further end of the 'canal' in 1764, as part of the evening entertainment for Princess Amelia.[72] The 'ship' at Stowe (Figure 6.12), described in the Benton Seeley guidebook as a 'Model Man of War in all her Rigging', was laid up during the winter, as accounts in the 1750s and '60s testify, whilst the smaller boats were stored in various summer houses and temples.[73] Ships of this size were not common, as the lakes on which to sail them had to be large, though Lord Bute had a similar ship at Luton Hoo.[74] This was on an irregular lake of *c*. 13 h, which was a substantial size, though rather narrow. Francis Dashwood also had a ship with two masts on his lake at West Wycombe.[75] Again, this was a lake of reasonable size – 4.6 h. Being a hybrid lake, it was roughly rectangular in shape, and more suited to sailing than Bute's lake. Another lake used for boating was the Upper Lake at Newstead Abbey. The 5th Lord Byron (not the poet) expanded it in the 1740s, and it was also more suited to sailing, being an

FIGURE 6.12 *The three masted ship on the Eleven Acre Lake at Stowe* by Chatelain and Bickham, 1753. *Courtesy of Stowe School/SHPT.*

irregular lake of *c.* 9.3 h. He had at least two sailing boats and indulged in naumachia, going as far as to build fortifications on the lake shore from which to fire guns. Naumachia were also enjoyed at Wotton, the second seat of the Grenvilles and, as Felus says, "Its vast extent and the firm evidence of the battery begs the question of whether its scale was determined by the desire to play such aquatic games."[76] Sir Francis Dashwood certainly engaged in mock battles on his lake at West Wycombe, one of his captains being slightly injured in one such battle. The cumulative evidence of images and texts suggests that pursuits such as pleasure trips, angling, and naumachia led to canals and lakes being enlarged, and was probably also a factor influencing their construction in the first place, from the 1730s onwards, gaining momentum in the 1750s and '60s. The fact that Bute was having his ship constructed – a first rate man of war – whilst Brown was creating the lake at Luton Hoo in 1766 also supports the theory that some lakes were made or enlarged for this purpose.[77]

One of the latest landscape paintings of boats on ornamental lakes is Constable's painting of Wivenhoe, 1816 (Figure 5.27). As noted earlier, pictures of houses in landscapes declined in numbers towards the end of the eighteenth century, so less information is available from this source. However, it seems that the passion for rowing or sailing still endured: in 1811 Jane Austen describes a pleasure party in *Sense and Sensibility* in which a sail on a noble piece of water

> was to form a great part of the morning's amusement; cold provisions were to be taken, open carriages only employed, and everything conducted in the usual style of a complete party of pleasure[78]

and, as we have seen earlier, George Eliot was looking forward to a row on the Serpentine in 1853.

Lakes as status symbols

Whilst the way in which parks were being used was changing, there was continuity in the form and function of the water in them. Irregular lakes having evolved from fishponds (*vivaria*), remained largely similar in shape to them in the eighteenth century and were also stocked with fish. It was this close relationship which gave lakes immense status, and that was a significant factor in their emergence.

The desire to display water appears to have been as strong in the eighteenth century as in the medieval period. Lakes did not derive their status directly from the freshwater fish, which was not as valuable as in the medieval period, but from the message they conveyed about the wealth of the owner, as someone who could afford not only the cost of making a lake but also to devote a large area of land to less than maximum productivity. The ostensible function changed from providing fish to providing aesthetic beauty and leisure opportunities. In fact, although fishponds retained their importance throughout the eighteenth century, lakes replaced them in terms of status, and possibly in terms of function to some extent as well.

The desire to display these status symbols – lakes – led to them occupying a more prominent position in the landscape in the second half of the eighteenth century. With the ha-ha being used to blend the gardens into the park, it became much easier to site the lake in view of the house, making it a very important element in the landscape. This created good opportunities to view the lake from the house, as well as to view the house from the park across the water. By the second half of the century, views of the lake as well as the house were being contrived on the approach drives – very reminiscent of the flanking fishponds of the Bishop of Ely at Somersham in the twelfth century. James

FIGURE 6.13 Chatsworth House, Derbyshire, with James Paine's bridge in the foreground. *Courtesy of Tom Williamson.*

Paine's bridge over Brown's river-lake at Chatsworth is a good example (Figure 6.13), and this was a device which Brown used frequently. The Lion Bridge at Burghley, and the (original) approach at Bowood are further examples.

In the early decades of the eighteenth century, lakes had frequently been positioned obliquely in relation to the house. This changed in the middle of the century, and placing the water in front of the house where possible became more common.[79] This also enabled the lake to reflect the house better: an important factor in the second half of the eighteenth century. The effect was to enhance the image of the house (Figure 5.32) and double its impact, just as lake-moats had done with castles. Garden buildings were similarly enhanced by being reflected in lakes, and this was particularly valuable where the house itself was not reflected, as at Studley Royal and Stourhead (Figure 6.14). There, the Pantheon, a substantial building, similar to the Pantheon in Rome, was constructed at a similar time as the lake, and fulfils the role of 'house', to some extent, being fully reflected in the lake. That this reflective element of lakes was a significant factor in their rising popularity in the second half of the eighteenth century is borne out by the many 'portraits' of houses in landscapes which showed the house reflected in the lake. The images published by W. Watts in 1779 in *The Seats of the Nobility and Gentry* frequently show the house reflected in the lake, sometimes even when it was impossible, as at Corsham Court, Wiltshire. The height of the house above the lake was the

FIGURE 6.14 The Pantheon reflected in the lake at Stourhead, Wiltshire.

significant factor in creating a reflection, and at Corsham the lake is quite some distance from the house (400 m) and also at a similar level to it, so not actually reflected in the water. As well as the aesthetic attractions of buildings reflected in lakes, the enhanced (double) image further emphasised the status of the owner, just as the lake-moat did at Bodiam.

Conclusion

The reasons for the emergence and evolution of irregular lakes are varied and complex. Although the different threads have been considered separately, these were actually interdependent, and woven together to produce the significant novelty of the ornamental lake. One factor which was conducive to this change was the political events of 1713–15. Men became free of the worry of war, as the Treaty of Utrecht put an end to the fighting in Europe. The death of Louis XIV saw the end of French domination in political and cultural affairs, and the death of Queen Anne ushered in regime change, with a new focus, in Britain. Whilst these last two factors did not have a direct bearing on the emergence of lakes, the changing cultural ambience did. French fashion in garden and landscape design began to seem increasingly like a sterile geometry of 'straight walks'.[80] People were receptive to new concepts and ideas, and this change in taste was fostered by men like Addison, Pope and Burlington. It led men to view their experiences on the Grand Tour, and the lakes they saw on the Continent, in a new light. There is no doubt that Grand Tourists were impressed with the Alpine lakes, and the villas and gardens bordering them. This, linked to the prevailing interest in Palladian concepts and the remains of Antiquity which the tourists saw, meant they wanted to replicate, in some measure, the landscapes they had seen on their travels. Only the elite could afford to do this, but they adopted the concept of large, irregular water and began to make 'lakes' in their landscapes, often with classical houses or garden buildings nearby. The bond of common experience which the Grand Tour provided cemented these concepts, and meant that these landscapes could be 'read' by any educated gentleman. Paintings of 'similar' scenes, often by Lorrain or Dughet, were brought back and reinforced the bond, with the added advantage that they displayed the owner's cultural sophistication.

Irregular lakes were a significant factor in precipitating the dissolution of the geometric landscape, as they did not fit easily into those designs, and geometry in landscapes began to dissolve. The desire for leisure activities, particularly boating, made lakes increasingly popular, and the increasing involvement of women in these activities may have enhanced that popularity. This in turn influenced the positioning of lakes nearer the house, which made them more accessible to women, and the rest of the family.

Lastly, as in the medieval period, there was a desire to display water, making it visible from the house, and as visitors approached. By the second half of the eighteenth century, the usual formula was to route approaches through the park across the lake at some point, ideally giving views of the house across the water. The great status this conferred was very important, as the lake clearly displayed the owner's wealth, just as the medieval *vivaria* had done. Men also wanted to see their houses reflected in their lakes. This was perhaps the ultimate status symbol, along with a yacht or several, as it doubled the impact of the house, and often signalled that a renowned landscape improver had been employed.

Notes

1   T. Williamson, personal communication, May 2015.
2   Ibid., February 2017.
3   Mowl, *Gloucestershire*, op. cit., p 70.
4   Ibid., p 68.
5   Ibid., p 70. Reproduced in Bishop, op. cit., p 202.
6   Bishop, op. cit., pp 69–70.
7   T. Nugent, *The Grand Tour* (London: S. Birt et al, 1749).
8   James, op. cit., p 187.
9   Ibid., pp 230–231.
10 Ibid., p 228.
11 T. Williamson, personal communication, December 2016.
12 James, op. cit., p 202.
13 Bishop, op. cit., p 207.
14 K. Woodbridge, 'Henry Hoare's Paradise', *The Art Bulletin*, Vol. 47, No. 1 (March, 1965) p 4.
15 Charles Montagu, later Lord Halifax, "arranged a Treasury grant of £200 allowing Addison to make an extended stay on the continent. The idea was that he should take advantage of his travel abroad to learn languages and equip himself for a diplomatic career". in Pat Barker, ODNB, op. cit.
16 J. Addison, *Remarks on Several Parts of Italy*, in R. Hurd, *The Works of the Right Honourable Joseph Addison, A New Edition, with Notes by Richard Hurd D. D. Lord Bishop of Worcester*, Vol. II (London: T. Cadell & W. Davis, 1811) p 174.
17 Ibid., p 170.
18 Ibid., p 145.
19 Addison remarked, "At one side of the walks you have the near prospect of the Alps, which are broken into so many steps and precipices, they fill the mind with an agreeable kind of horror, and form one of the most irregular mis-shapen scenes in the world." Addison, ibid., p 174.
20 J. Dixon-Hunt & P. Willis, eds., *The Genius of the Place* (London: Paul Elek, 1976) p 138.
21 Joseph Addison, *The Spectator*, No. 37, 12th April, 1711, quoted in Dixon-Hunt & Willis, op. cit., p 140–141.
22 Joseph Addison, *The Spectator*, No. 414, 25th June, 1712, quoted in Dixon-Hunt & Willis, op. cit., p 141.
23 Ibid., p 141.
24 Switzer, *Ichnographia*, op. cit., Vol. III, pp vi, xiv.
25 Kingston had a picture by Salvator Rosa in his London house when he died: I. Grundy, *Lady Mary Wortley Montagu* (Oxford: Oxford University Press, 2001) p 16 fn. We do not know if he made the Grand Tour, though his father did, and his friend, Carlisle, certainly did. He also had a copy of Addison's *Remarks* in his library.
26 Dixon-Hunt & Willis, op. cit., p 187.
27 A. Pope, *Essay* in *The Guardian* 1713, quoted in Dixon-Hunt & Willis, op. cit., p 205.
28 A. Pope, *Letter to Martha Blount c.* 1724, quoted in Dixon-Hunt & Willis, op. cit., p 209.

29 Dixon Hunt & Willis, op. cit., p 204.

30 T. Mowl & B. Earnshaw, *An Insular Rococo: Architecture, Politics and Society in Ireland and England, 1710–1770* (London: Reaktion Books, 1999) p 66.

31 Nugent, op. cit., p 165.

32 Ibid., p 175. N. B. This is much bigger now owing to the construction of a large reservoir in the 1960s.

33 Ibid., p 178.

34 North, op. cit., pp 72–73.

35 He is not riding, and the dogs are pointers, not hounds.

36 K. Felus, *'Beautiful Objects and Agreeable Retreats': Uses of Garden Buildings in England, 1720–1820*, unpublished PhD thesis, Bristol, 2009. p 149.

37 North, op. cit., p 72.

38 Lady Mary Pierrepont in a letter of 25th September, 1711, to Philippa Mundy, quoted in R. Halsband, ed., *The Complete Letters of Lady Mary Wortley Montagu*, Vol. I, p 110. She was 22. The following year, she eloped with Edward Wortley Montagu.

39 Lady Mary and her friends would have been riding side-saddle. Later in her diaries, she confesses to riding astride, when living in Italy.

40 Lady Mary Pierrepont in a letter of August, 1725 to her sister, Lady Mar, Halsband, op. cit., Vol. II, p 54.

41 D. Birley, *Sport and the Making of Britain* (Manchester: Manchester University Press, 1993) p 103; A. Pope in *The Guardian* 1713, quoted in D. Birley, ibid., p 106.

42 H. Fielding, *The History of Tom Jones, A Foundling* (London: Murray's Book Sales, c. 1960) p 158.

43 Bishop, op. cit., p 219.

44 Ibid.

45 Bishop, op. cit., Chapter 4 and the Image Database.

46 It is not being suggested that no women went into parks before this time, but that it became acceptable then, and was possibly encouraged.

47 T. Williamson, personal communication, November 2015.

48 Girouard, *Life in the English Country House*, op. cit., p 190.

49 K. Felus, *Boating, Driving and Dining in the Creations of Brown*: lecture at Compton Verney, 24.6.14.

50 Ibid.

51 P. Henderson, 'A Shared Passion: The Cecils and their Gardens', *Patronage, Culture and Power: The Early Cecils*, ed. P. Croft (New Haven, CT: Paul Mellon Centre BA, 2002) p 99.

52 Switzer, *Ichnographia*, op. cit., Vol. III, p 120.

53 Bishop, op. cit., Chapter 4 and the Image Database (Appendix II).

54 A few women played cricket towards the end of the eighteenth century, notoriously the Countess of Derby. See D. Underdown, *Start of Play: Cricket and Culture in Eighteenth Century England* (London: Allen Lane, The Penguin Press, 2000) illustration 7.

55 Felus, unpublished PhD thesis, op. cit., p 21.

56 J. Whitney, *The Genteel Recreation Or, the Pleasure of Angling, A Poem with a Dialogue between Piscator and Corydon* 1700, in the 'Dedication'.

57 Felus, K. 'Boats and Boating in the Designed Landscape', *Garden History*, Vol. 34, No. 1 (Summer, 2006), pp 22–46.

58 T. Mowl, 'Rococo and Later Landscaping at Longleat', *Garden History*, Vol. 23, No. 1 (Summer, 1995) p 59.

59 Harris, op. cit., p 126, painting by Thomas Smith.

60 Bishop, op. cit., p 226.

61 Ibid., p 227.

62 Felus, 'Boats and Boating', op. cit., p 26.

63 Ibid., p 27.

64 Ibid., p 28.

65 Approximately a third had boats.

66 Reproduced in Spring, op. cit.
67 Lennox-Boyd volume of Wright's drawings, no. 55, quoted in Felus, 'Boats and Boating', op. cit., p 35.
68 Her sister, Elizabeth, was married to Admiral Anson, who circumnavigated the world in 1744; he was a brother of Thomas Anson, owner of Shugborough.
69 Felus, 'Boats and Boating', op. cit., p 36.
70 29 July, 1766; BLAS, Lucas Papers, L 30/21/2, quoted in Felus, 'Boats and Boating', op. cit., p 39.
71 Thomas Robinson to Frederick Robinson (1st September, 1790); WDRO, Morley Papers, 1259/1/219, quoted in Felus, 'Boats and Boating', op. cit., p 39.
72 Felus, 'Boats and Boating', op. cit., p 28. Princess Amelia was a daughter of George II.
73 Felus, 'Boats and Boating', op. cit., p 28.
74 Felus, unpublished PhD thesis, op. cit., p 152.
75 Felus, 'Boats and Boating', op. cit., p 39.
76 Ibid., p 30.
77 Ibid., p 25.
78 Jane Austen, *Sense and Sensibility: A Novel. By a Lady*, 3 Vols. (London: Penguin Classics, 1986) pp 91–92.
79 Bishop, op. cit., p 235.
80 Mentioned in Coke's poem, quoted in James, op. cit., p 228.

7

LAKES IN THE NINETEENTH CENTURY

Developments in the evolution of lakes are examined in two main sections in this chapter: the first and second halves of the nineteenth century, beginning with what happened to lakes themselves, and then moving on to the people who designed them.

Lake chronology

Before looking at Repton and the protagonists of the Picturesque at the end of the eighteenth century and the beginning of the nineteenth century, it is useful to have an overview of what was happening to lakes themselves. This will provide a framework in which to analyse any changes which occurred. Lakes which were altered or increased in size were not considered to be new, so a lake made by Brown, but altered by Repton was counted once, for example. Trends in the eighteenth century have already been discussed and Figure 7.1 shows that lake numbers, which were starting to fall in 1800, continued to fall throughout the nineteenth century, except for the 1860s. The usual 'health warnings' about the data should be borne in mind, and the graph should be regarded as an indication of trends, rather than finite statistics. However, there is still a marked decline in the numbers of lakes from *c.* 1815 onwards, and this period included the tithe maps of the 1840s and the First Edition 6" OS maps, so more means of gathering data were available, and higher numbers might have been expected. The peak in the 1860s is almost certainly a result of an anomaly in the data. If, for example, a lake appears on a First Edition 6" OS map of 1880 but not on a tithe map of 1840 then, lacking any other information, a mid-point date of 1860 was created.[1] Similarly, the 'bulge' in the 1820s is likely to be caused by being the mid-point between the OS drawings of the 1800s and the tithe maps of the 1840s. As noted previously, the significant dip in lake numbers in the 1780s mirrors the dip in numbers of Parliamentary enclosures at

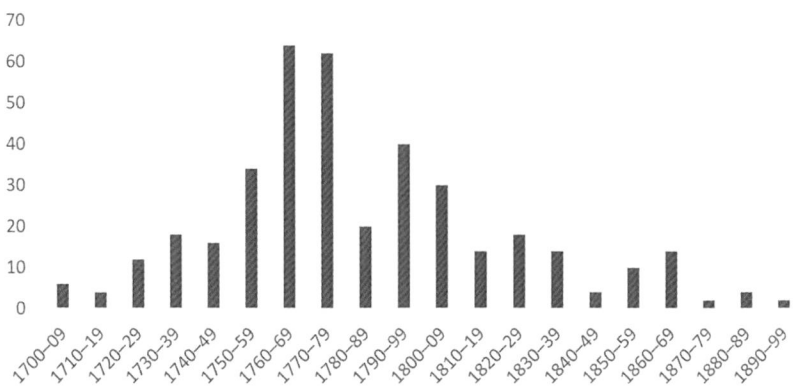

FIGURE 7.1 A chart of the numbers of all types of lakes, 1700–1899.

that time, which points towards economic and political causes. However, there is no obvious explanation for that, although it is worth mentioning general factors such as the Land Tax, the Navigation Acts, and disruption to trade, all of which had more impact in times of war, notably the American War of Independence (1775–83) and the Revolutionary and Napoleonic Wars (1792–1815).[2] The agricultural depression of the 1870s may also have had a similar effect: given that the First Edition 6" OS maps yield more information for that period, higher numbers of lakes might have been expected. Despite being a rather 'blunt instrument', a statistical analysis of this kind shows not only that lake making peaked in the 1760s and '70s, and but also that there was a hitherto unsuspected and significant fall in lake numbers in the nineteenth century, and the reasons for this will be discussed later.

Lake size and parks

From 1730–99, the most usual size for an ornamental lake was *c.* 2–4 h, though lakes of *c.* 5–7 h were common.[3] Brown's lakes were consistently larger than average, at *c.* 7 h, which probably reflects the fact that his clients were frequently the wealthiest men, with the largest estates. A small handful of lakes were over 30 h. The largest was Virginia Water, Surrey, which was over 47 h. However, in the period 1800–99, the most usual lake size was *c.* 1–2 h, with an average size of *c.* 3 h, and larger lakes were not common.[4] As it has not been possible to study all the ornamental lakes in England, these statistics should be regarded as indications of what was happening to lakes. However, the general conclusion must be that, as well as declining in numbers in the nineteenth century, lakes also decreased in size.

The reasons for that decline are complex, and not particularly clear. One factor may have been what was happening to parks in the nineteenth century, though there is little national information on which to base an evaluation. Data for Norfolk and Suffolk show that the trend in the nineteenth century was a decline in new large parks, and an increase in smaller ones.[5] In Norfolk, approximately 100 new parks were created from 1820–99, and of these only three had lakes. As Tom Williamson points out, the term 'park' is a rather slippery one, but the relevant point is how much land owners had available in which to make lakes. As those areas of land were smaller in new parks in the nineteenth century, it seems rational to suggest that lakes would also be generally smaller.[6] Another factor, also pointed out by Williamson, was that the best locations for parks in Norfolk, especially large ones, had been occupied by the end of the eighteenth century, so fewer suitable locations for large lakes remained, especially as these new parks tended to be located on dry interfluves where lake-making would be challenging.[7]

In some instances, a park was created in the nineteenth century, then expanded, and then a lake was made. This happened at Dauntsey Place, in Wiltshire (Figure 7.11). On the OS drawing of 1813, there was a small area of gardens around the house, and a collection of farms. By 1884, these farms were designated as 'park' on the OS map, but more importantly, Idover Farm to the west of the house had been taken into the park and an irregular lake of 1.2 h made. This process was not new; it had happened in the eighteenth century, and it continued to occur in the nineteenth century.

In general terms, out of *c.* 50 new lakes made in England in the period 1820–99, approximately two-thirds were made in existing parks, a small proportion being a second lake. Just under one third of new lakes was made in new parks.[8] It is noticeable that in these new parks, the majority of lakes were positioned directly in front of the main 'garden' façade of the house on a parallel axis. This may have been because space was relatively restricted in smaller parks. As well as being small – usually *c.* 1.5 h – lakes in new nineteenth-century parks were predominantly irregular. Even in existing parks approximately two-thirds of the lakes were 2 h or less in size, and presumably this number was slightly greater because existing parks tended to be larger. The relative heights and distances between house and lake remained dependent on topography, but the lakes in new parks were often somewhat closer to the house than eighteenth-century lakes, probably because those parks were generally smaller.

Bearwood, Berkshire, where two lakes were made in the early nineteenth century, was on the cusp of this change. Both were irregular, and the relationship of the house to the lakes belongs to the previous era, as they were set obliquely to the house. However, the larger lake was more like an eighteenth-century lake in size, being *c.* 18 h. It does not appear at all on the OS drawing of 1809, but it had been enlarged by 1819–20, probably by W. S. Gilpin. We have a description of the landscape by J. C. Loudon in *The Gardener's Magazine* in 1833, although he does not mention Gilpin. He does comment on the water, saying the "puny piece of water, ridiculously placed on the side of a steep bank" has been extended to nearly 30 acres (12 h), with other pieces of water, and altogether they "are so disposed, and disguised by plantations, relatively to each other, that the difference of their levels is scarcely ever observed".[9] In terms of shape though this lake was very much like the new, spreading shape which was coming into fashion, as we shall see.

Changes in lake shape

As well as declining in numbers and size, lakes also changed in shape. Whereas, in the eighteenth century, the majority of lakes could be characterised as elongated, lakes which were made in the nineteenth century could be characterised as spreading.

Figure 7.2 highlights these differences and adds significantly to our knowledge of lakes in the nineteenth century. Even when taking the topography into

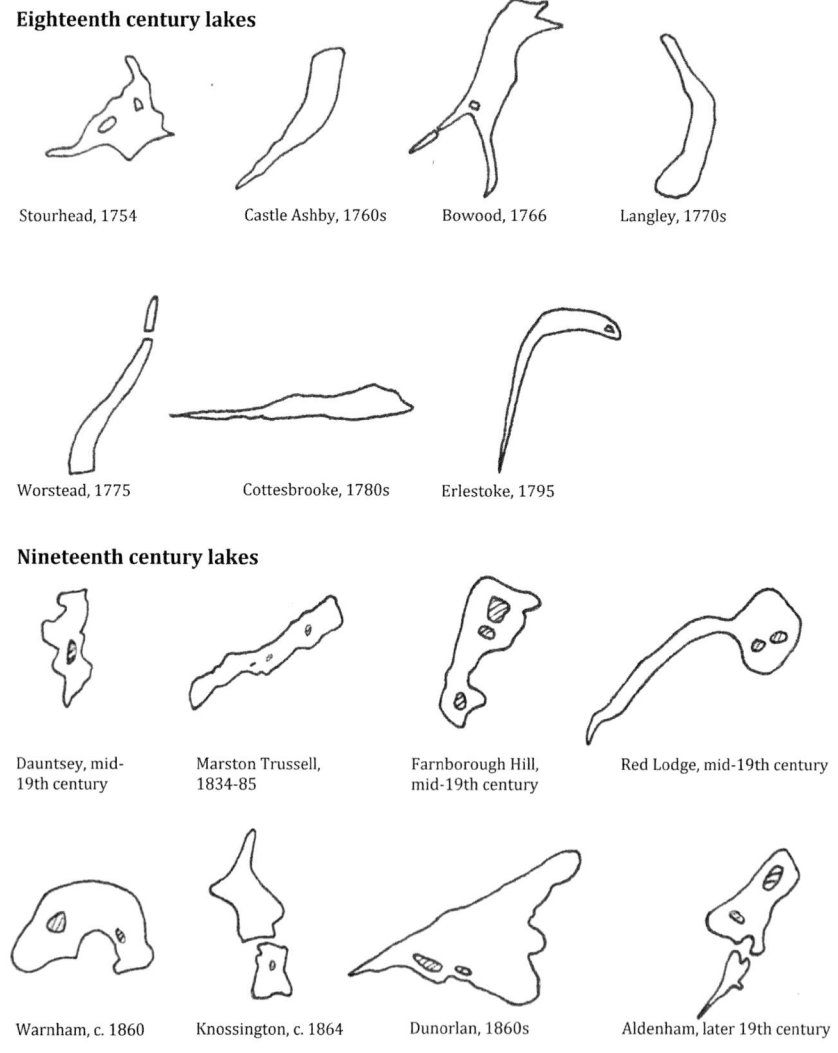

Eighteenth century lakes

Stourhead, 1754   Castle Ashby, 1760s   Bowood, 1766   Langley, 1770s

Worstead, 1775   Cottesbrooke, 1780s   Erlestoke, 1795

Nineteenth century lakes

Dauntsey, mid-19th century   Marston Trussell, 1834-85   Farnborough Hill, mid-19th century   Red Lodge, mid-19th century

Warnham, c. 1860   Knossington, c. 1864   Dunorlan, 1860s   Aldenham, later 19th century

FIGURE 7.2 Lake shapes in the eighteenth and nineteenth centuries.

consideration, it appears that this spreading shape was favoured over the more elongated form of the eighteenth century. The increase in the numbers of islands is also noticeable. An eighteenth-century lake might, if large, have one island, but a nineteenth century lake would usually have at least two, if not several. It is tempting to suggest that this spreading shape was a result of lakes being mostly made in fairly flat areas. However, the lake at Dunorlan, just outside Royal Tunbridge Wells, was probably made in the 1820s, in a valley below the house, and could as easily have been made more elongated, rather than a somewhat bulbous triangle (Figure 7.3).

Similarly, at Rood Ashton, Wiltshire, a park of *c*. 150 h, a bulbous dumpling of a lake was made at some time between 1838 and 1885. It would have been easier to make a more elongated, horned lake, like Stourhead, letting the water flow back up the two stream valleys (Figure 7.4). Instead, the outline was formed into pillow-like cusps, with rectangular projections in several places, and two islands were made. At places such as Trentham, Staffordshire, Longleat and Fonthill, in Wiltshire, where large, elongated lakes were made in the eighteenth century, spreading lakes were made further out in the park in the nineteenth century

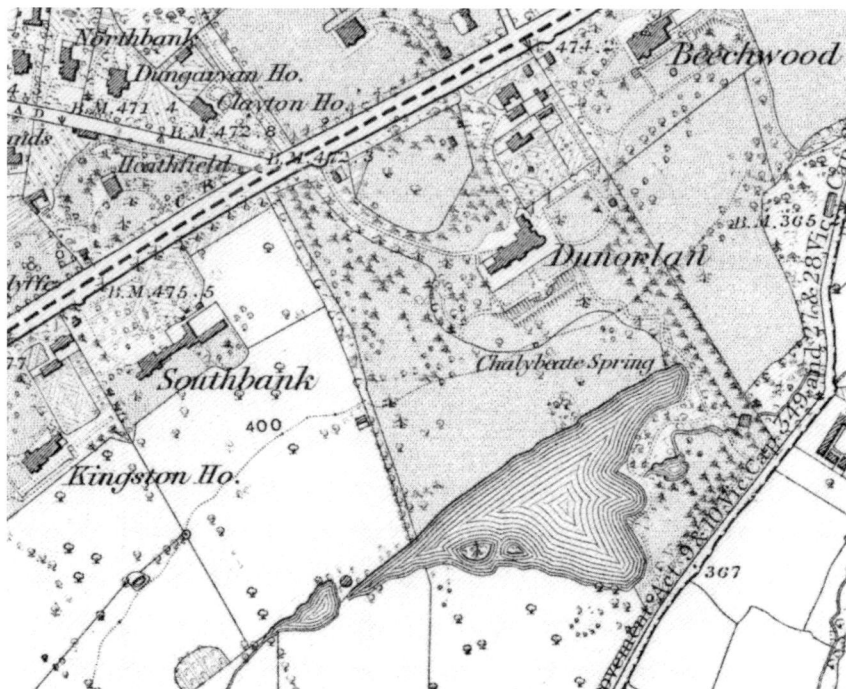

FIGURE 7.3 Dunorlan Park, Kent, OS 6" map, 1873. *Reproduced by permission of the National Library of Scotland.*

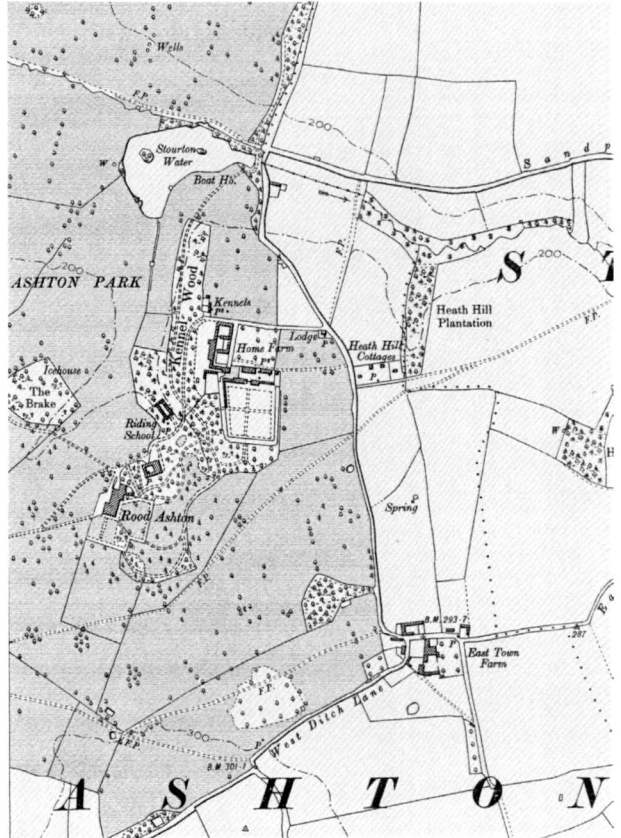

FIGURE 7.4 Rood Ashton Park, Wiltshire, OS 6" map, 1885. *Reproduced by permission of the National Library of Scotland.*

(Figure 7.5).[10] At Fonthill, the new 'abbey' was started in *c.* 1796, but the new lake does not appear to have been started until after *c.* 1808 (OS drawing, 1808). At Trentham, Black Lake was made by 1836, and Shearwater Lake at Longleat in the 1790s.[11] The rationale behind this change in shape would appear to be the desire to increase the perimeter area of the lake. This would agree with the Picturesque precepts discussed later. The decline of carriage drives in parks in the nineteenth century also points to a greater interest in walking rather than driving in parks, although evidence for activities in parks is not as robust as for the eighteenth century.[12] With the increase in turnpike roads at the end of the eighteenth century, and the expansion of a national road network, carriage driving was becoming a common-place necessity, and was no longer a novelty to be enjoyed by the elite. There was also less scope for carriage drives in smaller parks. Edward Kemp explored how to achieve the optimum shape for these lakes (spreading) in the mid-nineteenth century, in

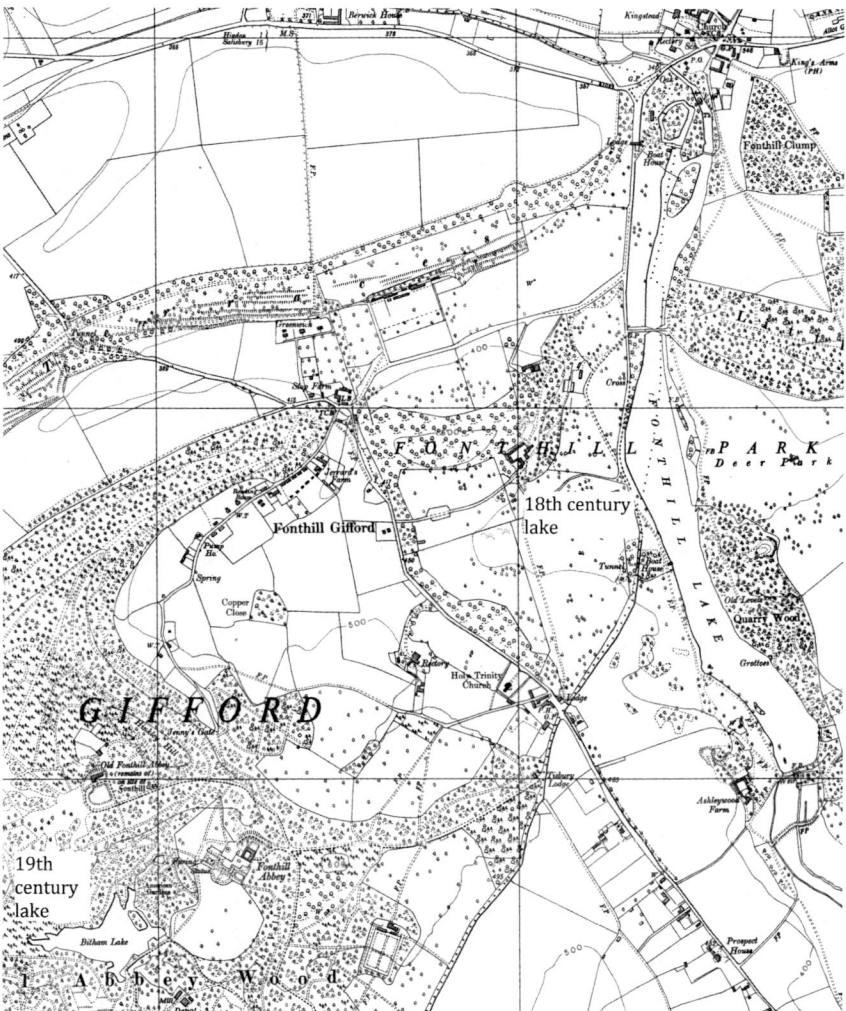

FIGURE 7.5 Fonthill, Wiltshire: Bitham Lake, OS 6" map, pre-1930–61. *Reproduced by permission of the National Library of Scotland.*

his book *How to Lay Out A Garden* 1858.[13] These changes to lake shapes and margins did not begin to appear until the 1820s and '30s, and will be explored in more detail below.

Roughly half of new lakes had boathouses in the nineteenth century. However, it was not until the production of the First Edition 6" OS maps that boathouses were routinely marked, so it is difficult to gauge when they first became popular. The OS drawings of the early nineteenth century are too small scale to show them, so it is only the presence of other data which reveals them before *c.* 1880. They sometimes appear on tithe maps but, "There is a huge variation in the detail included and quality of tithe maps".[14] Thus, the evidence for the lack of boat-houses before the 1880s is partly negative, as there are few illustrations and diary entries which include them before that time. Given the use of garden buildings for storing boats which we have come across at Stowe, and the boathouses mentioned by Felus, such as Kedleston, in the eighteenth century, it seems possible that in the nineteenth century relatively simple, purpose-built boathouses began to appear, perhaps largely for rowing boats.

First half of the nineteenth century

Humphry Repton 1752–1818

Having looked at the chronology of lakes in the nineteenth century, and the main changes to size, shape and position which occurred, we will turn to the practitio-ners responsible for making them, and the aesthetic principles which may or may not have influenced them. The leading landscape designer of the late eighteenth and early nineteenth centuries was Humphry Repton. His landscape career coin-cided with the national downward trend in the numbers of lakes in this period, as Figure 7.1 shows. Repton grew up in Norwich, and having spent some time in The Netherlands, and failed in business, he decided in 1788 to adopt the profes-sion of landscape gardener.[15] After several small commissions, he began work for the Duke of Portland at Welbeck, and the Duke's patronage launched his career. Perhaps Repton's best talent was as an artist and his method of working, as is well-known, was to produce plans and accompanying explanations and illustrations, usually bound in a Red Book. The first Red Book, for Welbeck, was produced in 1789. Repton claimed to have produced some 400 Red Books.[16] Steven Daniels makes the point:

> Repton saw the profession of landscape gardening not just as a way of mak-ing money but as an opportunity to mix with landed society.[17]

A further factor in Repton's career was his keenness, at the start of his career, to maintain Brown's reputation, which brought him into conflict with the proponents of the Picturesque ethos, Sir Uvedale Price and Richard Payne Knight.[18] These factors – his artistic talent, social aspirations and champion-ship of Brown's work – are the key to Repton's career as a landscaper. After a buoyant start with the Welbeck commission, Repton designed for a number of aristocratic clients, such as the Duke of Bedford at Woburn (1804), where he produced an irregular lake, but unlike Brown, he rarely obtained a remit for whole large parks, or even a substantial portion of them. He did design some whole 'estates' but they tended to be small, and in the later part of his life, the focus was on pleasure grounds.[19] However, the perception that these were mainly for minor landowners is false, according to Williamson, as around a quarter of his commissions in the period after 1804 were for the nobility.[20] Also unlike Brown, he did not design and build; he did not offer a complete package. This meant that his designs were liable to be misinterpreted, or altered by a foreman on the spot in favour of an easier alternative. Daniels makes reference to Repton's non-attendance at Woburn, where William Adam bewailed the lack of any overall direction.[21] Undoubtedly, Repton was an artistic man and perhaps not very practical; he was more at home designing pleasure grounds. His focus on them was underpinned by his concept that "the pleasure grounds were to partake more of the character of the house than of the park" and their extent and complexity should reflect its size and importance,[22] and reflects the beginnings of the change in the early nineteenth century of a shift in focus away from the park, and towards gardens.

Repton was particularly concerned with approaches. His success with them appears to have been recognised by contemporaries, as Jane Austen famously assures us in *Mansfield Park*. The stolid Mr. Rushworth, modelled on Austen's cousin, the Rev. Thomas Leigh and his house, Stoneleigh Abbey,[23] actually becomes animated when describing Repton's improvements at his friend's house, Compton: "The approach *now* is one of the finest things in the country. You see the house in the most surprising manner".[24] Repton was also very interested in the impact of the foreground, which was a painterly approach to landscape more in sympathy with the ideals of the Picturesque.[25] He was adept at framing a view, but there is little sense of major innovation in his designs. Rather, he was content to work within the Brown mode, with minor variations of his own. His contribution to landscape style was a largely a shift in emphasis, towards formal gardens, not one of innovation and change in the park.[26]

Repton did not design very many lakes, as Table 9 shows. At Welbeck, he possibly deepened an existing lake (Gouldsmeadow), but the chronology of the lakes

TABLE 9 Repton's works with water

Garden	County	Date	Feature	Size (h)
Welbeck Abbey	Nottinghamshire	c. 1790	Altered a lake	?
Thoresby Hall	Nottinghamshire	c. 1793	River-lake, cascade	4
Corsham Court	Wiltshire	1795–7	Irregular lake	6
Panshanger	Hertfordshire	By 1805	River-lake	4
Tewin Water/Cole Green	Hertfordshire	1799–1805	River-lake	4
Woburn Abbey	Bedfordshire	1804	Irregular lake, and one lake altered	5
Bayham Abbey	Kent	1799–1814	Irregular lake, ascribed to Repton	6
Barningham	Norfolk	1795–1816	Altered a lake	2
Langold	Yorkshire	1806	Lake extended	2

Source: Data extracted from the Landscape Database.

there is not precisely known. At Panshanger (*c.* 4 h), Culford (8 h) and possibly Bayham (*c.* 6 h) he was responsible for the lakes, and at Woburn he made the formal pool to the west of the house into an irregular lake, now *c.* 5 h. At Thoresby, he deformalized the canals, creating a river-lake with several weirs, and at Corsham Court, he appears to have enlarged a very small lake, or pond, made by Brown. At Longleat, he deepened and extended Brown's lake, whilst at Langold, he extended the lake in 1806.

Repton's work at Thoresby exemplifies the ideals of his approach to water. In *Observations* he recommended it being as natural as possible, especially in the case of rivers and cascades, and suggested using rocks to achieve a natural effect.[27] Brown made a plan for Thoresby, but it was not implemented as far as we know, though Repton drew his design for updating the old canals with reference to it (Figure 7.6), and on top of Francis Richardson's design for a formal cascade.[28] He indicated Brown's plan on it with dotted lines, and Brown's intended cascade.

Repton made the course of the River Meden much more informal, ponding it back with several weirs to make it wider, and introducing a new approach over a new bridge (Figures 7.7 and 7.8). It was his design for a cascade which was most noteworthy, and he describes it in *Observations*:

> In forming this cascade huge masses of rock were brought from the crags of Creswell, one in particular of many tons weight, with a large tree growing in its fissures; the water has been so conducted by concealed leaden pipes it appears to have forced itself through the ledges of the rocks.[29]

He disarms the critic by saying that if this is considered to be an artificial management of water, it is no less so than making an artificial lake in the first place. He coyly ends by saying he has not included an illustration because "the best reference is to the spot itself".[30] Fortunately, he did actually make a picture of it (Figure 7.9). He goes on to say, "A rapid stream, violently agitated, is one of the most interesting objects in nature".[31] He explicitly says that he aims to imitate the natural waterfall which occurs when a lake is made by rocks blocking the path of a river, which then tumbles over them with great fury. He excuses this artifice at Thoresby by pointing out that Creswell Crags are only a short distance away, in Derbyshire. In his preference for animated, natural water, he differed from Brown. He was much closer to the ideals of Price and Payne Knight, and at Thoresby he was "perhaps experimenting with his own interpretation of the new taste for the Picturesque".[32]

It is not easy to assess the actual extent of Repton's involvement with making lakes, but his work at Panshanger throws light on this.[33] His design for a lake there appears in his Red Book for Panshanger (Figure 7.10). It was a river-lake, and was constructed much as planned. The site was a flat river valley, as indicated on the plan, and Repton chose to construct the lake by using several weirs to pond back water, though only two are indicated on the plan. It is likely that the cascade mentioned in the accounts was actually one of these weirs. The plan for the northern end of the lake was modified slightly, as two more weirs were made to split the River Mimram and create a substantial island.

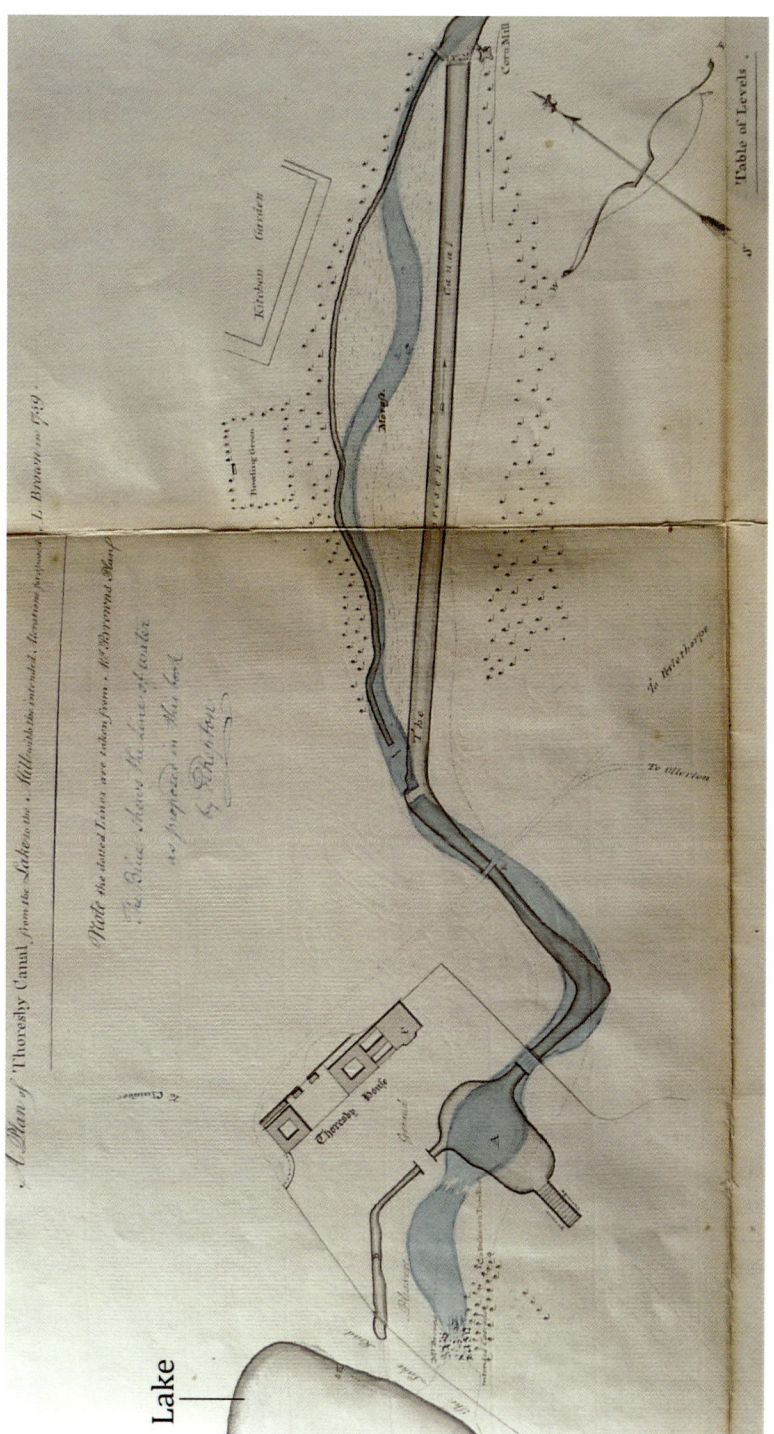

Lake

FIGURE 7.6 Repton's plan for Thoresby Hall, Nottinghamshire, in his 1791 Red Book for Thoresby. Repton's cascade is on the left, adjacent to the lake. *Courtesy of Gregor Pierrepont.*

FIGURE 7.7 Repton's 'slide' of the river and approach at Thoresby, Nottinghamshire, before improvement. *Courtesy of Gregor Pierrepont.*

FIGURE 7.8 Repton's 'slide' of the river and approach drive at Thoresby, Nottinghamshire, after improvement. *Courtesy of Gregor Pierrepont.*

Drawn by H.Repton Esqr   The CASCADE at THORESBY, NOTTINGHAMSHIRE.   Engraved by J.Peltro
                              Seat of Lord Viscount Newark.

FIGURE 7.9 The cascade at Thoresby Hall, designed by Repton, engraved by J. Peltro, 1801, after Repton. *Courtesy of Charles Boot, and the Nigel Temple Collection, The Gardens Trust.*

A ledger of payments by Dr. Thos. Pallett on behalf of Earl Cowper for the Piece of Water in the Park, in 1799, shows that a considerable amount of earth was moved:

> 'about 60 hands [were] at work' digging and 'wheeling out' the 'moory soil' and huge quantities of the underlying 'gravel soil' in the valley bottom. In August another twenty men joined the workforce and in September 110 men were employed. Work continued through the winter, spring and summer of 1800 and included widening the water to the south and north, 'wheeling and spreading earth' and 'altering and slopeing'. By September 1800 the 'Piece of Water' was nearing completion and Repton was on site to supervise the final levelling on the north and south sides of the water, on the island and below the cascade. The total cost of making the 'piece of Water below Pansanger' recorded at Michaelmas 1800 was £2,030 2s 1d.[34]

These accounts included "men attending Mr. Repton", and give considerable insight into how Repton operated.[35] His was clearly quite involved in the work, and the 'slopeing' mentioned is possibly indicated by lines on the plan on either side of the lake south of the middle island. It is clear from the accounts that a considerable amount of earth was moved in wheelbarrows. The straightening and widening involved spreading out the spoil suitably on the adjacent land, especially below the cascade (probably the weir by the lower island). There was considerable digging

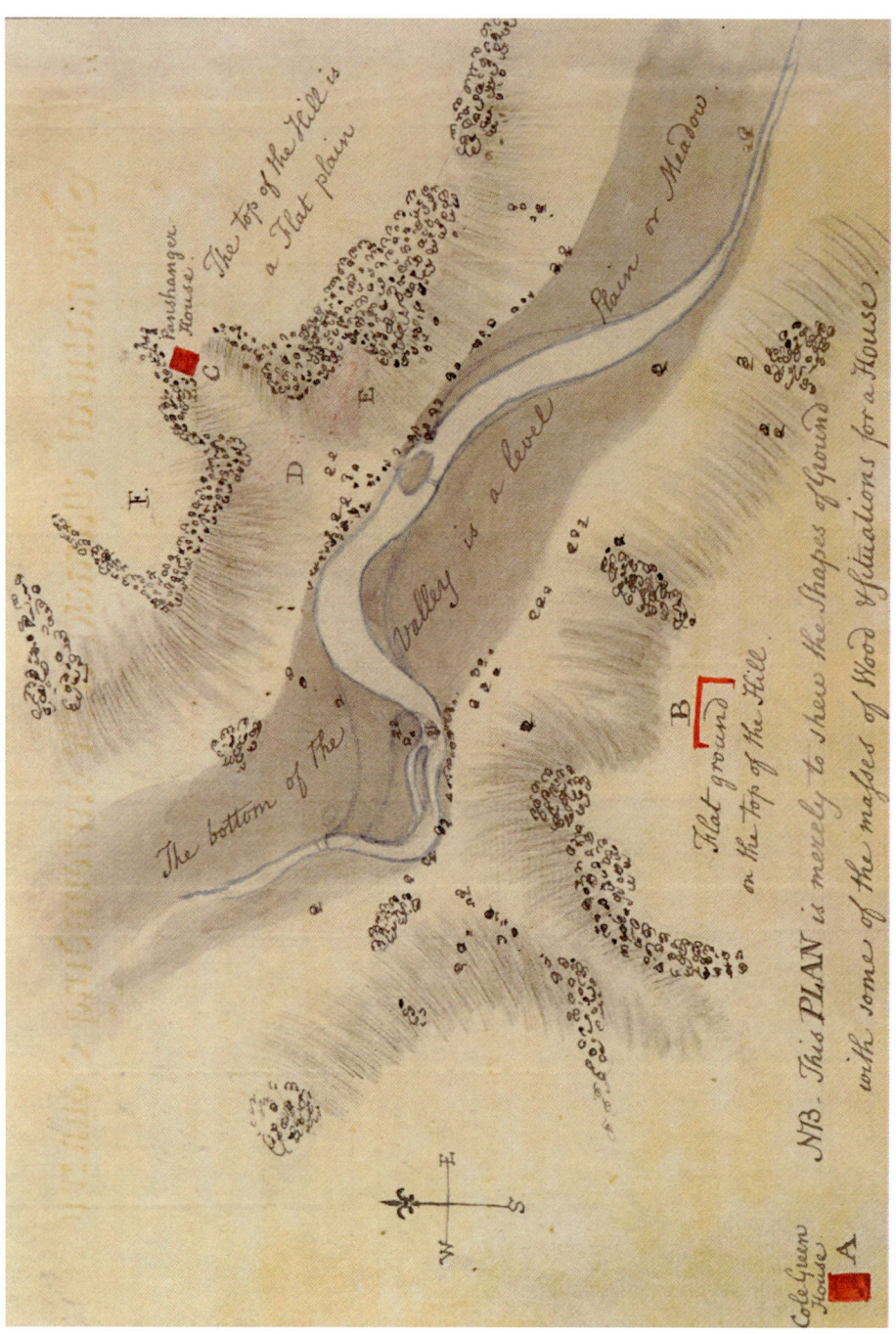

FIGURE 7.10 Humphry Repton's plan for a lake at Panshanger, Hertfordshire, 1799. *Reproduced by permission of HALS. Ref. DE/P/P21 No. 1.*

out above the 'engine house', although it is not clear where that was. Perhaps a device to deliver water to the mansion, it accounted for half the cost of making the lake, and may help to explain how expensive the lake was. That cost also illustrates how the owner's requirements might have an impact. A spreading, irregular lake could have been less expensive to make because it would have involved less earth moving. The likelihood is that a slim, sinuous lake was regarded as more fashionable in 1799. Another factor was that the firm of Mat^w Willcox & Co. was used to do the earth moving, rather than estate labour, which possibly increased the expense.

At Langold, which straddles the Yorkshire-Nottinghamshire border, it is difficult to assess Repton's competence in making lakes. He successfully extended an existing lake in 1806 for Gally Knight by constructing a weir.[36] However, a plan thought to be Repton's for extending this lake very significantly to the north was not successful. Work began in 1812, under the aegis of the landscaper, John Webb. By 1815, the lake was still not finished and was leaking seriously, and by *c.* 1819 the project was abandoned. On current OS maps (2020), it is labelled 'Dry lake'! After Gally returned from Europe in 1811, Repton did meet him again, and so seems to be linked with the new lake plans. Whilst the finger of blame may well be pointed at Webb rather than Repton, it is difficult to envisage Brown in a similar position. Either he would not have chosen that site (limestone deposits outcrop very nearby), or he would have solved the problem, as he did at Bowood and Harewood.

Although Repton's work at Panshanger gives considerable insight into how that lake was made, his significance in relation to lakes in general is questionable. He is associated with only a handful, and with most of these he was generally altering or extending existing lakes. At Bayham, it is not known whether his proposals for the lake were adopted, although a lake was made there at some point before 1841.[37] Repton did not make a new contribution in design terms, being content to work in the established mode. However, one place where he did show originality and flair was at Thoresby, and the cascade he made there is on a different level in terms of design, and much closer to Picturesque ideals.

The picturesque movement

Not only did Repton's career coincide with a decline in lake numbers and size, linked to a general decrease in park size and the fact that most large parks already had lakes, but it also coincided with a prevalent ambivalence about aesthetic style. Brown's death in 1783 emphasised this ambivalence, not just about who to consult in landscape design, but also which style to aim at.[38] Although Repton attempted to fill the gap, his apparent championship of Brown did not recommend him to patrons who were beginning to feel that Brown's style was old-fashioned. It was at this point that ideas about 'picturesque' landscape design began to be discussed. The main proponents were Sir Uvedale Price (1747–1829) and Richard Payne Knight (1750–1824), seconded by William Gilpin (1762–1843). They debated many nuances between them, but the nub of their ideas was that *beauty* was smooth, flowing, tranquil, tame, and that the *picturesque* was rough, varied, exciting, turbulent. In terms of water, this translated into rushing streams, rivers and cascades. They eschewed Brown's smooth lawns flowing down to placid lakes (as they saw them). Their contributions were chiefly through their publications. Price published his *Essay on the Picturesque* in 1794, and was answered by Payne Knight in *The Landscape: a Didactic Poem, Addressed to Uvedale Price* in the same year. Gilpin published a series of *Observations* on various areas of Britain in the 1780s, which were illustrated commentaries on the regions he visited, and *Three Essays* in 1792, including *On Picturesque Beauty*, which was more analytical, on the nature of picturesque beauty, picturesque travel, and on the sketching of landscape, together with a poem on landscape painting. Gilpin's work was complemented in the field of literature by William Wordsworth, whose work did much to popularise the Lake District, just at the time when the Continent became less attractive to British visitors. These three men were not landscape designers themselves. Price did design his own landscape at Foxley, Herefordshire, as did Payne Knight at Downton, but their significance lies in the intellectual debate which they fostered about picturesque ideals and the sterility of Brown's landscapes, as they perceived them.

In his design for Thoresby, Repton seems to embrace fully Picturesque ideals – imitating the crashing and roar of a natural cascade, with 'Nature's Bridge' extending over the torrent. He considered this to be one of his most successful creations, which demonstrated his love of romantic scenery.[39] In *Observations*, he recommends taking nature as a model: "the highest perfection is, to imitate nature so judiciously, that the interference of art shall never be detected".[40] In this regard, Repton agreed with Price and Payne Knight. However, his next statement perhaps explains why he did not attempt to imitate picturesque nature more often: "her wildest features are seldom within the range of man's habitation. The rugged paths of alpine scenery will not be daily trod by the foot of affluence".[41] In other words, the scope for such imitation is limited – suitable sites and suitably large pockets. As Jacques points out, by *c.* 1791, Repton was finding that the practicalities of landscape designing interfered with the aim of trying to imitate a painting, and this led to a rupture between him and Price, who had initially treated him as a friend, recommending him to possible patrons.[42] He was castigated by Price and Payne Knight for his departure from their tenets, and for his championship of Brown, and the controversy was at its height in 1794–5.[43] In fact, Repton's ideas on the treatment of rivers and cascades, notably at Thoresby, were close to theirs in some respects; Price valued the utility of the agricultural context, for example buildings such as mills, whilst Knight favoured a surprising degree of formality in the house surroundings, including terraces.[44] In this respect, they did not differ so much from Repton.

The Picturesque debate raged in the 1790s and early 1800s, but then subsided, to be replaced by a multiplicity of styles and,

> The growing diversification of style was accompanied by a revival of interest in the garden at the expense of the park. Indian, Italian, French and cottage styles of gardening appeared at a bewildering rate.[45]

Although all four men, including William Gilpin, wrote about their ideas, Repton was the only one who designed widely. However, with their writings, and Gilpin's pictures, they changed the taste of landscape design from that which had prevailed in Brown's time, to one which was more concerned with variety and, in terms of water, movement and noise.

An assessment of the impact of the Picturesque ideas on ornamental water depends on the definition of that term. If Price's or Payne Knight's definitions of 'picturesque' are adhered to, the assessment has to be that their impact was small. However, if the Picturesque is defined as a movement concerned with detail, especially in the foreground, and water which was animated – babbling streams and cascades – as opposed to still and reflective, like Brown's, then it was a significant movement in its influence on lakes in the nineteenth century, although the effects were not apparent until later in the century.

J. C. Loudon and W. S. Gilpin

Changes in the shapes of lakes in the eighteenth century did not depend on topography but were driven by changes in aesthetics and park size, and the two men who were linked with this change, through their influential writings, were John Claudius Loudon (1783–1843) and William Sawrey Gilpin (1761/2–1843).[46] Loudon published *An Encyclopaedia of Gardening* in 1822, and Gilpin published *Practical Hints Upon Landscape Gardening: with Some Remarks on Domestic Architecture as Connected with Scenery* in 1832. Both these treatises laid down how lakes should be made, and how the margins should be treated, and it is evident from Loudon's illustrations that if his instructions were followed, lakes like those in Figure 7.11 would be the result. Loudon was a Scot who began working part-time in 1794 as a nurseryman and landscape gardener at Dalry, Ayrshire. He had settled in London by 1803, and by 1804 he was executing commissions for the Duchess of Brunswick and others in the London area and in Scotland.[47] His career as a designer was chiefly connected with public spaces – cemeteries at Histon (Cambridgeshire), Bath Abbey, Southampton Old Cemetery, a public garden (1839) and an arboretum (1840), both in Derby.

Loudon's chief contribution to landscape design was his writings: *An Encyclopaedia of Gardening* in 1822, *The Encyclopaedia of Agriculture* in 1825, *The Gardener's Magazine* from 1826, *The Magazine of Natural History* from 1828, and *Arboretum et Fruticetum Britannicum* in 1835–8. As well as writing, he travelled on the Continent, as far as Moscow, in 1813–14, and again in 1840. In his *Encyclopaedia of Gardening* the essence of what he recommended was rushing water, with a highly varied course and varied sound, with 'naturally' planted margins and carefully positioned islands, as well as 'natural' cascades to add noise and excitement. He felt the reflective, mirror-like qualities of water were not much sought after, and distinguished between still water (lakes) and running water (rivers and rills), recommending the latter.[48] However, he made the point that formal gardens require formal water features, and these should be as lavish as the general scale of the gardens they were for. He placed great emphasis on 'natural' water looking as natural as possible, though

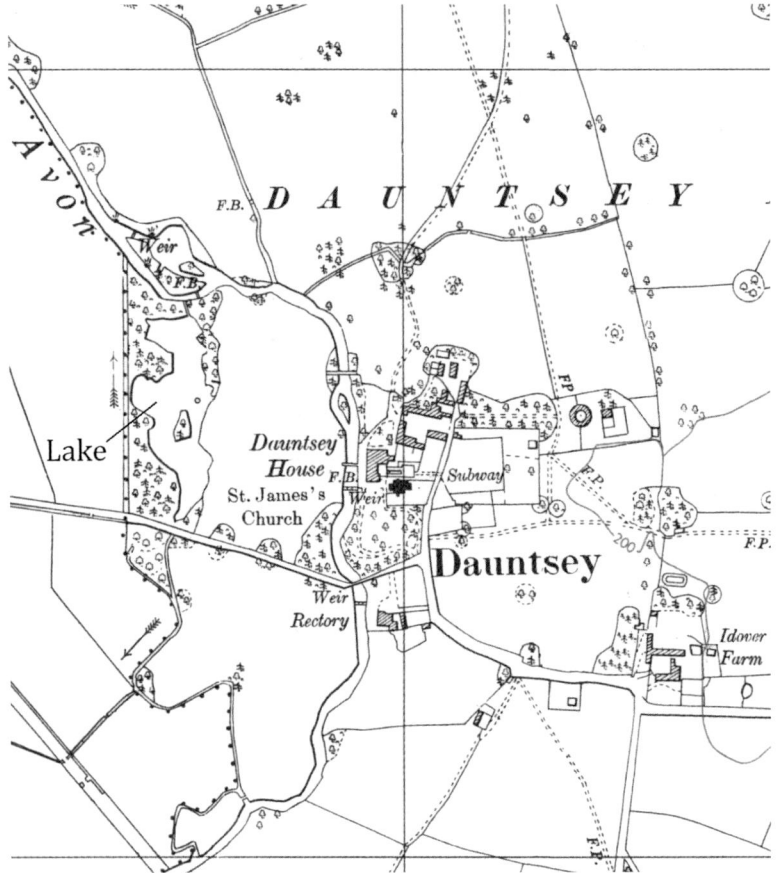

FIGURE 7.11 Dauntsey lake, Wiltshire, possibly made *c.* 1865, OS 1:10,000 map, pre-1930–59. *Reproduced by permission of the National Library of Scotland.*

artificial mountain streams were out of the question as being too difficult to simulate successfully.[49] Figure 7.12 shows what Loudon had in mind for lakes, and he describes them like this:

> The outline of the plan of the lake is to be varied by the contrasted position of bays, inlets and smaller indentations, on the same principles as we suggested for varying a mass of wood. To the irregularity of outlines so produced, islands and *aits* (fig. 697) may be added on the same principle.[50]

He recommended studying the natural situations of lakes and rivers, and copying them to achieve a convincing result. Studying the stones and rocks, as well as the

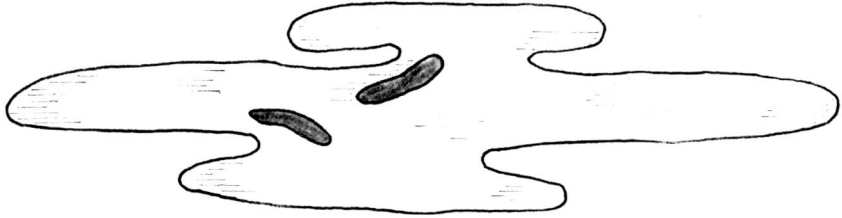

FIGURE 7.12 A diagram of an ideal lake shape, by J. C. Loudon, based on his designs in *An Encyclopaedia of Gardening* 1825.

natural trees and shrubs which occur on the margins of natural lakes and rivers, was the key:

> The *marginal banks* of water in nature, are tame or bold, gravelly or sedgy, stony or rocky, according to the character of the surrounding ground. Art, therefore, must imitate each in its proper place, not always by a studious picturesque arrangement, but by excavating the ground-work, planting the trees and shrubs, and leaving the rest to the motion of the waves of the water. After the effects of one winter, stones or gravel may be deposited in spots suitable for stony or gravelly shores. But to enter into this, and many other circumstances of the imitation of lakes, would exceed the proper limits.[51]

It is interesting that he regards the landscaping of lakes as being outside his remit. Quite why, is not obvious. Despite this, he continues to give advice on them. He is concerned that rushes and 'aquatics' (marginal plants) should not extend too far into the water as that would give a marshy appearance. He recommends significant plantings of trees by lakes, not 'sparingly or indiscriminately scattered around the margin' but to reflect shapes, colour, light and shade, and to 'relieve the brilliancy of the water'.[52] Islands are held to be 'the greatest ornaments of lakes'. He points out that if they do not link visually with other islands or the shore, forming part of a prominence or recess, they will appear quite unnatural. Likewise, centrally placed islands should also be avoided. What he says about puddling lakes is ambiguous. His advice is "to arrange by puddling and under-draining, that a marshy appearance may not surround the lake".[53] He did not actually make any lakes, as far as we know, and this statement has a 'catch all' ring about it. He did advise upon the outline of the lake at Harewood, which had been created by Brown. Illustrations of 1806 show that he proposed an extensive alteration of the shores to give a broken, rocky outline, but any works which took place were restricted to fairly minor irregularities introduced along the north-east shore.[54] His chief interest seems to have been in streams and rivers, in which the ideal was to be improving one, rather than making a new one, and he gives advice for straightening the course occasionally, to increase the speed, or undercutting the banks to increase the swirling of the water, and hence the noise.[55] He distinguished between formal and naturalistic cascades, and included instructions for constructing them.

Loudon's publications were also important in disseminating current thinking on horticultural advances, for example, in hot-house construction and use. His focus was on individual specimens of plants placed in an orderly fashion, usually on a symmetrical axis.[56] He introduced the term 'gardenesque' in the *Gardener's Magazine* in Dec., 1832, and it was subsequently used to "describe a style of garden layout characterised by rampant eclecticism and lack of artistic unity" and it became associated with exotic planting.[57] In terms of lakes, however, Loudon's writings appear to have served, at least in part, as a blueprint for the spreading, irregular lakes which proliferated in the nineteenth century, from the 1820s onwards.

Although William Sawrey Gilpin perhaps carried out a greater number of practical projects than Loudon, typically designing terraces and pleasure grounds, altering approach drives, advising on pinetums, or planting in parks, no lakes can be attributed to him, though there were planned alterations at Wolterton (Figure 7.14).[58] He became a friend of Uvedale Price, and held that a landscape was akin to a painting, favouring the ideas of intricacy, variety and connection.[59] His *Practical Hints Upon Landscape Gardening* 1835, is a good summary of the direction in which landscape design was developing in the mid-nineteenth century. He inveighs against the 'baldness' of Brown's landscapes, the insipid 'easy sweep' of his approaches and the regularity of his clumps. He has a completely different opinion about views: he does not want the *approaching* visitor to see the best views; these should be seen from the windows of the house.[60] This was the complete opposite of Brown's ethos. In one respect, his ideas rolled back landscape style by over a century. He very much favoured an architectural division between the 'dress ground' and the park: "a sunk fence I hold to be wholly irreconcilable to a shadow of taste."[61] In order to connect the house with its surroundings (the gardens, not the park), Gilpin wanted, "the principle of an architectural foreground to be established" (Figure 7.13), and typically he suggested terrace walks, parterre gardens, and balustrades. This would increase the separation of house and lake, of course. His work in the 1820s and 1830s led the way in re-introducing formality into the garden, good examples being Gorhambury, Hertfordshire, Wolterton Hall, Norfolk, and Sudbury Hall, Derbyshire.[62]

Gilpin's theories on making lakes were similar to Loudon's. Concealing the dam was his first concern, followed by the creation of a 'natural' variation in the shoreline, which should not appear straight at any point.[63] He advised that one should not stake out the shape of a lake or pool too exactly, but rather place stakes within the expected shape, so that the water could make its own, natural outline. He also advocated leaving some parts of the banks as they were when

FIGURE 7.13 W. S. Gilpin's illustration of the architectural separation of gardens and park at Heanton, Devon, in *Practical Hints Upon Landscape Gardening* 1835. *Courtesy of Historic England.*

broken by pickaxes, which looked natural, and to avoid a "'hanging level', as the workmen call it".[64] He suggested avoiding a drive, or walk, across the dam if possible as this reveals where the lake finishes, and also avoiding having a walk which goes all around a piece of water as that also betrays its extent. Planting should touch upon the edges of the water not follow it, as this would make it seem bigger.[65] In terms of size, this was a more economical way of achieving what Brown set out to do when he concealed the ends of his lakes by making them turn out of sight. The implication of these recommendations is that the lakes in question were not large, and that walking around them, not driving, was the usual activity. The treatment of the lake margins should be adapted to the situation: they should be wild and broken in wild country, gentler in undulating country. He thought it was essential to connect the woods adorning the banks of lakes with the rest of the woodland in the area. This is in direct contrast with Brown's approach. His (Brown's) aim was to keep the lake margins clear and uncluttered, so that the beauty of the water, with its reflective qualities, could be fully appreciated.

Gilpin makes an important distinction between trying to create a lake and an artificial river, pointing out correctly that it is easier to make a 'natural looking' lake than to imitate a river:

> The difficulties of concealing the extremities of the artificial river, so as to impress the idea of continuity, will be considerable, even under the most favourable circumstances.[66]

This is an indication that the impact of Picturesque ideas was a focus on animated water, whilst acknowledging the difficulties that that entailed. He went on to recommend islands as a good way of concealing or distracting attention from the dam or end of a lake, but the number and size should depend on the situation, and they should not be regular in shape or height.[67] He favoured 'lower growths' and 'fern' for islands, and exposing tree roots, with stones, at the edges of lakes, as this would be 'picturesque'. If possible, trees should be encouraged to lean over banks, which may have to be raised to achieve this effect.[68] There is some evidence of this roughening of edges in his plans for the lake at Wolterton (Figure 7.14), where he suggested altering and extending the lake towards the north-east in one plan, and to the north-west in another, and the intention to construct an island.[69]

A further recommendation in *Practical Hints* is a boathouse or fishing cottage on the lake shore.[70] It is possible that, just as Whateley's positive comments on islands in his *Observations* of 1770 may well have been responsible for the popularity of islands towards the end of the eighteenth century, this comment by Gilpin may have brought the boathouse a significant measure of popularity in the mid-nineteenth century. They are not mentioned by Whateley, and do not occur in many of Brown's plans, so it would seem likely that they increased in popularity in the nineteenth century, perhaps as a result of Gilpin's writing. There were some elaborate structures in the eighteenth century which incorporated boathouses, such as the fishing pavilions at Kedleston and Enville, but these were banqueting houses as well as boathouses.

FIGURE 7.14 W. S. Gilpin's late 1820s sketch of proposed changes to the south of Wolterton Hall, Norfolk. *Courtesy of Lord Walpole and Norfolk Record Office. Ref. WLP 10/104/1.*

Mid-nineteenth century

In the middle and later parts of the nineteenth century, the work of men like Edward Kemp, the Pulham firm, Charles Barry, Anthony Salvin and his brother-in-law, William Andrews Nesfield, is noteworthy in different ways. One thing which is immediately obvious is that, in terms of water, there is no great name among them similar in stature to Brown, or even Emes. Perhaps only James Pulham II approaches that, with the cascades he made. As the number of new lakes being made declined significantly in the nineteenth century, this is not surprising. Some lakes in the new public parks were made, but the rationale for making lakes in public parks, as opposed to private parks, was completely different as they did not have a mansion as a focus (except where a traditional house and park was converted to public use), so they are only touched upon here.

Edward Kemp

Edward Kemp (1817–1891) was born in Surrey, and apprenticed to Joseph Paxton at Chatsworth in the 1830s.[71] From 1843, he superintended the development of the public park at Birkenhead (Figure 7.15), which Paxton had designed (1842–5), and remained head gardener there for forty years. As Paxton withdrew from the project in 1845,[72] Kemp's contribution was significant:

> Birkenhead Park was drained by the creation of two lakes, and the spoil
> was artfully contrived to look like low hills around the lakes, which, with

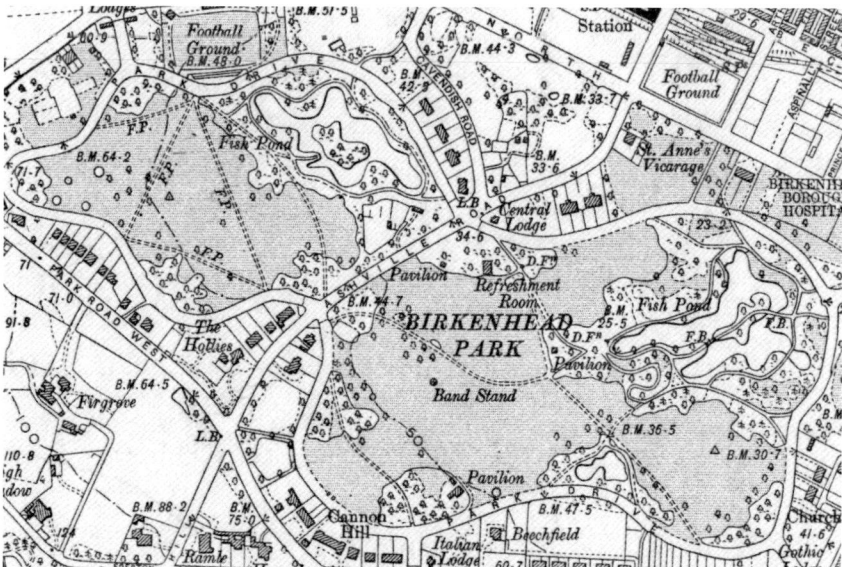

FIGURE 7.15 Birkenhead Park, Liverpool, OS 6" map, 1897. The lakes are 1.7 h and 1.2 h. *Reproduced by permission of the National Library of Scotland.*

their islands, helped to provide views across the water. The hills were made craggy with stone from the excavation of the lakes, and trees were planted to separate the views; all the additions gave a feeling of the park as a 'natural' phenomenon.[73]

Table 10 shows Kemp's main works. His commissions included flower gardens as well as public parks and cemeteries, and his publications included *How to Lay Out a Small Garden* 1850. As the title suggests, Kemp was not generally dealing with landscape parks, but with gardens of 1.2–12 h (3 to 30 acres). He mentions geometric shaped basins as being the most appropriate water features for formal gardens, and he repeats much of the traditional wisdom about designing ponds and lakes and making dams. He maintains that ponds should be lined with puddled clay, and so should 'lakes', unless they are on clay.[74] Although he talks of 'lakes', he does appear to mean ponds, as he describes the water as 'stagnant'. However, in practice he did not make more than a handful of small lakes. Like Gilpin, he thought the ends of lakes should be disguised, and that islands should be in proportion to the size of a lake, and not used in very small lakes.

Both the irregular 'lake' designs illustrated by Kemp in *How to Lay Out a Small Garden* are small – *c*. 0.5 h (Figure 7.16), and the lakes he designed in public parks – Hesketh, Southport, Stanley, Liverpool, and Saltwell, Gateshead – were usually 1–1.5 h and were mainly on flattish ground.[75] Thornton (Stanacres), Cheshire, is a typical spreading Kemp pond. He says the site for this small park (40 h) was chosen in 1850, and had abandoned marl pits, which he decided to convert into a small 'lake' (0.5 h). He adds:

> These pits . . . are always filled with clear water, and often with Water-lilies and other pleasing aquatic plants. In this instance too, as is usual, they were accompanied by a number of rugged old Oaks, of stunted growth; and picturesque masses of Thorns, Furze and other brushwood clothed the banks between them.[76]

TABLE 10 Edward Kemp's works with water

Garden	County	Date	Feature	Size (h)
Garswood (New Hall)	Lancashire	By 1838	Geometric pond	0.4
Birkenhead Public Park	Cheshire	1843–7	Irregular lakes	1.7, 1.2
Thornton (Stanacres)	Cheshire	1850	Irregular pond	0.5
Hesketh Public Park	Lancashire	1864	Irregular lake	1
Stanley Public Park	Lancashire	1868	Irregular lake	1.5
Newsham Public Park	Cheshire	1865	Irregular lake	2
Saltwell Public Park	Northumberland	1876	Irregular lake	1.6

Source: Data extracted from the Landscape Database.

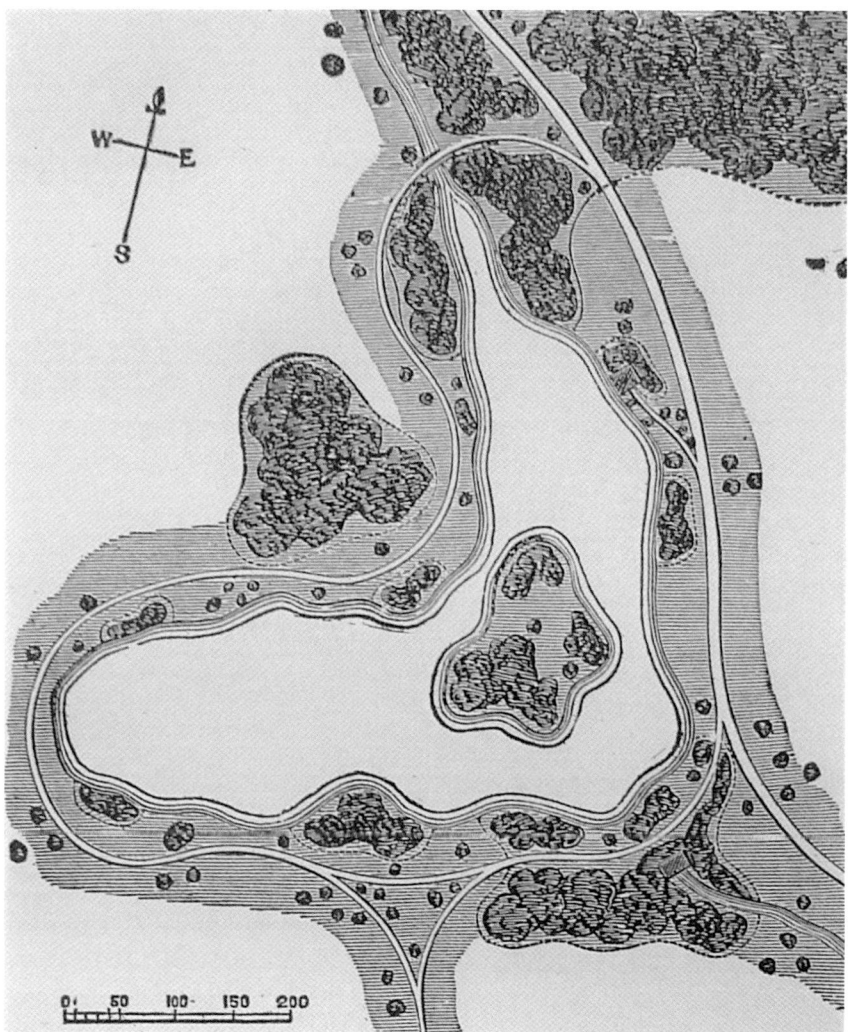

FIGURE 7.16 A design for a pond by Edward Kemp, in *How to Lay Out a Small Garden* 1850. *Courtesy of Historic England.*

It is difficult to tell from Kemp's plan whether his 'lake' had a strip of clear turf around it, as he recommended (later), or not. It appears so, but this could also be the pitching which he recommends for edges.[77] What he says about planting lake margins is interesting:

> Smoothness and softness in the finish of the banks around the water should be a leading feature, and the grass should slope down, more or less gently, to the very edge of the water so that there should be no hard line of earth between them.[78]

This is in direct contrast to what Loudon says in his *Encyclopaedia* of 1822, and perhaps marks the definitive change from Picturesque ideas to high Victorian precepts.[79] Even where plantations come down to the water, Kemp recommends a strip of turf so that the water is not washing against bare earth anywhere.[80] He maintains that lake margins should be either planted or 'mounded', and that too many trees too near a lake make the water dull and unreflective, as well as concealing the water. He favours a scattering of specimen trees: weeping willows and birches, cut-leaved alder, swamp cypress, liquidambar, tamarisk. His recommendations for planting mounds and lake margins differ in detail though not substance from Gilpin's, but are still a contrast to Brown's ideas of keeping lake margins clear and uncluttered, with a sweeping gradient, rather than mounds. Perhaps the difference is more one of scale rather than attitude, however. On dams, Kemp favours planting shrubs, rather than trees, whereas Brown generally used trees. He does not discuss approaches over water at all. In general, the tenor of Kemp's work tends towards a 'garden style' rather than a landscape style, even in his public parks. It is domestic rather than park-like, and his designs and writings about ornamental water reflect this, as does his design at New Garswood, which included formal gardens leading, via steps, down to a formal pond.

The impact of the Picturesque

The significance of the Picturesque debate, especially in terms of ornamental water, arguably had little to do with Picturesque principles themselves. Rather, the fact that the debate occurred signified that people wanted a change in style. Popular opinion had been sated with the Brownian English Landscape style, and wanted a fresh look. The Picturesque did not have an impact on lakes initially, but subsequently translated into a desire for more animation in the landscape, especially relating to water, with a greater focus on detail in the foreground, as we have seen W. S. Gilpin and Loudon recommending. Lakes continued to be made throughout the nineteenth century, though not in large numbers, and the generally smaller park size, combined with the aesthetics of the Picturesque, albeit much muted and modified, had a direct effect on lakes. In response to the smaller park size, the aim was to increase their *apparent* size, and their spreading shape arose from the indentations to increase their perimeters, allowing more planting on or near the margins, to make them seem larger than they were. The response to the Picturesque aesthetic was the emphasis on water seeming 'natural', with margins often being made deliberately rough, and being planted with native shrubs common to river banks. More importantly, Picturesque aesthetics led to cascades becoming popular. They rushed and tumbled in an animated way, and could easily be accommodated in the smaller parks, or added to existing lakes. Although apparently 'natural', they were actually stage managed, particularly in the later part of the century, with carefully positioned rocks and boulders, and 'picturesque' planting amongst them and alongside. It is doubtful whether Price and Knight would have recognised their ideals in these landscapes at first glance.

Cascades and the Pulham firm

Cascades in the latter part of the nineteenth century are inextricably linked with the firm of James Pulham and Co. and their synthetic rock, Pulhamite. Initially, in the 1820s and '30s, James Pulham I made architectural ornaments from a type of cement – Portland Cement – pioneered by the firm of William Lockwood. In the 1840s, his son, James Pulham II, began producing a synthetic rock, which became known as Pulhamite, and the firm became extremely skilled in producing such good imitations of natural rocks that they appeared to be real. It was the combination of the desirability of the picturesque elements of water tumbling over 'natural' looking rocky cascades and the Pulhams' ability to supply 'rocks' to suit any situation which led to the proliferation of cascades from the 1850s, peaking in the 1880s.[81]

The criteria for dating cascades are the same as for dating lakes: where known, the date of creation is given (usually the commencement of work), but where there is a span of years, even decades, a mid-point date is given. There are two other points about cascades. They are harder to identify on maps, and data is more plentiful in the later nineteenth century, so those dates tend to be firmer, as with lakes. Such cascades were often linked to pools, or a series of pools, or sometimes lakes. As with lakes, cascades require some definition. The main criterion is whether they are naturalistic or geometric in form, and in the nineteenth century, they were exclusively naturalistic. Another aspect, which is not addressed in any detail here, is what exactly a cascade is. Does water rushing over a weir constitute a cascade, for example? How high does it need to be a 'real' cascade? At Panshanger, it appears that Repton's 'cascade' was one of the weirs that he used to create the lake.

The flexibility of Pulhamite was the key to its success: 'rocks' of almost any type could be reproduced and cascades or cliffs or caves, for example, could be created which blended in with the location and appeared to be 'natural'. It enabled people who did not have access to supplies of real rock, such as Repton had used at Thoresby, to have similar features in their landscapes which were just as convincing. Over the course of the latter part of the nineteenth century, the Pulhams created numerous rocky landscapes, many with cascades and features such as 'boat-caves', or cliff walks as at Bawdsey, Suffolk, and Madeira Walk at Ramsgate, Kent. James Pulham II was the driving force in the company in the nineteenth century (Table 11).

Just as ornamental lakes were an essential feature of the eighteenth-century landscape, so 'natural' tumbling water, and preferably a series of ponds and cascades, became the signature ornamental water feature of the second half of the nineteenth century. Where landscapes were sufficiently large, and of suitable topography, lakes with cascades were made, or cascades were added to existing lakes. At Dunorlan,

TABLE 11 Works by James Pulham II

Garden	County	Date	Feature	Size (h)
Bayfordbury	Hertfordshire	1845–6	Rock garden & 'lake'	0.8
Highnam	Gloucestershire	1847–9	Informal cascade	
Berry Hill	Buckinghamshire	1856–60	Informal cascades & pond	0.7
Fonthill Abbey	Wiltshire	1859–60	Informal cascade	
Orchardleigh Park	Somerset	1862–3	Informal cascades	
Rendcomb	Gloucestershire	c. 1865	Informal cascade, irregular lake	2.5
Castle Donington	Leicestershire	1866–7	Cascade, Informal canal	
Audley End	Essex	1867	Ponds	
Bearwood Park	Berkshire	1865–71	Rock garden	
Dunsdale	Yorkshire	1866–73	Informal canal	
Lockinge House	Oxfordshire	1864–71	Informal cascade	
High Leigh	Hertfordshire	1871	Informal cascade	
Hutton Hall	Yorkshire	1868–74	Informal cascades	
Titsey Place	Surrey	1871	Informal cascade	
Abberley Hall	Worcestershire	1867–80s	Informal cascades	
Smithills Hall	Lancashire	1873–5	Informal cascade	
Welcome Hall	Warwickshire	1875	Cascade	
Sandringham	Norfolk	By 1877	Informal cascades, boat-cave	
Brogyntyn Hall	Shropshire	1870s	Informal cascades	
Bearwood Park	Berkshire	1879–85	Informal cascade	
Winterbourne	Devon	1884	Informal cascades	
Sheffield Place	Sussex	1882–5	Informal cascades, lake	3
Dumbleton	Gloucestershire	1880s	Boat-cave, pond,	0.5
Waddesdon Manor	Buckinghamshire	1881–92	Water gardens	
Holly Hill	Hampshire	c. 1892	Informal cascades	
Buckhurst Park/ Stoneland	Sussex	1890s	Informal cascade	

Source: Data extracted from the Landscape Database; much of this information is drawn from Hitching *Rock Landscapes: The Pulham Legacy. Rock Gardens, Grottoes, Ferneries, Follies Fountains & Garden Ornaments* (Woodbridge: Antique Collectors' Club, 2010).

Royal Tunbridge Wells, for instance, the lake was made in the early nineteenth century, then extended to Robert Marnock's design in the 1860s, with cascades built by James Pulham II (Figure 7.17).

One important factor which makes cascades different to other water features is that their 'footprint' is relatively small because they are predominantly vertical features. This means that they can be included in relatively small landscapes, with the scope for using ponds or pools rather than lakes in many instances. Like lakes,

FIGURE 7.17 A cascade by James Pulham II, Dunorlan Park, Kent. *Courtesy of Kent Archive and Local History Service and Claude Hitching.*

however, cascades are not always mentioned in accounts of a landscape, and accurate dates are also uncommon.

James Pulham II's work at Highnam Court, Gloucestershire (1847–9 and 1851–62) is the earliest surviving complete Pulham rock garden.[82] Pulham made a series of small pools and cascades in the boggy area between the upper and lower lakes west of the house, uniting them with a water garden complete with a 'gorge' and grottoes. The 'stream' is spanned by a Japanese style bridge and a rocky outcrop forms an island. The rock features were made from a combination of Pulhamite and York stone. At Sheffield Park, James Pulham II used sandstone from local quarries to construct an artificial cascade between the Ten Foot Pond (or First Lake) and Second Lake (Figure 7.18) from 1882–85. The list of works by the Pulham firm compiled by Claude Hitching is detailed and includes other notable places such as Audley End, Sandringham, Waddesdon Manor, Buckingham Palace and Bawdsey Manor.[83] By no means were the Pulhams restricted to water features: at Buckingham Palace a bridge and rocky banks and mounds were made. An innovative feature which emerged from the

FIGURE 7.18 The Pulham cascade at Sheffield Park, Sussex, 1988. *Courtesy of Claude Hitching.*

Pulhams' construction of artificial banks and islands was the boat-cave: a boat-house constructed in the (artificial) rocky banks of a lake as a cave. The 'bulge' in cascade numbers in the 1870s and '80s also coincides with the Pulham firm at its peak, and it seems likely that their skill and ingenuity fuelled the fashion for ponds and cascades in the second half of the nineteenth century. The adaptability of cascades also contributed to their success: an animated though low cascade falling into a small pond could have considerable appeal, and would fit into quite restricted gardens.

Separation of house and lake

In the mid-nineteenth century, the trend was for the big estates to focus on constructing lodges and gateways on the edges of parks, and terraces around the house. In terms of ornamental water though, little happened after *c.* 1820, although fountains and formal garden basins came back into fashion later in the century. In large established parks, having built lodges and gatehouses, the general pattern was to build terraces around the house, with formal parterres, and sometimes conservatories, in the middle of the century, and the lakes remained largely unchanged.

In order to understand the role of lakes at this time, it is necessary to understand the changing relationship between the house and the landscape. The effect of
building terraces was to segregate the house from its surroundings. This meant that
the lake, which in Brown's time had been connected seamlessly with the house by
lawns, was now at one remove. It could perhaps be viewed better from the raised
terraces, but a person casually stepping out of the house would have to be tempted
beyond the flowerbeds and gravel paths to reach it. This segregation actually began
to develop early in the century, with the use of 'transparent fences' by Repton,
and Gilpin's emphasis on an architectural division between gardens and the park
exemplifies this (Figure 7.13).[84] Eventually, it had the effect of isolating the lake in
the park. In Figure 7.19, the women appear to be imprisoned in the gardens, and
flowers in pots are being arranged on the lawn.

With the emphasis on 'unimproved' landscape which the Picturesque movement advocated, there was an increasing tension between 'house' and 'landscape',
which these images hint at, and Figures 7.20 and 7.21 make more explicit. The
terraces act as a barricade between the house and the park. Although there may
have been little real difference between how parks looked in 1745 and 1845, these
paintings demonstrate that perceptions of the park had changed, and the suggestion
is that within the balustrades or terrace walls everything is ordered, whilst beyond
that is a wildness, or at least an 'unknownness', which might be characterised as
Picturesque. A comparison with Brown's landscape at Benham Park (Figure 7.22)

FIGURE 7.19 Harrow Manor, Middlesex, by John Glover, *c.* 1820.

FIGURE 7.20 The lake at Kimberley Hall, Norfolk, seen from the terrace, 2014. Possibly designed by Nesfield, the terrace was originally gravelled.

FIGURE 7.21 Knostrop Hall, Yorkshire, by J. A. Grimshaw, 1870.

FIGURE 7.22 Brown's landscape at Benham Park, Berkshire, 2015.

makes this change of perception even clearer, with some far-reaching implications for lakes: they were becoming isolated in the park in a way which is reminiscent of their fishpond (*vivaria*) ancestors.

Salvin, Nesfield and Barry

The people who were employed by the wealthy to make those terraces and parterres were Anthony Salvin and his brother-in-law and later business partner, William Andrews Nesfield, and Charles Barry. However, there is evidence that these three men did engage with the landscape beyond the terraces in different ways, though how much this was initiated by the clients and how much by the designers is difficult to assess.

Salvin (1799–1881) is mainly noted for his houses, and significant works were Harlaxton Manor (1835–43), Keele (1854–60) and Thoresby Hall (1864–76). At Harlaxton, a new lake (2.3 h) was created to flank the approach to the new house from the north. Though it is not known who was responsible for the lake, Salvin appears to have been responsible for the bridge taking the main approach directly across the lake.[85] The lake and the bridge considerably enhanced the approach to the house and, with the land rising, the hall appeared dramatically above. Ensuring that the approach route had water flanking it is again reminiscent of Somersham, or Raglan or Staunton Harold, and highlights that passing across water on the approach to the residence continued to be important in the nineteenth century, for those wealthy enough to command it.

Perhaps Salvin's main contribution, in terms of lakes, was an indirect one: he frequently recommended Nesfield (1793–1881), his relative and business partner,[86] to design the parterres for his terraces. This aspect of Nesfield's work is well-known, based on the 'bedding out' which was pioneered in the 1820s and '30s by such men as John Caie at Bedford Lodge, Kensington, George Fleming at Trentham, Staffordshire, and Donald Beaton at Shrubland, Suffolk.[87] What is less well known is the extent of his accomplishments, which included:

> fashioning lakes, fountains and cascades; positioning new houses, lodges and gateways; screening railway lines; planting trees and avenues; . . . laying out terraces and balustrade walkways.[88]

At Crewe Hall, Cheshire, where he worked from 1840 to 1860, he linked the house to the existing lake with his design of terraces and parterres (Figure 7.23), according to his guiding principle that, "the formal area around the house should merge gradually with the natural area in the far distance, with a transitional zone in between."[89] The lake no longer exists there as the dam burst in 1941.[90] At Crewe, as elsewhere, it is often difficult to establish exactly which elements Salvin

FIGURE 7.23 W. A. Nesfield's parterre at Crewe Hall, Cheshire, *c.* 1870. *Charles Latham / Country Life Picture Library*.

and Nesfield were responsible for. Through Salvin's recommendation, Nesfield produced a similar plan for Sudbury Hall, Cheshire, in 1852 (Figure 7.24).[91] Like Crewe, the site was fairly flat, with a 7-h lake (eighteenth century), and Nesfield's aim was to re-orientate the existing axis of the gardens, which was parallel to the house, to extend longitudinally away from the house, and so provide a longer perspective view, directly engaging with the lake, and marrying the house to the lake. However, this was not executed.

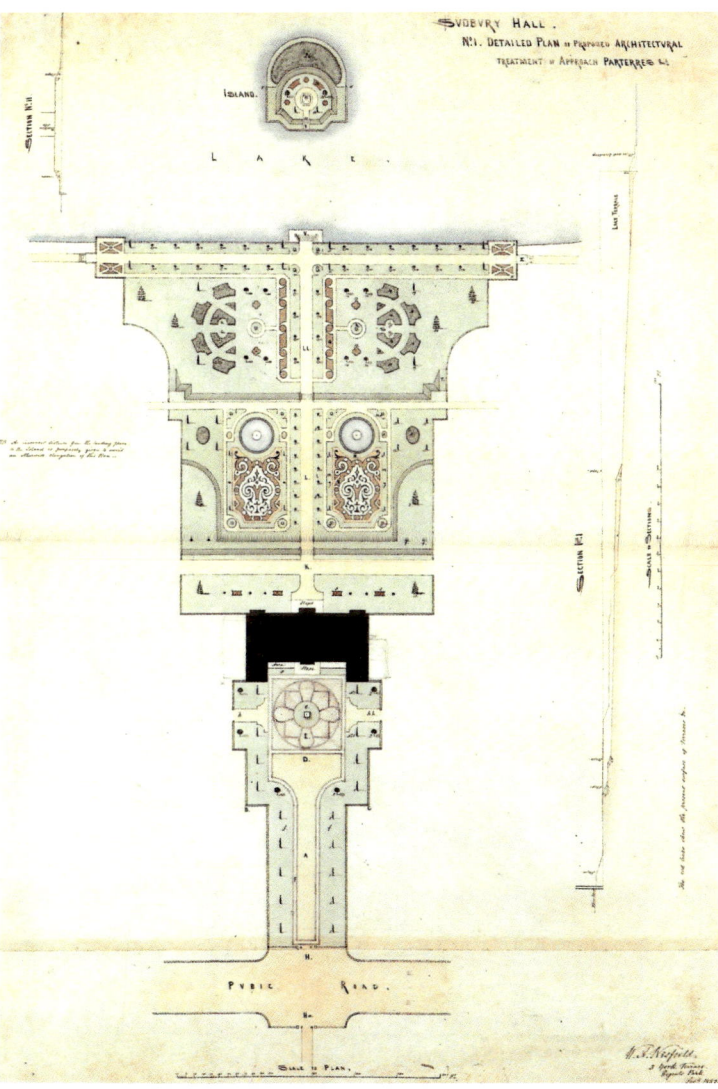

FIGURE 7.24 A garden plan for Sudbury Hall, Cheshire, by W. A. Nesfield, 1852. North is towards the lower right.

Nesfield worked extensively at Castle Howard from 1849 to the mid-1860s, beginning with a commission to alter the shoreline and island of the Great Lake (north of the house).[92] By 1850–1, an enormous balustraded parterre to the south of the house was being constructed, along with the Atlas Fountain. This, and the Prince of Wales Fountain in the South Lake, were gravity fed from the reservoir in Wray Wood (which tended to leak), which itself was supplied with water pumped up by a steam engine from Coneysthorpe village, a mile to the north.[93] South Lake was adjacent to the parterre, and another part of Nesfield's remit was to link it with the New River. He did this by creating the Temple Pool with a cascade between the two, although Christopher Ridgway suggests the idea itself may have been Henderson's (the agent). Nesfield also altered the shape of South Lake, making it more formal. The rationale behind the scheme was to create the illusion, looking from the house towards the Mausoleum, of a continuous sheet of water. Nesfield added an additional pool, and two cascades to the concept. As Christopher Ridgway points out,

> The effect of adding these formal features (both additional pools repeated the symmetry of the South Lake) was to extend the geometrical region begun with the parterre south of the house and continued by the South Lake, which itself received additional formal embellishments in 1864.[94]

This created the *trompe l'oeil* effect which Henderson had predicted, the changes in level being hidden from a viewer near the house, with the Mausoleum, nearly a mile away, forming part of the vista (Figure 7.25).

FIGURE 7.25 View from the south parterre at Castle Howard over South Lake to the Mausoleum, 2017.

However, fashion was beginning to change as early as the 1870s, and people were beginning to take note of the ideas of William Robinson, as promulgated in his book *The Wild Garden* 1870. By 1894, Rosalind, 9th Countess of Carlisle, had removed Nesfield's parterres and gravel walks, and was busy trying to soften the geometric outlines of the South Lake. (This was at least the third outline for this lake shore.) Trees and shrubs were also planted above the lake which eventually obliterated the full impact of the planned vista. It has been a general view that Nesfield was primarily a creator of intricate parterre designs, but his work at Castle Howard belies this. His ability to fashion water features, and the engagement with lakes in his parterre designs deserves more recognition.

Sir Charles Barry (1795–1860) had a similar impact to Salvin on houses and landscapes. Remembered in the twentieth-century for his design of the Houses of Parliament, his work most widely known today (2020) is probably his remodelling of Highclere Castle. His signature was the Italianate style, his taste having been formed during an extensive tour of Europe and the Near East in 1817–20.[95] Like Salvin, his commissions often included formal gardens, with terraces and parterres, and extended into the landscape with the design of lodges, gates and obelisks. At Bowood, for example, he designed the Golden Gates, and the Lansdowne Monument which stands on Cherhill Down. Like Salvin, Barry did not design any lakes but at two places, he designed terraces which bridged the gap between house and lake: Trentham and Clumber, and at a third, Harewood, he remodelled the house in the 1840s, and his Italianate gardens provided views over the lake (Figure 7.26).

FIGURE 7.26 Charles Barry's terrace overlooking the lake at Harewood, Yorkshire, 2019. *Copyright Trevor Nicholson/Harewood House Trust.*

As with Salvin, it can be difficult to disentangle Nesfield's work from Barry's, and it is possible that Nesfield did the planting plans there. The influence of the Italian lakes can be clearly detected, in the juxtaposition of terraces and water. Despite this, there is still a sense of separation from the lake. How do you reach it from the house? The OS map of 1891–2 (Figure 7.27) suggests the route was a very indirect one, either around the stables, or possibly along the eastern boundary. As today, there was a clear vista between trees to the lake. There were steps from the terraces down to a grass rampart but the conclusion has to be that the lake was primarily for looking at, at least initially, and further effort was required to actually reach it. That may have depended on the 3rd Earl of Harewood's taste, because at Trentham and Clumber, the gardens designed by Barry extended to the lake edge. The other factor may have been that the lake at Harewood is further from the house, and considerably lower, so the cost of a similar scheme would have been greater.

The gardens at Trentham are well-known today, with their planting by Piet Oudolf and Tom Stuart Smith, which follows the outlines of Barry's parterres. Barry's gardens were large, and occupied the whole area between the house and the lake, which is a flat site. Barry conceived his parterres as a transitional zone taking the visitor from the formality of the house, via the 'outside formality' of the gardens to the lake.

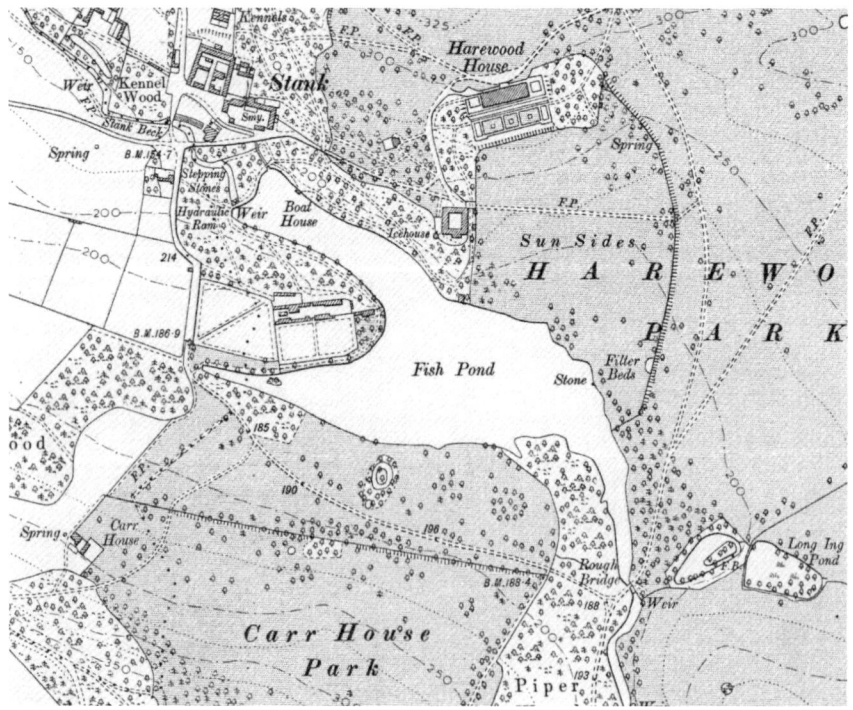

FIGURE 7.27 Harewood, Yorkshire, OS 6" map, 1906–9. *Reproduced by permission of the National Library of Scotland.*

Visible in Figure 5.22, is the straight edge that Brown's lake has at this end, which complemented Barry's gardens, and which was the remainder of the semi-geometric lake of *c.* 1727 (Figure 5.23). Here, there is no doubt that the gardens are meant to link the house and lake, and a small bastion jutting into the lake emphasises this. Like Nesfield's design at Sudbury, Barry's is on an axis stretching away from the house and draws the eye and the visitor from house to lake.[96] At Clumber, the house was much closer to the lake, and the axis was parallel to the house. The 1884 First Edition OS map suggests that this parterre was made very much as indicated in the plan, with one or two minor alterations and, again, Nesfield's name is linked with this work. In 1837, he designed a battery jutting into the lake, and may have been responsible for the actual terraces.[97] The house was rebuilt in 1879, by Charles Barry – junior, which illustrates how complex the situation is when assessing the works of Salvin, Nesfield and Barry; W. E. Nesfield also worked on some of these sites.

What emerges from this is that designers such as Nesfield, Salvin and Barry, were capable of dealing with many aspects of landscape, and that they were alert to opportunities for engaging with ornamental water. There remains a sense of uneasiness, or at least ambivalence, about the role of that water. Clearly, some patrons embraced it, but it would appear that others did not, preferring to remain safely on their terraces. At Bowood, for example, where terraces were constructed in the early to mid-nineteenth century, there was no attempt to link them with the lake, and at Sudbury, where the topography was relatively flat, Nesfield's striking plan to link the house with the lake was not adopted (Figure 7.24).[98]

As well as acting as a barrier between house and park, extended terraces and parterres also provided scope for ladies' talents in planting flowers, and Jane Loudon's publications testify to the increasing practical interest of women in the subject. The continuing interest in collecting plants from overseas fostered the interest in conservatories, which could be palatial (Chatsworth, Kew, Enville, Tatton Park), and 'American' Gardens, which consisted of primarily exotic trees, mainly from North America, as at Fonthill, Wiltshire.

Terraces clearly affected the way in which lakes were approached from the house, but having negotiated the terraces, the approaches to any lake which was at a distance from the house remained similar to the eighteenth century: paths led obliquely from house to lake, usually winding around it. However, there is some evidence that the main approach to the house in new parks was less likely to involve the lake, which was somewhat isolated in the park in the case of larger parks. This is certainly the case with Bearwood, Berkshire, Red Lodge, Wiltshire, and Felthorpe in Norfolk. The decline in circuit drives in the nineteenth century further contributed to this isolation of the lake.[99] It would also be a mistake to think 'only terraces' in relation to the nineteenth century. They were substantial and architectural, and could be extensive, as at Holkham and Trentham. Figure 7.28 gives a flavour of them. These were gardens in their own right. It must also be acknowledged that the juxtaposition of highly formal gardens with informal lakes has a jarring note. However, this may be what the Victorians, or their designers, perceived that they had seen in Italy. Certainly, they revelled in the extensive bedding schemes which

FIGURE 7.28 Holkham Hall, Norfolk, showing the terraces by W. A. Nesfield and W. Burn. *Copyright Keiron Tovell Photography.*

the new technology of heated glass houses made possible, as well as the increasing general wealth of patrons, at least up to the 1870s.

In summary, lakes in the mid-nineteenth century suffered from a tension between the house and the landscape, whereas Brown and his contemporaries had been happy to let the park come right up to the house. In the context of the mid- and later nineteenth century, the advent of the 'moral' house may be relevant.[100] There was a dawning feeling that there was a proper place for everything. In house planning, this was evinced by rooms primarily for men or women: drawing rooms and morning rooms for ladies, libraries and smoking rooms for men. Externally, these spawned the terraces with flower beds, and parterres below, or sometimes a conservatory, and the main sections of the parterres would be aligned on the windows and doors of the principal rooms.[101] In landscapes, this translated into a sense that the 'polite' place to be, especially for women, was on the terraces, or in the pleasure ground, not in the park, and pleasure grounds were becoming more 'garden-like', as at Alton Towers, for example.[102] Men, of course, continued hunt- ing, shooting and fishing in the park. The apparent proliferation of boathouses suggests they also rowed on the generally smaller lakes, fishing perhaps, and possibly the more intrepid women joined them, genteelly rowing on lakes near the house, but there are few paintings of them. The 'portrait landscape' painting is virtually absent after the 1820s, and there are very few images of people in parks after this time, suggesting that patrons did not commission paintings of them, or activities in them. Instead, in the 1850s, Adveno Brooke portrayed women in gardens, especially on terraces. This, and the later nineteenth-century pre-Raphaelite focus on the female form, implies that parks were no longer of interest.

Post script: the late nineteenth century

Very few lakes were made in the later nineteenth century. From 1870–99, possibly half a dozen were made, and two were altered. In comparison, around 28 lakes were made in the preceding 30 years, which itself is not a large number when compared with the same decades of the eighteenth century.[103] By 1900, the age of the motor car was dawning, to be closely followed by that of the plane, and a different focus was developing, which did not involve the park or lakes.

Conclusion

Several significant points have emerged from this survey of lakes in the nineteenth century. Numbers declined from *c.* 1815 onwards, and lakes also became generally smaller, probably because new parks were also smaller on average. They were, of necessity, often in front of the house and to be walked around by those who could be tempted beyond the new terraces, rather than driven around.

The shapes of lakes also changed, becoming spreading, whereas they had generally been elongated in the eighteenth century. This appears to have been a response to the aesthetics of the Picturesque movement as, combined with the 'roughing up' of lake margins, it produced more varied edges, with more scope for 'surprises', and the impression that a walk around such a lake was longer than it actually was. Islands also became more popular, fulfilling a similar role of creating interest and making a lake seem bigger. Boathouses, even boat-caves, became popular as well.

This survey also shows that in the large, established parks there was little significant change in lakes after *c.* 1820. Margins might be altered slightly, or planted differently, the rhododendrons and laurels on the lake shores at Stourhead being a case in point, planted in the early nineteenth century by Richard Colt Hoare, but these were changes in detail rather than substance. This kind of planting, plus the increased indentation of edges and the mounding of banks, may have made lakes look more 'naturalistic' to nineteenth-century viewers on the ground.

The aesthetics of the Picturesque movement led, eventually, to a desire for animated water. Ponds in series with cascades became popular, especially in smaller parks, the Pulham firm being instrumental in providing them. Some were made in larger parks: cascades and ponds at Sheffield Park, for example. Cascades were also added to existing dams, and this may be the case with the two cascades at Bowood, as there is no indication of either on a 1789 plan for managing the plantations of the estate. This plan is very detailed in many respects, even showing the buildings of the menagerie in a clearing on the north side of the small, western lake (Figure 5.20).

There is no doubt that the large, established parks focussed on lodges and gateways in the earlier part of the century, and elaborate terraces and formal gardens in the later decades, at the expense of water. In the lakes which were made, boathouses were popular. The distance of lakes from the house, and relative height, remained dependent on factors such as topography and park size. Other interests also came to the fore, particularly large conservatories, and the bedding schemes which they facilitated.

However, probably one factor in the decline of lakes was stronger than the rest: the novelty of lakes had worn off – they had been 'done'. A lake was no longer deemed to be a vital ingredient of the ornamental landscape, but depended largely on personal taste, as opposed to prevailing fashion. A new park might well have an irregular lake, but it did not have to be large because it was no longer seen as the main statement. This was particularly pertinent after *c.* 1820.

In the eighteenth century, the English Landscape style predominated, primarily driven by Brown's 'design and build' practice, but in terms of lakes, the nineteenth century did not see a similar practitioner. Repton did not have the business acumen of Brown, preferring to concentrate on designing rather than building, and he did not develop anything new in terms of lakes.[104] There was also a developing sense that experts were required – architects, engineers – whereas Brown and his contemporaries were happy to provide the whole 'package': house, lake, landscape, planting, drives. Consequently, Brown was able to 'roll out' his style of house and landscape, whereas the process began to become more fragmented in the nineteenth century, and no single style predominated. Most significant was the increasing separation between gardens and park, which tended to leave lakes isolated in the park, whilst gardens and their appurtenances gained in importance.

Notes

1 The date range 1840–1880 would then be entered in the Landscape Database in the column called 'Dating'. It would not be possible to sort that column sensibly by date, however.
2 Total taxation increased sharply towards the end of the eighteenth century, according to J. V. Beckett & M. Turner in 'Taxation and Economic Growth in Eighteenth-Century England', *The Economic History Review* New Series, Vol. 43, No. 3 (August, 1990) p 384. Possibly, wars had some kind of delayed effect so that the events of the later 1770s had most impact in the 1780s, a decade which was largely war free, with the effects of that peace translating into the 1790s.
3 Bishop, op. cit., Table 30, p 297.
4 As before, Cheshire lakes were largely excluded because of the effects of salt extraction.
5 T. Williamson, *Suffolk's Gardens and Parks: Designed Landscapes from the Tudors to the Victorians* (Macclesfield: Windgather Press, 2000) p 109 and P. Wade-Martins, ed., *An Historical Atlas of Norfolk* (Norwich: Norfolk Museums Service, 1993) p 111.
6 Wade-Martins, ibid., p 110.
7 T. Williamson, personal communication, December 2014.
8 Bishop, op. cit., Table 33, p 300.
9 J. C. Loudon, *The Gardener's Magazine and Register of Rural and Domestic Improvement*, Vol. 9 (1833) p 679.
10 In the case of Fonthill, a different part of the park became the focus, as the new house was built there.
11 Historic England listing: Longleat Park and Garden.
12 T. Williamson, personal communication, June 2016.
13 E. Kemp, *How To Lay Out A Garden* (Massachusetts, 1858) p 295.

14 Steven Cable, Remote Enquiries Duty Officer, National Archives, Kew, personal communication, February 2016.
15 S. Daniels, entry for H. Repton in ODNB, op. cit.
16 Ibid.
17 Ibid.
18 S. Daniels, *Humphry Repton: Landscape Gardening and the Geography of Georgian England* (London: Yale University Press, 2000) p 104.
19 T. Williamson, personal communication, March 2017.
20 T. Williamson, *Humphry Repton: Landscape Design in an Age of Revolution* (London: Reaction Books, 2020) p 199.
21 Daniels, *Humphry Repton*, op. cit., p 177.
22 Williamson, *Repton*, op. cit., p 199.
23 Historic England listing: Stoneleigh Park and Garden.
24 J. Austen, *Mansfield Park* (London: Penguin Books, ed. 1996) p 46.
25 Williamson, *Suffolk's Gardens*, op. cit., p 92.
26 T. Williamson, personal communication, June 2017.
27 H. Repton, *Observations on the Theory and Practice of Landscape Gardening* (London, 1805), pp 32 and 38.
28 Williamson & Brown, op. cit., p 116.
29 Repton, op. cit., fn., p 38.
30 Ibid., fn., p 38.
31 Ibid., p 38.
32 J. Finch, 'Three Men in a Boat: Biographies and Narrative in the Historic Landscape', *Landscape Research*, Vol. 33, No. 5 (October, 2008) pp. 511–530, p 521.
33 I am indebted to Tom Williamson and Anne Rowe for drawing my attention to this.
34 T. Williamson & S. Flood, eds., *Humphry Repton in Hertfordshire* (Hatfield: Hertfordshire Publications, 2018) p 123.
35 Accounts ledger for 1798–1811 in Hertfordshire Archives and Local Studies office, ref. HALS D/EP/EA23/2 fol. 180.
36 P. Eyres & K. Lynch, *On the Spot: The Yorkshire Red Books of Humphry Repton, Landscape Gardener* (Huddersfield: New Arcadian Press, 2018) p 91.
37 Historic England listing: Bayham Abbey Park and Garden.
38 Jacques, op. cit., p 181.
39 A. Gore & G. Carter, eds., *Humphry Repton's Memoirs* (Norwich: Michael Russell, 2005) p 162, quoted in Daniels, *Humphry Repton*, op. cit., p 161.
40 Repton, op. cit., p 38.
41 Ibid., p 39.
42 Jacques, op. cit., p 146.
43 Daniels, *Humphry Repton*, op. cit., p 104.
44 J. Appleton, 'Some Thoughts on the Geology of the Picturesque', *Journal of Garden History*, Vol. 6, No. 3 (1986) p 280.
45 Jacques, op. cit., p 183.
46 W. S. Gilpin was the nephew of William Gilpin.
47 B. Elliott, entry for J. C. Loudon, ODNB.
48 J. C. Loudon, *An Encyclopaedia of Gardening* (London, 1825) p 1010.
49 Ibid., p 1010.
50 Ibid., p 1011.
51 Ibid.
52 Ibid.
53 Ibid.
54 Historic England listing: Harewood Park and Garden.
55 Loudon, op. cit., p 111.
56 H. Leathlean, 'From Gardenesque to Home Landscape: The Garden Journalism of Henry Noel Humphreys', *Garden History*, Vol. 23, No. 2 (Winter, 1995) p 176.
57 Elliot, op. cit.

58 He did formulate proposals for altering the existing lake at Wolterton: T. Williamson, personal communication, June 2017.
59 S. Piebenga, entry for W. S. Gilpin, ODNB.
60 Gilpin, op. cit., p 20.
61 Ibid., p 85.
62 Piebenga, op. cit.
63 Gilpin, op. cit., pp 153–154.
64 Ibid., p 155.
65 Ibid., p 157.
66 Ibid., p 161.
67 Ibid., p 164.
68 Ibid., p 170.
69 T. Williamson, *The Archaeology of the Landscape Park: Garden Design in Norfolk, England, c. 1680–1840*, BAR British Series 267 (Oxford: Archaeopress, 1998) p 214.
70 Gilpin, op. cit., p 162.
71 J. Waymark, entry for Edward Kemp, ODNB.
72 Parks & Gardens UK entry: Birkenhead Park.
73 Waymark, op. cit.
74 Kemp, op. cit., p 297.
75 Ibid., pp 301–302.
76 Ibid., p 301.
77 Ibid., p 305.
78 Ibid., p 297.
79 Loudon, op. cit., p 1011.
80 Kemp, op. cit., pp 297–300.
81 Bishop, op. cit., Table 37, p 329.
82 C. Hitching, *Rock Landscapes: The Pulham Legacy. Rock Gardens, Grottoes, Ferneries, Follies Fountains & Garden Ornaments* (Woodbridge: Antique Collectors' Club, 2010) p 61.
83 Hitching, op. cit., p 291.
84 Gilpin, op. cit., p 39.
85 Historic England listing: Harlaxton Manor Park and Garden.
86 S. Evans, 'William Andrews Nesfield: An Introduction to His Life and Work', *William Andrews Nesfield, Victorian Landscape Architect: Papers from the Bicentenary Conference, King's Manor, York, 1994*, ed. C. Ridgway (York: University of York, 1996) p 3.
87 G. Jellicoe, S. Jellicoe, P. Goode, & M. Lancaster, eds., *The Oxford Companion to Gardens* (Oxford: Oxford University Press, 1986) p 42.
88 Evans, op. cit., p 7.
89 Ibid., p 8: from "numerous reports" for various properties written by W. A. Nesfield.
90 Historic England listing: Crewe Hall Park and Garden.
91 Jackson-Stops, op. cit., pp 133–135.
92 C. Ridgway, 'Design and Restoration at Castle Howard', in C. Ridgway, op. cit., pp 39–54.
93 Ibid., p 44.
94 Ibid., p 47.
95 M. Girouard, 'Charles Barry: A Centenary Assessment', *Country Life* 13th October, 1960.
96 Jackson-Stops, op. cit., p 137.
97 Historic England listing: Clumber Park and Garden.
98 Jackson-Stops, op. cit., p 135.
99 T. Williamson, personal communication, June 2017.
100 Girouard, op. cit., p 276.
101 Evans, op. cit., p 8.
102 Mowl & Barre, op. cit., p 222.
103 These numbers will change as new data emerge, but the trend is clear.
104 Daniels, *Humphry Repton*, op. cit., p 1.

8

CONCLUSION

For a visitor to a late eighteenth-century landscape, the most striking feature apart from the house, would often have been the lake. They were frequently very large, so it is all the more surprising that they have attracted little attention from garden historians. Perhaps this is in line with today's tendency to assume that lakes are a natural part of the landscape – that they have always been there. This was not the case, as we have seen, and this study has illuminated a neglected aspect of garden history.

When we think of a lake, we tend to imagine a body of water with informal, curving edges, which might occur naturally, and England seems to be plentifully supplied with them. However, ornamental lakes like these did not occur in England before *c.* 1720, and the word 'lake' was not normally used to describe them until the late eighteenth century. The only natural lakes in England are in the Lake District, although all sorts of other water could be seen in landscapes before 1720: large fishponds (*vivaria*), moats, millponds, hammer ponds, and small fishponds. By the late seventeenth century, there were also some geometric pieces of ornamental water. As well as the fact that virtually all the lakes we see are man-made, another important concept is the difference between a pond and a lake. Size might seem to be the distinguishing feature, and a minimum of 1 h has been used to define a lake, but the most significant feature is that a lake is constantly replenished by a spring or river, whilst a pond usually relies on run-off and rainwater to fill it. Superficially, this fact may seem unimportant but actually it is vital, because it governs where these features can be sited in landscapes, and how they are made. Lakes have to be sited where the suitable water source is, whereas ponds can be situated almost anywhere, providing they are lined (usually with puddled clay).

One major aim has been to establish a chronology of ornamental lakes as a basis for finding out why and in what form they first appeared, and how they evolved. Very little has been written about lakes in the past or the present, and Roger North and Stephen Switzer stand out as historical sources. Much of the information, therefore, has been drawn from maps, plans and images (mainly landscape paintings) and archival documents, and a statistical approach was adopted. In order to make lakes more amenable to analysis, they were grouped into categories according to various characteristics. This meant that they could be counted in various ways – according to date, size, location or designer, for example. The main categories are irregular and geometric lakes, with various sub-categories. Two new sub-categories in particular emerged: the river-lake, and the hybrid lake. Their names suggest their origins and, indeed, the river-lake is made from a river, by widening it to look plausibly lake-like, using weirs, and the hybrid lake is a cross between a geometric lake (straight-sided) and an irregular (informal) lake. Another point to emerge was the difference between a dam and a weir. A dam holds the water back completely, whilst a weir has water flowing over it. Why is this important in a history of lakes? The answer lies in the complexity of them, and therefore the expense of making them. A dam is much more complex to make, and will usually retain a much greater body of water than a weir, so weirs enabled lakes to be made relatively inexpensively and convincingly, and required a smaller area of parkland.

The first irregular lake was made in *c*. 1719, and numbers began to increase in the following decades, peaking in the 1760s and '70s. 'River-lakes' did not come into existence until *c*. 1750, but were relatively popular, accounting for about 10 percent of lakes in the eighteenth century. The earliest ones are attributable to Brown, and he may well have invented them. They continued to be made in small numbers throughout the nineteenth century. Hybrid lakes were made throughout the eighteenth and nineteenth centuries. Initially, this may have been because a geometric lake was desired but was too expensive to make. Subsequently, they may have occurred because geometric lakes were updated to become somewhat irregular, to conform to changing fashion. Their flexibility ensured their enduring popularity.

There was a marked dip in the numbers of lakes being made in the 1780s, which mirrored the dip in numbers of Parliamentary enclosure acts. The reasons for this dip are not clear cut but were almost certainly related to factors such as the economic costs of the American War of Independence, and the threat of war with France. Despite recovering to some extent in the 1790s, the general trend in lake numbers was downwards after this point, and this decline increased in the early decades of the nineteenth century, especially after *c*. 1820.

The polemics of the Picturesque did not appear to have a direct effect on ornamental water other than producing a preference for animated water, with the mirror-like, Brownian lake being regarded as dull. In the middle decades of the nineteenth century, cascades became very popular, with the Pulham firm having a dominant role. By 1900, very few lakes were being made, with estate owners turning their focus towards other elements such as lodges, terraces and conservatories. Occasionally, a second lake was made, possibly instead of building a conservatory. Lakes changed in size and shape, becoming generally smaller, and spreading in shape, rather than elongated as they had been in the eighteenth century. They were also more likely to be positioned in front of houses, probably because many new parks in the nineteenth century were smaller than their predecessors, and this also affected lake size, 2 h or smaller being typical. Walking round lakes, rather than carriage driving, became more common, and this also had an impact on lake shapes and planting. More indentations of the shoreline, and planting to disguise the limits of the lake, became common, as this made walking around them more interesting, resulting in the spreading shape.

Although virtually nothing is known about the designers of geometric lakes in the early eighteenth century, in the succeeding decades (1720s–'40s) it was largely the owners of elite landscapes who were the impetus behind the making of the new irregular or hybrid lakes. Predictably, the owners of large estates led the way, but it is important to recognise the contribution of Sir John Vanbrugh. His work as an architect is well-known, but the evidence suggests that it was he who introduced the concept of an irregular lake to clients such as the 3rd Earl of Carlisle and possibly the 1st Duke of Kingston. By the 1750s, when irregular lakes were becoming fashionable, the situation had changed and it became usual to employ an 'improver' to design the lake, and often the house and accompanying landscape. Whilst Brown made around one quarter of the lakes (irregular and river-lakes) in the eighteenth century, most lakes were made by unknown people. These would probably have been estate owners in conjunction with their own workers, or local experts. A small percentage of lakes – about 10 percent – was made by men such as William Emes, Nathanial Richmond, Richard Woods and later, Edward Kemp. These men were clearly competent at the engineering side of making lakes as there is little evidence of skilled men at a national level who specialised in making dams, although this view may be owing to the difficulties of identifying them in archives. In the later nineteenth century, there was a tendency for specialisms to arise, with more of a distinction between architects and landscapers. Edward Kemp came into the latter category.

Lakes evolved for a variety of reasons and the factors responsible for their appearance were complex and intertwined. One of the key factors was the influence of Italy and the experience of the Grand Tour. Men such as Carlisle of Castle Howard, Coke of Holkham, Burlington of Chiswick and Londesborough, Cobham of Stowe (all members of the Kit-Cat Club) made the Grand Tour, which exposed them to other styles of landscapes and other types of topography, notably lakes in the Alps, and in Italy. These men made the earliest lakes, and in some instances were directed by Vanbrugh. The importance of this foreign travel was that the configuration of gardens and landscapes bordered by large bodies of water made an impact on tourists, who began to want to replicate these views at home. The irregular lake was a vital ingredient in attempting to construct a landscape reminiscent of those scenes. Being visible from the house, and having plantations adjacent to some parts of the lake, were two other important ingredients. Addison had remarked on plantations, made by the monks at Ripaille, bordering Lake Geneva, with vistas cut through plantations to the lake, and this happily coincided with the existing popularity of geometric plantations on large estates in England at that time. Although by no means mountainous, it may well be that Coke felt that he had achieved something similar with the tree planting on the steep eastern banks of his lake at Holkham (Figure 8.1). Burlington's patronage of Castell, which led to the publication of the influential *Villas of the Ancients* in 1728, was a secondary way in which the influence of Italy was disseminated throughout the elite. The fact that Addison published his *Remarks* in 1701–3, with Vanbrugh, Kingston, Carlisle and Burlington making irregular lakes in the 1720s, highlights the fact that these lakes pre-dated Brown by several decades. Although Brown had a distinctive style, and transformed the style of irregular lakes, he did not 'invent' them.

A subsidiary aspect of the Grand Tour was the paintings which tourists brought back, often by Lorrain, Poussin, Rosa, Dughet, and their imitators. Men like Coke at Holkham collected works by Lorrain, as did Hoare with Dughet. Typically, these artists portrayed landscapes (often rural), with water of some sort, and classical buildings, or their remains. Although these paintings, and the subsequent engravings of them, often depicted lake-like areas of water, or natural lakes, it seems that they imprinted the formula of 'irregular landscape plus water plus classical architecture' on the minds of many people. The paintings cemented the Italian experience for the returned tourists and some of them, notably Hoare at Stourhead, aimed to produce landscapes which reflected these scenes and experiences. The paintings also served to announce to visitors the provenance of those landscapes.

A further factor in the emergence of irregular lakes was the development in the way parks, and bodies of water in particular, were used There is good evidence,

FIGURE 8.1 View across the lake to Holkham Hall, Norfolk, 2013. The house can just be seen at the end of the lake.

from both images and texts, that the desire to row, and later sail, was instrumental in lakes being made, and also increasing in size later in the eighteenth century. The 1726 painting of Thoresby Park provides some of the clearest and earliest evidence, depicting several luxurious rowing boats on the lake. The incidence of pictures showing people boating increased after this, becoming significant in the 1750s, and sailing boats also became more common at this time. This is supported by Rigaud's pictures of boats at Stowe, for example, and textual accounts such as of Jemima, Marchioness Grey's entertainments on boats at Wrest, and her extension of the canals there. Naumachia also became popular from the 1740s onwards, with the 5th Lord Byron's activities at Newstead, and Sir Francis Dashwood's at West Wycombe, being noteworthy. Entertainments such as these would have been very unsatisfactory on the generally small and unexciting geometric lakes of the early eighteenth century.

An increase in the use of parks by both men and women, but particularly women, may also have contributed to the development of lakes. Boating was one such use but the increase in the number of paintings showing women doing other things, such as carriage driving, from the 1750s onwards, and later walking and sketching in the park (usually by the lake), strongly suggests that carriage drives and circuit walks developed at least partly in response to women wanting to do these things. The tendency to place lakes in front of the house in the latter part of the eighteenth century, may have been partly driven by the rationale of making lakes more accessible for women. In the nineteenth century, writers such as W. S. Gilpin, Kemp and Loudon were very much talking in terms of walks around the lake, which, if new, was likely to be smaller than in the eighteenth century.

Once the concept of an ornamental lake had become established, lakes were undoubtedly used as status symbols, to display the wealth of a fashion-conscious owner, with bigger being better, in this context. Part of this display was the role of lakes in relation to the house. In the second half of the eighteenth century, it became desirable for the lake to reflect the house, thereby enhancing both lake and house considerably. Secondly, routeing the main approach to the house across the lake was held to add considerably to the impact on the visitor, and Brown was particularly adept at this. Approaching a residence over water seems to have had an enduring importance. This was recognised in relation to the medieval period, where fishponds often flanked elite residences, or where the approach to a castle was over a lake-moat. Simply crossing an ordinary moat conferred a *cachet*. Excursions into psychology are not the remit of this work, but it does appear that, for over a millennium, crossing water to reach a residence satisfied a deep and enduring psychological need, and marked the boundary between the exterior, public space and the interior, private space. A third aspect of lakes as status symbols was their fish stocks. Again, this refers back to the medieval era, when fishponds were systems for producing high status, freshwater fish. Whilst freshwater fish did not have the same high value in the eighteenth century, it was still a valuable 'crop', and most garden water features, as well as lakes, were stocked. However, while some of this 'second hand' status still attached to fish production, it was mainly the expense of constructing a large area of water, as much for aesthetic and leisure reasons, which meant that lakes conferred considerable status.

The study of the topography as it related to lakes has led to an understanding of the factors which governed the siting of different types of lakes. As might be expected, geometric lakes were usually sited on fairly flat land because this meant that less earth would have to be moved to make the straight sides. This in turn would reduce the expense of making them. Hybrid lakes, with their requirement of two straight sides, were also easier to make in fairly flat areas. Because irregular lakes did not require a lot of earth-moving to make them, although this might be done for aesthetic reasons, they could be made anywhere that there was a river or stream. The topography affected the general shape of the lake though: a spreading lake would occur in flatter areas, and a narrower, linear lake in deeper valleys. River-lakes were found to be quite site specific, requiring a good water course, though one which was not too 'steep', so they were usually made in suitable river valleys. As many residences were situated near water sources, it was often possible to construct a lake of some description nearby.

One of the significant findings of this research is that irregular lakes were first made in *new* geometric landscapes, albeit ones which were becoming 'unbalanced', before landscapes became informal in style. Whilst to some extent styles of ornamental water are related to landscape styles, hitherto it has been assumed that lakes formed part of a 'package'; that they appeared because landscape style had changed. It was assumed that geometric landscapes fell out of fashion, to be replaced by informal landscapes, and that irregular lakes then developed. This was not the case. There is evidence that landscapes were already beginning to become less symmetrical and more unbalanced by 1700. This was partly because they were getting bigger, a process begun by Le Nôtre in laying out Vaux-le-Vicomte and Versailles. Lakes intensified that 'unbalancing' process. Once ornamental water became irregular, the landscape had to change around it. Because lakes were irregular, they would not fit easily into geometric landscapes, which were linear, whereas ornamental canals had fitted neatly into those landscapes. The only way to fit a 'lake' into a geometric landscape without unbalancing it was to make it rectangular, and put it in place of a parterre. This was usually only practicable on a relatively flat site with an adjacent water source, as at Boughton and Welford. At places such as Thoresby, Holkham and Castle Howard, owners had estates large enough to make a lake and to retain significant geometric plantations with vistas and rides, but eventually even those landscapes became much less formal. These early lakes were off-set from the house, and further contributed to the unbalancing of the landscape.

Another significant finding was that lakes were not usually lined (with puddled clay). There is relatively little primary evidence for how eighteenth-century lakes were constructed, but it has been a widely held view that lakes were lined, based on the knowledge, chiefly gained from North and Switzer, that ponds were lined with puddled clay. Confusion has arisen over this subject because they are both talking about small ponds – store ponds for fish, or garden ponds. The use of clay in the making of dams – a clay core 'wall' – has led to further confusion and the assumption that the entire lake was made with a clay lining. Whilst some lakes may have been lined, or partly so, most were not. In North's account of making larger fishponds, which were much like ornamental lakes, he recommends using any clay in the bottom of them to make the dam with.

The story of lakes does not stop in 1900. Although very few ornamental lakes have been made since then, reservoirs could be regarded as the post script. These began to appear in the late eighteenth century when they were made to supply early canals, and in the early nineteenth century, to supply power to industry and water to towns. From the mid-nineteenth century, they were also made to supply the railways. Regardless of their size, these reservoirs were constructed in the same way as ornamental lakes, although Robert Thom, a successful reservoir builder, seems to have been alone in using a clay 'mantle' on the upstream side of his dams, instead of a clay core wall. He built Loch Thom in 1827, to supply power and water to the town of Greenock. The dam was 18 m high (Bowood is 4 m) and *c.* 155 h in area. His technique has stood the test of time, however, as his reservoir still supplies the town with water, and today also provides angling, walking and wildlife study facilities. So, like the lakes of the eighteenth century, these bodies of water have more than one function.

This story of ornamental lakes shows that there is a clear narrative for water in the landscape. The large fishponds of the medieval era were made to produce fish, and were also appreciated for their status and beauty by those who saw them. Crucially, they provided the technical blueprint for the construction of orna-mental lakes. Those lakes also contained fish, but were primarily about status and ornament. In turn, the knowledge of how to build large ornamental lakes in the eighteenth century enabled engineers to build the reservoirs which powered the Industrial Revolution of the nineteenth century. Like those eighteenth-century lakes, today's reservoirs are also leisure facilities, and one of the newest is Roadford Lake (1989), which supplies water to North Devon and is an arena for activities ranging from sailing to archery.

Lakes have a very deceptive air of simplicity, but this exploration of ornamental water has brought to light the many complexities which lie beneath the surface. Their construction could be fraught with difficulty, and their surfaces were vari-ously required to be calm and reflective, then animated and Picturesque. One of the main attractions of water is that it is always changing, never static, and the story of ornamental lakes is also one of change, from their beginnings in *vivaria*, via formal water features, to informal lakes. Their evolution reflects general changes in society, just as many eighteenth century lakes mirrored the mansions of their owners.

Appendices

APPENDIX I

A chronological table of semi-geometric lakes[1]

Place	County	Dating	Lake Size (h)
Trentham	Staffordshire	1695–1700	8.8
Raynham	Norfolk	By 1724	9.2
Blenheim Palace	Oxfordshire	1716–24	*c.* 3
Studley Royal	Yorkshire	1720s–30s	*c.* 2
Hyde Park	Middlesex	1730	12.5, 3.7, 2.9
Wolterton	Norfolk	*c.* 1727	*c.* 4
Compton Verney	Warwickshire	By 1736	*c.* 3
Deene	Northamptonshire	After 1746	1, 1
Barton Abbey	Oxfordshire	By 1848	1.2

APPENDIX II

A chronological table of river-lakes[2]

Place	County	Dating	Lake Size (h)	Designer
Latimer	Buckinghamshire	1750s	4.4	Brown
Peper Harrow	Surrey	1757–8	c. 3	Brown
Stratfield Saye	Hampshire	c. 1757	4.7, 2.6	Brown
Horton	Northamptonshire	c. 1760	1.8	–
Belhus	Essex	1753–63	c. 1.5	Brown
Chatsworth	Derbyshire	c. 1763	10	Brown
Audley End	Essex	By 1764	3	Brown
Cranford Hall, Cranford St. Andrew	Northamptonshire	1760s?	1	–
Ranston	Dorset	By 1774	1.1	–
Shortgrove	Essex	1753–70s	1.7	Brown
Wycombe Abbey	Buckinghamshire	1760s	3	Brown
Brocket Hall	Hertfordshire	By 1768	7.5	–
Broadlands	Hampshire	c. 1768	c. 3.5	Brown
Kedleston	Derbyshire	1765–74	13.7	W. Emes?
Ramsbury	Wiltshire	By 1775	5.2	–
Youngsbury	Hertfordshire	c. 1770	c. 1.9	Brown
Olantigh	Kent	1760s?	c. 2	–
Badger Hall/Badger Dingle	Shropshire	c. 1780	1.5, 0.8	W. Emes
Costessy	Norfolk	1775–95	3	–
Netheravon	Wiltshire	1755?	5.8	T. Wright
Woodhall	Hertfordshire	c. 1780	3	–
Cottesbrooke	Northamptonshire	c. 1780–1	2	–

Place	County	Dating	Lake Size (h)	Designer
Bramshill	Hampshire	By 1799	3.4	–
Pinkney	Wiltshire	1773–1813	1	–
Brockhall	Northamptonshire	1775–1813	1.6	–
Thoresby	Nottinghamshire	1794	4	Repton
Panshanger	Hertfordshire	1800	4.2	Repton
Tewin	Hertfordshire	1795–1805	4	Repton
Estcourt	Gloucestershire	By 1813	3.6+3.6	–
Easton Grey	Wiltshire	*c.* 1800?	2.8	–
Draycot	Wiltshire	1784–1813	1.4	–
Hatfield	Hertfordshire	By 1805	5.9	–
Lexham	Norfolk	1795–1817	1.7	–
Quidenham	Norfolk	*c.* 1810	2	–
Swerford	Oxfordshire	*c.* 1815	1.4	–
Riddlesworth	Norfolk	1795–1840	5	–
Bayfield	Norfolk	1816–39	3	–
Tackley Park	Oxfordshire	1815–39	1.4	–
Lynford	Norfolk	1821–42	1.7	–
Toddington	Gloucestershire	*c.* 1849	2.5	–
Easton Neston	Northamptonshire	1814–82	1	–
Heydon	Norfolk	1841–85	1	–

Notes

1 Bishop, op. cit., data extracted from the Landscape Database.
2 Ibid.

BIBLIOGRAPHY

Addison, J. 'Remarks on the Several Parts of Italy etc. in the Years 1701, 1702, 1703', in Hurd, R., ed., *The Works of the Right Honourable Joseph Addison*, Vol. 2 (London, 1811)

———. '*The Tatler*, No. 161, 18–20 April, 1710, *The Spectator*, No. 37, 12 April, 1711, *The Spectator*, No. 414, 28 June, 1711, *The Spectator*, No. 477, 6 September, 1712', in Dixon-Hunt, J. & Willis, P., eds., *The Genius of the Place* (London: The MIT Press, 1976)

Allibone, J. *Anthony Salvin: Pioneer of Gothic Revival Architecture, 1799–1881* (Colombia: Missouri University Press, 1987)

Andrews, M. 'Theobalds Palace: The Gardens and Park', *Garden History*, Vol. 21, No. 2 (Winter, 1993), pp. 129–149

Aston, M., ed. *Medieval Fish, Fisheries and Fishponds in England*, BAR British Series 182 (i) (Oxford: Archaeopress, 1988)

Atkyns, Sir Robert. *The Ancient & Present State of Glostershire* (London, 1712)

Aubrey, J. *The Natural History of Wiltshire* (Teddington: The Echo Library, 2006)

Austen, J. *Mansfield Park* (London: Penguin Books, 1996)

Bacon, Sir Francis, in Spedding, J., Ellis, J. R. & Heath, D., eds. *The Works of Francis Bacon: Literary and Professional Works*, Vol. I, 1625 (London, 1861)

Badeslade, J., Rocque, J., Woolfe, J. & Gandon, J. *Vitruvius Brittanicus. Second Series, Volume the Fourth, 1739 & Vols. 4 & 5, 1767–71* (New York: Dover Publications, 2009)

Bapasola, J. *The Finest View in England: The Landscape and Gardens at Blenheim Palace* (Oxford: Blenheim Palace, 2009)

Barnett, J. & Williamson, T. *Chatsworth: A Landscape History* (Macclesfield: Windgather Press, 2005)

Batey, M. & Lambert, D. *The English Garden Tour: A View into the Past* (London: John Murray, 1990)

Beard, G. *The Work of John Vanbrugh* (London: B. T. Batsford, 1986)

Beckett, J. V. & Turner, M. 'Taxation and Economic Growth in Eighteenth-Century England', *The Economic History Review* New Series, Vol. 43, No. 3 (August, 1990), pp. 377–403

Berners, Juliana & Worde, Wynkyn de. *A Treatise of Fysshynge Wyth an Angle; Being a Facsimile Reproduction of the First Book on the Subject of Fishing Printed in England by Wynkyn de Worde at Westminster in 1496*. With an introduction by M. C. Watkins (1880)

Binnie, G. M. *Early Dam Builders in Britain* (London: Thomas Telford, 1987)

Birley, D. *Sport and the Making of Britain* (Manchester: Manchester University Press, 1993)

Black, J. *The British Abroad: The Grand Tour in the Eighteenth Century* (Stroud: Sutton Publishing, 1997)

Blomefield, F. *An Essay Towards a Topographical History of the County of Norfolk*, Vol. 2 (London, 1805)

Bowden, M., MacKay, D. & Topping, P., eds. *From Caithness to Cornwall: Some Aspects of Field Archaeology*, BAR British Series 209, 1989 (Oxford: Archaeopress, 1989)

Britton, J. *Beauties of Wiltshire*, Vol. 2, 1801 (London, 1801)

Brooks, E. St. J. *Sir Christopher Hatton: Queen Elizabeth's Favourite* (London: Jonathan Cape, 1946)

Brown, D. *Nathaniel Richmond (1724–1784): Gentleman Improver*. Unpublished PhD thesis, University of East Anglia, 2000

Brown, D. & Williamson, T. *Lancelot Brown and the Capability Men: Landscape Revolution in the Eighteenth Century* (London: Reaktion Books, 2016)

Brown, J. *Lancelot 'Capability' Brown: The Omnipotent Magician 1716–1783* (London: Pimlico, 2012)

Campbell, C. *Vitruvius Britannicus, or the British Architect*, Vol. 3, 1725 (New York: Dover Publications, 2007)

Cannadine, D., gen. ed. *Oxford Dictionary of National Biography* (Oxford: Oxford University Press), online accessed 2016.

Carmi Parsons, J. *Eleanor of Castile: Queen and Society in Thirteenth Century England* (London: Macmillan Press, 1994)

Chauncy, Sir Henry. *The Historical Antiquities of Hertfordshire* 1700 (Bishop's Stortford, 1826)

Cheney, C. S., et al. *The Physical Properties of Minor Aquifers in England and Wales*, in Hydrogeology Group Technical Report WD/00/04 Environment Agency R&D Publication 68 (Nottingham: British Geological Survey, 2000)

Churchyard, Thomas. *The Worthines of Wales* (London, 1587)

Clarke, G., et al. *Stowe Landscape Gardens: A Comprehensive Guide* (Swindon: National Trust Enterprises, 1997, rev. 2005)

Colvin, H. gen. ed. *The History of the King's Works*, Vols. 1–4 (London: HMSO, 1962–83); Vol. 5, 1660–1782 (London: HMSO, 1976); Vol. 6, 1782–1851 (London: HMSO, 1973)

Cook, H. & Williamson, T., eds. *Water Management in the English Landscape: Field, Marsh and Meadow* (Edinburgh: Edinburgh University Press, 1999)

Cornforth, J. 'The History of the Boughton Landscape', *Country Life* (11th March, 1971)

———. 'The Making of the Bowood Landscape', *Country Life* (7th September, 1972)

Cowell, F. *Richard Woods (1715–1793): Master of the Pleasure Garden* (Woodbridge, Suffolk: Boydell Press, 2009)

Creighton, O. *Designs Upon the Land: Elite Landscapes of the Middle Ages* (Woodbridge: Boydell Press, 2009)

Crosby, B. 'Private Concerts on Land and Water: The Musical Activities of the Sharp Family, *c.* 1750–*c.* 1790', *Royal Musical Association Research Chronicle*, Vol. 34, No. 1 (January, 2001), pp. 1–118

Currie, C. 'The Early History of the Carp and Its Economic Significance in England', *The Agricultural History Review*, Vol. 39, No. 2 (1991), pp. 97–107

———. 'Fishponds as Garden Features, *c.* 1550–1750', *Garden History*, Vol. 18, No. 1 (Spring, 1990), pp. 22–46

Dalton, C. '"He That . . . Doth Not Master the Human Figure": Sir John Vanbrugh and the Vitruvian Landscape', *Garden History*, Vol. 37, No. 1 (Summer, 2009), pp. 3–37

————. *Sir John Vanbrugh and the Vitruvian Landscape* (Abingdon: Routledge, 2012)

Daniels, S. *Humphry Repton: Landscape Gardening and the Geography of Georgian England* (London: Yale University Press, 2000)

Defoe, Daniel. *A Tour Through the Whole Island of Great Britain by a Gentleman*, Vols. I–III, 1742 (London, 1742)

Dennys, John. *The Secrets of Angling . . .* (London, 1613)

Dixon Hunt, J. *Garden and Grove: The Italian Renaissance Garden in the English Imagination: 1600–1750* (London: Dent and Sons, 1986)

Dixon Hunt, J. & Willis, P., eds. *The Genius of the Place* (London: Paul Elek, 1976)

Downes, K. *Sir John Vanbrugh* (London: Sidgwick and Jackson, 1987)

Eburne, A. 'The Passion of Sir Thomas Tresham: New Light on the Gardens and Lodge at Lyveden', *Garden History*, Vol. 36, No. 1 (Spring, 2008), pp. 114–134

Eiche, S. 'Prince Henry's Richmond: The Project by Constantino de' Servi', *Apollo* (November, 1998), pp. 10–14

Elliott, B. *Victorian Gardens* (London: B. T. Batsford, 1986)

Evans, S. 'William Andrews Nesfield: An Introduction to His Life and Work' in C. Ridgway, ed. *William Andrews Nesfield, Victorian Landscape Architect: Papers from the Bicentenary Conference, King's Manor, York, 1994* (York: University of York, 1996)

Everson, P. '"Delightfully Surrounded With Woods and Ponds": Field Evidence for Medieval Gardens in England', in *There By Design: Field Archaeology in Parks and Gardens*, BAR British Series 267, 1998 (Oxford: Archaeopress, 1998)

————. 'The Gardens of Campden House, Chipping Campden, Gloucestershire', *Garden History*, Vol. 17, No. 2 (Autumn, 1989), pp. 109–121

————. *Survey of Quarrendon, Aylesbury Vale, Bucks* (Unpublished: English Heritage, 1999)

Eyres, P. & Lynch, K. *On the Spot: The Yorkshire Red Books of Humphry Repton, Landscape Gardener* (Huddersfield: New Arcadian Press, 2018)

Feluś, K. 'Boats and Boating in the Designed Landscape', *Garden History*, Vol. 34, No. 1 (Summer, 2006), pp. 22–46

Feluś, K. *'Beautiful Objects and Agreeable Retreats': Uses of Garden Buildings in England, 1720–1820* Unpublished PhD thesis, Bristol, 2009

Feluś, K. *The Secret Life of the Georgian Garden* (London: I. B. Tauris, 2016)

Fiennes, C. *Through England on a Side Saddle: In the Time of William and Mary* (New York: Cambridge University Press, 2010, 1888 ed.)

Finch, J. 'Three Men in a Boat: Biographies and Narrative in the Historic Landscape', *Landscape Research*, Vol. 33, No. 5 (October, 2008), pp. 511–530

Finnegan, R. *Letters From Abroad: The Grand Tour Correspondence of Richard Pococke and Jeremiah Milles. Vol. 1: Letters From the Continent (1733–34)* (Piltown, Ireland: Pococke Press, 2011)

Floud, R. *An Economic History of the English Garden* (London: Allen Lane, 2019)

Fowler, A. *The Country House Poem* (Edinburgh: Edinburgh University Press, 1994)

Gard, R., ed. *The Observant Traveller: Diaries of Travel in England, Wales and Scotland in the County Archives* (London: HMSO, 1989)

Garnett, O. *Kedleston Hall* (Swindon: National Trust Enterprises, 1999, rev. 2009)

Gilpin, W. *Essay on Picturesque Beauty* (London, 1792)

————. *Observations on the River Wye, and Several Parts of South Wales etc. 1771* (London, 1789)

Gilpin, W. S. *Practical Hints Upon Landscape Gardening With Some Remarks on Domestic Architecture as Connected With Scenery* (London, 1832)

Girouard, M. 'Charles Barry: A Centenary Assessment', *Country Life* (13th October, 1960)

———. *Life in the English Country House: A Social and Architectural History* (Harmondsworth: Penguin Books, 1980)

Gregory, J., Spooner, S. & Williamson, T. *Lancelot 'Capability' Brown: A Research Impact Review Prepared for English Heritage by the Landscape Group, University of East Anglia, Research Report Series No. 50–2013* (English Heritage: ISSN 0246–98002 (Online), 2013)

Grundy, I. *Lady Mary Wortley Montagu* (Oxford: Oxford University Press, 2001)

Gwyn, P. *The King's Cardinal: The Rise and Fall of Thomas Wolsey* (London: Barry & Jenkins, 1990)

Hale, Thomas. *A Compleat Body of Husbandry*, Vol. II, 1756 (London: 2nd ed., 1758)

Halsband, R., ed. *The Complete Letters of Lady Mary Wortley Montagu*, Vols. I–III (Oxford: Oxford University Press, 1965)

Harris, F. 'The Manuscripts of John Evelyn's "Elysium Britannicum"', *Garden History*, Vol. 25, No. 2 (Winter, 1997), pp. 131–137

Harris, J. *The Artist and the Country House* (London: Philip Wilson Publishers, 1979)

———. 'The Prideaux Collection of Topographical Drawings', *Architectural History*, Vol. 7 (1964), pp. 17, 19–21, 23–39, 41–108

———. 'Thoresby House, Nottinghamshire', *Architectural History*, Vol. 4 (1961), pp. 11–19; 'Thoresby Concluded', *Architectural History*, Vol. 6 (1963), pp. 103–105

Harvey, J. *Medieval Gardens* (London: B. T. Batsford, 1981)

Henderson, P. 'Sir Francis Bacon's Water Gardens at Gorhambury', *Garden History*, Vol. 20, No. 2 (Autumn, 1992), pp. 116–131

———. 'A Shared Passion: The Cecils and their Gardens', in Croft, P., ed., *Patronage, Culture and Power: The Early Cecils* (New Haven, CT: Paul Mellon Centre BA, 2002), p. 99

———. *The Tudor House and Garden* (London: Yale University Press, 2005)

Hentzner, Paul. *Travels in England During the Reign of Queen Elizabeth* (London, 1892)

Hewlett, H., ed. *Improvements in Reservoir Construction, Operation and Maintenance* (London: Thomas Telford Publishing, 2006)

Hinde, T. *Capability Brown: The Story of a Master Gardener* (London: Century Hutchinson, 1986)

Hitching, C. *Rock Landscapes: The Pulham Legacy. Rock Gardens, Grottoes, Ferneries, Follies Fountains & Garden Ornaments* (Woodbridge: Antique Collectors' Club, 2010)

Holt, R. *The Mills of Medieval England* (Oxford: Basil Blackwell, 1988)

Hudson, D. *Holland House in Kensington* (London: Peter Davies, 1967)

Jackson-Stops, G. *An English Arcadia 1600–1990* (London: National Trust Enterprises, 1991)

Jacques, D. *Georgian Gardens: The Reign of Nature* (London: B. T. Batsford, 1983)

———. 'Garden Design in the Mid-Seventeenth Century', *Architectural History*, Vol. 44 (Essays in Architectural History Presented to John Newman, 2001), pp. 365–376

Jacques, D. & Van der Horst, A. J. *The Gardens of William and Mary* (Bromley: Christopher Helm, 1988)

James, C. W. *Chief Justice Coke: His Family and Descendants at Holkham* (London: Country Life, 1929)

Jeffery, S. 'The Formal Gardens at Moor Park [Herts.] in the Seventeenth and Early Eighteenth Centuries', *Garden History*, Vol. 42, No. 2 (2014), pp. 157–177

Kemp, E. *How to Lay Out a Garden* (1851 and 1858 ed.)

Kenyon, J. R. *Raglan Castle* (Cardiff: Cadw, Welsh Historic Monuments, rev. ed. 2003)

Knyff, L. & Kip, J. *Britannia Illustrata (Nouveau Theatre de Grande Bretagne, 1708)*, eds. Harris, J. & Jackson-Stops, G. (Bungay: Paradigm Press, 1984)

Laird, M. *The Flowering of the English Landscape Garden: English Pleasure Grounds 1720–1800* (Philadelphia: University of Pennsylvania Press, 1999)

Landsberg, S. *The Medieval Garden* (London: The British Museum Press, no date)

Laneham, R. *Laneham's Letter Describing the Magnificent Pageants Presented Before Kenilworth Castle in 1575* (Philadelphia, 1822)

Langdon, J. *Mills in the Medieval Economy: England 1300–1450* (Oxford: Oxford University Press, 2004)

Langley, B. *New Principles of Gardening* (London, 1728)

La Rochefoucauld, F. & A. *Tour of England, 1785*, ed. Scarfe, N. (Woodbridge: Boydell Press, 1995)

Larsen, R. 'For Want of a Good Fortune: Elite Single Women's Experiences in Yorkshire, 1730–1860', *Women's History Review*, Vol. 16, No. 3 (2007), pp. 387–401

Lasdun, S. *The English Park: Royal, Private & Public* (London: Andre Deutsch, 1991)

Lawrence, John. *The Modern Land Steward in Which the Duties and Functions of Stewardship Are Considered and Explained . . .* (London: 2nd ed., 1806)

Leathlean, H. 'From Gardenesque to Home Landscape: The Garden Journalism of Henry Noel Humphreys', *Garden History*, Vol. 23, No. 2 (Winter, 1995)

Leslie, M. 'An English Landscape Before the "English Landscape Garden"?', *Journal of Garden History*, Vol. 13, No. 1–2 (1993)

Liddiard, R. *Castles in Context* (Macclesfield: Windgather Press, 2005)

Liddiard, R. & Williamson, T. 'There by Design? Some Reflections on Medieval Elite Landscapes', *The Archaeological Journal*, Vol. 165, No. 1 (2008)

Loudon, J. C. *An Encyclopaedia of Gardening* (London: 3rd ed., 1825 and 4th ed., 1826)

Macky, J. *A Journey Through England: In Familiar Letters*, Vols. 1 & 2 (London, 1724)

MacNair, A. & Williamson, T. *William Faden and Norfolk's 18th-Century Landscape* (Oxford: Windgather Press, 2010)

Markham, Gervaise. *Country Contentments, or the Husbandman's Recreation 1615* (London: 2nd ed., 1631)

Marsden, J. *Stowe Landscape Gardens: A Comprehensive Guide* (Swindon: National Trust Enterprises, 1997)

Marshall, Rev. C. *A Plain and Easy Introduction to the Knowledge and Practice of Gardening, With Hints on Fishponds c. 1797* (London: 5th ed., 1813)

Martin, A. 'Theobalds Palace: The Gardens and Park', *Garden History*, Vol. 21, No. 2 (Winter, 1993), pp. 129–149

McGarvie, M. *Notes Towards a History of Gardening at Marston House 1660–1905* (Frome: Frome Local History Group, 1987)

Milledge, J. 'Hadham Hall and the Capel Family', *Hertfordshire Garden History*, Vol. II (Spring, 2012), pp. 60–85

Moore, A. W. *Norfolk and the Grand Tour: Eighteenth-century Travellers Abroad and Their Souvenirs* (Norwich: Norfolk Museums Service, 1985)

Mowl, T. *Gentlemen and Players: Gardeners of the English Landscape* (Stroud: Sutton Publishing, 2000)

———. 'Rococo and Later Landscaping at Longleat', *Garden History*, Vol. 23, No. 1 (Summer, 1995), pp. 56–66

———. 'John Drapentier's Views of the Gentry Gardens of Hertfordshire', *Garden History*, Vol. 29, No. 2 (Winter, 2001)

———. *William Kent: Architect, Designer, Opportunist* (London: Random House, ed. 2007)

Mowl, T. & Earnshaw, B. *Architecture Without Kings: The Rise of Puritan Classicism Under Cromwell* (Manchester: Manchester University Press, 1995)

———. 'Inigo Jones Restored', *Country Life* (30th January, 1992)

———. *An Insular Rococo: Architecture, Politics and Society in Ireland and England, 1710–1770* (London: Reaktion Books, 1999)

Mowl, T., et al. *The Historic Gardens of England*, series

National Trust. *Stowe Landscape Gardens* (Swindon: National Trust Enterprises, 1997, rev. 2005)

National Trust Guidebook (John Newman et al.). *Blickling Hall* (Swindon: National Trust Enterprises, 1997)

Neave, D. & Turnbull, D. *Landscaped Parks and Gardens of East Yorkshire* (Unknown: Georgian Society for East Yorkshire, 1992)

North, Roger. *A Discourse on Fish and Fishponds* (London, 1714)

Nugent, Thomas. *The Grand Tour*, Vol. III (London, 1749)

Phibbs, J. 'A List of Landscapes That Have Been Attributed to Lancelot "Capability Brown"', *Garden History*, Vol. 41, No. 2 (2013), pp. 244–277

———. 'The Use of Plants in Eighteenth Century Gardens', *Garden History*, Vol. 38, No. 1 (2010)

Phillips, J. & Burnett, N. 'The Chronology and Layout of Francis Carew's Garden at Beddington', *Garden History*, Vol. 33, No. 2 (Autumn, 2005), pp. 155–188

Piggott, S. *William Stukeley: An Eighteenth-Century Antiquary* (London: Thames and Hudson, ed. 1985)

Pococke, Bishop Richard. *The Travels Through England of Dr. Richard Pococke, 1750–57 2 Vols.*, Vol. 41 & 42, ed. Cartwright, J. (London, 1888)

Rackham, O. *The History of the Countryside* (London: Phoenix, ed. 2000)

RCHME. *An Inventory of the Historical Monuments in the County of Northampton*, Vol. II *Archaeological Sites in Central Northamptonshire* (London: HMSO, 1979)

———. *An Inventory of the Historical Monuments in the County of Northampton*, Vol. III *Archaeological Sites in North-West Northamptonshire* (London: HMSO, 1975–84)

———. *An Inventory of the Historical Monuments in the County of Northampton*, Vol. IV *Archaeological Sites in South-West Northamptonshire* (London: HMSO, 1975–84)

RCHMW. *Glamorgan: The Early Castles*, Vol. III (London: HMSO, 1991)

Renn, D. *Caerphilly Castle* (Cardiff: Cadw, Welsh Historic Monuments, rev. ed. 1997)

Repton, H. *Observations on the Theory and Practice of Landscape Gardening* (London: J. Taylor, 1805)

Repton, H. & Repton, J. A. *Fragments on the Theory and Practice of Landscape Gardening* (London: J. Taylor, 1816)

Rickard, C., Day, R. & Purseglove, J. *River Weirs: Good Practice Guide R&D Publication W5B-023/ HQP* (Bristol: Environment Agency, 2003)

Ridgway, C., ed. *William Andrews Nesfield, Victorian Landscape Architect: Papers From the Bicentenary Conference, King's Manor, York, 1994* (York: University of York, 1996)

Ridgway, C. & Williams, R., eds. *Sir John Vanbrugh and Landscape Architecture in Baroque England 1690–1730* (Stroud: Sutton Publishing, 2000)

Roberts, E. 'The Bishop of Winchester's Fishponds in Hampshire, 1150–1400: Their Development, Function and Management', *The Proceedings of the Hampshire Field Club and Archaeological Society*, Vol. 42 (August, 1986)

Roberts, J. 'Cusworth Park: The Making of an Eighteenth-Century Designed Landscape', *Landscape History*, Vol. 21 (1999), pp. 77–93

———. 'Well Temper'd Clay': Constructing Water Features in the Landscape Park', *Garden History*, Vol. 29, No. 1 (2001), pp. 12–28

Robinson, J. M. *Temples of Delight: Stowe Landscape Gardens* (Andover: Pitkin Pictorials, 1994)

Rowe, A., ed. *Hertfordshire Garden History: A Miscellany* (Hatfield: Hertfordshire Publications, 2009)

Rowe, A. & Williamson, T. *Hertfordshire: A Landscape History* (Hatfield: Hertfordshire Publications, 2013)

Rowell, C. *Petworth: The People and the Place* (Swindon: National Trust Enterprises, 2012)

Saumarez Smith, C. *The Building of Castle Howard* (London: Random House, 1990)

Shields, S. *Moving Heaven and Earth: Capability Brown's Gift of Landscape* (London: Unicorn Publishing, 2016)

———. '"Mr Brown Engineer": Lancelot Brown's Early Work at Grimsthorpe Castle and Stowe', *Garden History*, Vol. 34, No. 2 (Winter, 2006), pp. 174–191

Simon, R. & Smart, A. *The Art of Cricket* (London: Martin Secker and Warburg, 1983)

Skempton, A. W. *A Biographical Dictionary of Civil Engineers in Great Britain and Ireland . . . 1500–1830* (London: Thomas Telford, 2002)

Smellie, W., ed. *Encyclopaedia Britannica* (Edinburgh: C. Macfarquhar and A. Bell, 1768–71)

Smith, N. *A History of Dams* (London: Peter Davies, 1971)

Spring, D., ed. *Hertfordshire Garden History*, Vol. II (Hatfield: University of Hertfordshire Press, 2012)

Stevens, D. *Humphry Repton: Landscape Gardening and the Geography of Georgian England 1752–1818* (London: Yale University Press, 1999)

Strong, R. *The Artist & the Garden* (London: Yale University Press, 2000)

———. *The Renaissance Garden in England* (London: Thames & Hudson, 1998)

Stroud, D. *Capability Brown* (London: Faber & Faber, 1975)

Stukeley, W. *The Family Memoirs of the Rev. William Stukeley MD and the Antiquarian and Other Correspondence of Roger and Samuel Gale etc* (London & Edinburgh, 1882)

Sutton, J. *Materializing Space at an Early Modern Prodigy House: The Cecils at Theobalds, 1564–1607* (Aldershot: Ashgate Publishing, 2004)

Swift, J., ed. *The Works of Sir William Temple in Two Volumes*, including *'Upon the Gardens of Epicurus'* (London, 1731)

Switzer, Stephen. *Ichnographia Rustica, or the Nobleman, Gentlemen, and Gardener's Recreation*, Vols. I, II & III, 1718 (London, 1718)

———. *An Introduction to a General System of Hydrostaticks & Hydraulics* 1729 (London, 1729)

———. *A Universal System of Water & Water-Works*, Vol. I (London: 1734)

Symes, M. 'Flintwork, Freedom and Fantasy: The Landscape at West Wycombe Park, Buckinghamshire', *Garden History*, Vol. 33, No. 1 (Summer, 2005), pp. 1–30

———. *Mr. Hamilton's Elysium: The Gardens of Painshill* (London: Frances Lincoln, 2010)

Symes, M. & Haynes, S. *Enville, Hagley, The Leasowes: Three Great Eighteenth-Century Gardens* (Bristol: Redcliffe Press, 2010)

Syson, L. *The Watermills of Britain* (Newton Abbot: David & Charles, 1980)

Taigel, A. & Williamson, T. *Know the Landscape: Parks and Gardens* (London: B. T. Batsford, 1993)

———. 'Some Early Geometric Gardens in Norfolk', *The Journal of Garden History*, Vol. 11, Nos. 1 & 2 (1991), pp. 1–111

Taverner, John. *Certaine Experiments Concerning Fish and Fruite Practised by Iohn Taverner, Gentleman* 1600 (London, 1600)

Taylor, C. 'From Recording to Recognition', in Pattison, Paul, ed., *There By Design: Field Archaeology in Parks and Gardens*, BAR British Series 267 (Oxford: Archaeopress, 1998)

———. 'Medieval Ornamental Landscapes', *Landscapes*, Vol. 1, No. 1 (April, 2000), pp. 38–55

———. *Parks and Gardens of Britain: A Landscape History From the Air* (Edinburgh: Edinburgh University Press, 1998)

———. 'Somersham Palace, Cambridgeshire: A Medieval Landscape for Pleasure?', in Bowden, M., MacKay, D. & Topping, P., eds., *From Caithness to Cornwall: Some Aspects of Field Archaeology*, BAR British Series 209 (Oxford: Archaeopress, 1998)

Taylor, C. & Whittle, E. 'The Early Seventeenth-Century Gardens of Tackley, Oxfordshire', *Garden History*, Vol. 22, No. 1 (Summer, 1994), p. 57

Temple, William. *The Works of Sir William Temple in Two Volumes, Including 'Upon the Gardens of Epicurus'* 1731 (London, 1731)

Thurley, S. *The Royal Palaces of Tudor England: Architecture and Court Life 1460–1547* (London: Yale University Press, 1993)

Till, E. C. 'The Development of the Park and Gardens at Burghley', *Garden History*, Vol. 19, No. 2 (Autumn, 1991), pp. 128–145

Turner, M. *English Parliamentary Enclosure* (Folkstone: Wm. Dawson and Sons, 1980)

Turner, R. *Capability Brown and the Eighteenth-Century Landscape* (Chichester: Phillimore & Co., 1999, 2nd ed.)

Underdown, D. *Start of Play: Cricket and Culture in Eighteenth Century England* (London: Allen Lane, 2000)

University of Manchester Archaeological Unit. *Northenden Weir: An Archaeological Desk-Based Assessment, June 2004 (26), Amended August 2004* https://www.gov.uk/government/uploads/system/uploads/attachment_data/file/290655/sw5b-023-hqp-e-e.pdf, accessed August 2014

Wade-Martins, P., ed. *An Historical Atlas of Norfolk* (Norwich: Norfolk Museums Service, 1993)

Waldstein, Baron, trans. G. W. Groos. *The Diary of Baron Waldstein* (London: Thames and Hudson, 1981)

Walpole, H., ed. J. Dixon-Hunt. *The History of the Modern Taste in Gardening*, text from *Anecdotes*, published in 1782 (New York: Ursus Press, 1995)

Walton, Izaak, ed. Brian Loughrey. *The Compleat Angler or the Contemplative Man's Recreation 1653* (Harmondsworth: Penguin Books, 1985)

Watts, W. *The Seats of the Nobility and Gentry in a Collection of the Most Interesting and Picturesque Views Engraved by W. Watts From Drawings by the Most Eminent Artists* (London, 1779)

Whateley, T. *Observations on Modern Gardening* (London: 2nd ed., 1770)

White, R., et al. *Holkham* (Arie & Ingrams, 2010)

Whitney, J. *The Genteel Recreation Or, the Pleasure of Angling, A Poem with a Dialogue between Piscator and Corydon 1700* (London, 1700)

Whittle, E. H. 'The Renaissance Gardens of Raglan Castle', *Garden History*, Vol. 17, No. 1 (1989), pp. 83–94

Williamson, T. *The Archaeology of the Landscape Park: Garden Design in Norfolk, England, c. 1680–1840*, BAR British Series 267 (Oxford: Archaeopress, 1998)

———. 'Fish, Fur and Feather: Man and Nature in the Post-Medieval Landscape', in Barker, K. & Darvill, T., eds., *Making English Landscapes: Changing Perspectives* (Oxford: Oxbow Books, 1997)

———. *Humphry Repton: Landscape Design in an Age of Revolution* (London: Reaktion Books, 2020)

———. *Polite Landscapes: Gardens & Society in Eighteenth-Century England* (Beauchamp, UK: Matador, 2012)

———. *Suffolk's Gardens & Parks: Designed Landscapes from the Tudors to the Victorians* (Macclesfield: Windgather Press, 2000)

Williamson, T. & Flood, S., eds. *Humphry Repton in Hertfordshire*, (Hatfield: University of Hertfordshire Press, 2018)

Williamson, T., Liddiard, R. & Partida, T. *Champion: The Making and Unmaking of the English Midland Landscape* (Liverpool: Liverpool University Press, 2013)

Willis, P. *Charles Bridgeman and the English Landscape Garden* (London: A. Zwemmer, 1997)

————. *Studies in Architecture, Vol XVII: Charles Bridgeman and the English Landscape*, in Blunt, A., Harris, J. & Hibbard, H., eds. (London: A. Zwemmer, 1977)

Wilson-North, R. 'Two Relict Gardens in Somerset: Their Changing Fortunes Through the 17th and 18th Centuries as Revealed by Field Evidence and Other Sources', in Pattison, Paul, ed., *There By Design: Field Archaeology in Parks and Gardens*, BAR British Series 267 (Oxford: Archaeopress, 1998)

Wirtemburg, Duke of Rathgeb, J. & others, trans. William Brenchley Rye. *England as Seen by Foreigners in the Days of Elizabeth and James the First 1592 & 1610* (London, 1865)

Woodbridge, K. 'Henry Hoare's Paradise', *The Art Bulletin*, Vol. 47, No. 1 (March, 1965), pp. 83–116

Woodside, E. 'Kenilworth, the Earl of Leicester's Pleasure Grounds Following Robert Laneham's Letter', *Garden History*, Vol. 27, No. 1, Tudor Gardens (Summer, 1999), pp. 127–144

Woodward, H. B. *The Geology of Water-supply* (London: Edward Arnold, 1910)

Worsley, G. *Classical Architecture in Britain: The Heroic Age* (London: Yale University Press, 1995)

Young, Arthur. *The Farmer's Tour Through the East of England*, Vol. II (London, 1771)

————. *A Six Months Tour Through the North of England*, Vol. IV (London, 1771)

————. *A Six Weeks Tour Through the Southern Counties of England and Wales* (London, 1769)

INDEX

Acton Court 19
Adam, Robert 165, 170–171
Adam, William 217
Addison, Joseph 76, 182, 186–188, 205, 255, 256
aesthetics 18
Alfred's Hall 181
Allen, Ralph 150
Alresford Pond 21
Alton Towers 248
Ancaster, 3rd Duke, Peregrine Bertie 107
Androuet Du Cerceau, Jacques 31
angling 43, 45, 192, 195, 196, 198, 202, 259
Anne, Queen of England, Scotland, and Ireland 195, 205
Arbury Park 112
Armstrong, John 68, 80, 162
Arundel, Thomas 32
Atkyns, Robert 15, 49, 56
Aubrey, John 39
Audley End House and Gardens 150, 237
Augustus, William, Duke of Cumberland 165
Austen, Jane 202, 217

Bacon, Anthony 38
Bacon, Francis 11, 36, 38–39, 45, 46, 76
Badeslade, J. 50, 52, 69
Badminton Estate 54, 88, 140
Badminton House 181
Barrett, Thomas Lennard, Lord Dacre 149, 171
Barry, Charles 231, 245–247

Bathurst, Allen, Lord Bathurst 181
Battersea Park 237
Bayham Abbey 218
Beachborough Hall 196
Beachborough Manor 90–91
Beale, John 47
Bearwood House 211, 247
Beaton, Donald 242
Beckley Park 27
Bedford, 6th Duke, John Russell 216
Bedford Lodge 242
Beeston Park 139
Belhus Park 121, 149, 171
Belton Park 179
Benham Park 239
Binnie, G. M. 11, 113
bird's-eye-view 13–15, 69
Birkenhead Park 118, 231–232
Bishop of Ely's Palace 18, 203, 241
Bishop of Lincoln's Palace 24
Bishop's Waltham 21
Black Lake 214
Blathwaite, William 52
Blenheim Palace 2, 54, 67–68, 82–84, 142, 147, 158, 162, 165
Blickling Estate 66, 94
Blickling Hall 27
Blomefield, Francis 90, 92
Blount, Martha 190
boat-caves 235
boathouses 215–216, 249
boating 198–202, 257
Bodiam Castle 22, 23, 24, 204

Bosworth, Battle of 29
Boughton Park 64, 69, 72, 258
Bowood House 28, 52, 66, 100, 101, 103, 129, 134, 142–146, 148, 155, 157, 162, 203, 245
Boydell, John 162
Bramshill Park 45, 115, 126–127
Bretby Hall 65, 114
Bridewell Palace 29
bridge-dams 151–157
Bridgeman, Charles 7, 68, 87, 135, 151
Bridgeman, Orlando 52
bridge-weirs 165, 171
Briggs, James 101
Britannia Illustrata (Knyff and Kip) 15, 47, 49, 50, 55, 56, 66, 72, 181
British Dam Society 113
Britton, John 172
Broads, The 128
broad water 2, 62, 119
Brocket Park 165
Brockman, Mr. 90–91
Brooke, Adveno 249
Brooke, Lord (Francis Greville, 1st Earl of Warwick) 134
Brown, David 11, 168
Brown, Lancelot 'Capability': clients 139; design of ornamental water, 146–158, 257; focus on house and lake 140, 142; lake at Belhus Park 121, 149; lake at Blenheim Palace 84, 142, 158–162; lake at Bowood House 101, 134, 142, 144–146, 148, 155, 157, 162, 165, 167, 203; lake at Chatsworth House 120–121, 142, 150, 203; lake at Croome Court 142, 155; lake at Longleat House 142, 143, 146–147; lake at Luton Hoo Estate 202; lake at Melton Constable Hall 148; lake at Trentham Park 146, 158–162; lake at Wakefield Lodge 134; lake at Wimpole Estate 2, 148; lake at Wotton House 152–155, 162; lakes at Petworth House 108–112, 148; landscape style 173–175, 256; manipulated landscapes 142–146; mill at Bowood House 28; naturalistic landscape style 43, 46, 97, 189, 250; plan for Packington Hall 136, 140, 142, 196; plan for Prior Park 150; plan for Thoresby Park 72; ponds at Burghley House 107, 142; as predominant lake-maker 11, 62, 93, 134–165; river-lake at Shortgrove Park 142, 150; siting of lakes 129, 131, 146–158; use of bridge-dams 151–157, 165; use of contemporary construction techniques 148; use of

dams 118; use of river-lakes 119–120, 142, 147–150, 162; use of split-level lakes 150–152; as water engineer 148
Buckingham, 3rd Duke, Edward Stafford 29
Buckingham Palace 237
Burghley, 1st Baron, William Cecil 22
Burghley House 32, 39, 94, 103, 107, 142, 162, 203
Burley-on-the-Hill 55
Burlington, 3rd Earl, Richard Boyle 70–72, 76, 84–85, 97, 183, 187, 190, 205, 255, 256
by-pass channel 10
Byron, 5th Baron, William Byron 257

Caerphilly Castle 22, 23, 24, 41
Cale, John 242
Campbell, Colen 15, 49, 72, 82, 86
canals 49, 52, 54–55, 92, 143, 179, 198
Cannon Hall 139
Carlisle, 3rd Earl, Charles Howard 76–79, 81–82, 97, 182, 187, 255, 256
Carlisle, 9th Countess, Rosalind Howard 245
Caroline of Brandenburg-Ansbach, Queen consort of Great Britain and Ireland 7, 87, 89, 92
carriage drives 142
cascades: Emes's work 172–173; formal and naturalistic 225–228; Nesfield's work 242, 244; Picturesque ideals and 224–225; popularity of 52, 234, 249, 254; Pulham firm and 231, 235–238; Repton's work 218, 224–225; water-carriage for 104
Case, John 101
Cassiobury Park 181
Castell, Robert 86, 188–190, 256
Castle Ashby 162
Castle Howard 2, 57, 69–70, 76–79, 81, 181, 184, 244, 255, 258
castle-palaces 29
Caus, Isaac de 49
Cecil, William, Lord Burghley 11, 22, 32, 33–36, 39, 41, 43–45, 196
Château de Blois 31
Château de Fontainebleau 31
Château de Gaillon 31
Chatsworth House 28, 46, 56, 119–120, 121, 142, 150, 181, 203
Chauncy, Henry 15, 49
Chester, William de 101
Chipping Camden Manor 36, 43
Chiswick House 84–85, 195, 255
Churchill, John, Duke of Marlborough 3, 68, 80, 162, 188

Churchill, Sarah, Duchess of Marlborough 2–3, 68, 80, 84
Churchyard, Thomas 41
circuit drives 134, 139, 140, 173, 247
Cirencester Abbey 20
Cirencester Park 181–182
Clare, Gilbert de 26
Claremont Estate 2, 66, 68, 86, 184, 189
Clark's Hill 157–158
Clavering-Cowper, Peter, Earl Cowper 221
clay walls 108, 110, 258–259
cliff walks 235
Clinton, Geoffrey de 25
clover 146
Clumber Park 165, 245, 247
Cobham, 1st Viscount, Richard Temple 110–111, 134, 135, 142, 255
Coke, Thomas 70, 76, 97, 182–184, 187, 188, 255, 256
Combs Hall 66
Compton Verney Estate 94
Constable, John 202
contour ponds 20
Cope's Castle 27, 35, 43, 45
Cope, Walter 43–45
Copley, Godfrey 2
Coppy Map 117
Corsham Court 204, 218
county maps 12
Coventry, 6th Earl, George Coventry 134
Cowdray House 29
Cowell, Fiona 11
Creighton, Oliver 17
Creswell Crags 218
Crewe Hall 242–243
Croome Court 134, 142, 155, 162
Culford Park 218
Cullings Manor 22, 27, 34, 43
Currie, Christopher 39
Cusworth Hall 106, 148
cut-off trench 102, 106, 108, 148
Cuttle Mill 28

Dalingridge, Edward 22
dams: basic construction 9–10, 100–114, 130–131, 258–259; bridge-dams 151–157; fishponds 19–21, 28; geometric lakes 114; of great length 162; irregular lakes 118–121; masking 172, 228–230, 234; semi-geometric lakes 114–117; sites for 125, 127, 129, 147–148; styles 168–170; weirs and 121, 254
Daniels, Steven 216–217
Danson Park 168
Darch, Richard 101

Dashwood, Francis 257
Dauntsey Place 211
Deeping Fen 107
deer parks 32
Defoe, Daniel 11, 79, 84
de Lacy, Henry 21
Denbigh, 5th Earl, William Feilding 134
Deplorable Mapp 41
design 18
Devonshire, 4th Duke, William Cavendish 120
Dézallier d'Argenville, Antoine Joseph 76
Ditchley Park 91–92
Dixon-Hunt, J. 188
Dogmersfield Park 195
Dormer, James, Colonel 146
drainage systems 129
Drake, William 168
Dudley, Robert, Earl of Leicester 32, 41
Dughet, Gaspard 184, 184, 205, 256
Dunham Massey 22
Dunorlan Park 213, 235–236
Dupérac, Étienne 32, 35
Dyrham Park 50–52, 150

Edward I, King of England 26
Egremont, 2nd Earl, Charles Wyndham 108, 146
Eleanor of Castile, Queen of England 19, 26
Eliot, George 91–92, 202
Elizabeth I, Queen of England and Ireland 32, 34, 35, 41
elongated lakes 212–213
Elvetham Park 32
embankment dams 19, 21
Emes, William 15, 87, 139, 155, 164, 165, 170–175, 255
Enstone Marvels 45
Enville Estate 104, 114, 230
Erlestoke Park 139, 171–175
Esher Palace 29
estate maps 12
Evelyn, John 47, 76
Eyre, Thomas 12–13

Felus, Kate 196, 216
Ferne Park 128
Ferrars, 14th Baron and 1st Earl Ferrars, Robert Shirley 55
fishponds (*vivaria*): as aesthetic water feature 38, 41, 43, 203; basic construction 100–113; for breeding fish 16, 27, 190, 259; Bretby Hall 114; construction 28; depth of water 119; dual purpose 35, 50–52, 257–259; geology 125; geometric lakes and 62; irregular lakes

and 93–94, 131, 202; Italian influence 30; Longleat House 143; medieval period 18–22; on OS maps 12; as part of landscape 32, 34, 35, 253; as proto-lake 29, 36; role of islands 45; status conferred by 24, 55
Fitzwilliam, William 29
Fleming, George 242
Flitcroft, Henry 164, 165
Fonthill Estate 119, 213–214
Fougeroux, Pierre Jacques 67
fountains 29, 34, 39, 49–52, 104, 238, 244
Framlingham Castle 23
French gardens 47
freshwater fish 18–22, 202, 257

Gamlinghay Park 65
garden buildings 204
gardenesque layout 228
gardens: Italian influence 29–36; water gardens 36–46
Garswood, New Hall, 232, 234
gender 192–202, 257
geology 113, 125–129
geometric fishponds 18
geometric lakes 3, 52, 62–66, 113, 114, 253–255, 257
geometric landscapes 15
George I, King of Great Britain and Ireland 76
George II, King of Great Britain and Ireland 191
George IV, King of the United Kingdom of Great Britain and Ireland 165
Gilpin, William 211, 223, 228–230, 234, 239, 257
giochi d'acqua 34, 45
Girouard, Mark 54, 140
Glencorse Reservoir 112–113
gloriette 26
Godfrey de Lucy, Bishop of Winchester 21
Golden Gates 245
Gorhambury House 32, 36, 228
Grafton, 2nd Duke, Charles Fitzroy 134
Grand Tour 76, 179, 182–191, 205, 255–256
great poole 26, 41, 108
great water 2, 62, 108
Grey, 2nd Marchioness, Jemima Yorke 199–201, 257
Grimsthorpe Castle 49, 84, 103, 106, 107, 110, 129, 148, 181
grottoes 45
groves 49

Grundy, John 103, 106–108, 110, 129, 131, 148, 164
Guernsey, 1st Baron, and 2nd Earl of Aylsford, Heneage Finch 136, 140

ha-ha 135, 192
Hale, Thomas 45
Halifax, 2nd Earl, George Montagu 119
hammer ponds 253
Hampton Court, Herefordshire 29, 31, 104
Harborne, John 43–45
Harewood House 148, 245
Harlaxton Manor 241
Harris, John 13
Hartlib, Samuel 47
Hatfield House 32, 39
Hatton, Christopher 35–36
Hawk Lake 171
Hawkstone Park 171
Haytley, Edward 195, 196, 198
Henry III, King of England 21, 101
Henry V, King of England 25
Henry VIII, King of England 29
Hentzner, Paul 34
Herstmonceux Castle 23
Hertford, 1st Earl, John Seymour 32
Hesketh Park 232
Hickes, Baptist 36, 43
Highclere Castle 129, 245
Highnam Court 237
Hill, Richard 171
Himley 164
Hitching, Claude 237
Hoare, Henry 94
Hoare, Henry, II, Henry the Magnificent 94, 184, 256
Hoare, Richard Colt 249
Holdenby House 32, 35–36
Holkham Hall 69–70, 86, 88, 181, 183–184, 186, 196, 255, 256, 258
Horton, Northamptonshire 29, 36
Horton Court, Gloucestershire 119
Houghton Hall 184
Houses of Parliament 245
Hughes, Andy 113
hunting 195
Hussey, Christopher 7, 92
hybrid lakes 5, 52, 115–118, 254, 258
Hyde Park 87, 89, 90

Ichnographia Rustica (Switzer) 11, 92, 104, 105–106
Idiver Farm 211
industrial reservoirs 112–113

informal lakes 39, 57, 119
informal ponds 52
Ingman, George 171
irregular lakes 5–9, 17, 57, 68–86, 93–94, 97, 114, 118–119, 129, 179, 182, 192, 202, 205, 211, 232, 254–258
islands 6, 45, 86, 230, 238
Italy: Grand Tour and 182–191; influence on water features 29–36, 45

James, C. W. 183
James I, King of England and Ireland 27
James Pulham and Co. 235–238, 249, 254
Jardine, James 112
Jefferys, Thomas 12–13
John of Waverley, Brother 21, 101
Jones, Inigo 46

Kedleston Park 121, 165, 170–171, 196, 216, 230
Keele Hall 241
Kemp, Edward 11, 118, 214–215, 231–234, 255, 257
Kenilworth Castle 19, 22, 23, 25, 32, 41
Kensington Gardens 87
Kent, William 28, 76, 86, 88, 97, 135, 189, 190
Kew 247
Kimberley Park 90, 119, 121, 128, 136, 175
Kimbolton Castle 54
Kingston, 1st Duke, Evelyn Pierrepont
Kip, Johannes 13, 15, 47, 49, 50
Kit Cat Club 76, 187
Knight, Gally 223
Knight, William 29–30
Knyff, Leonard 13, 15, 47, 49, 50
Kyrle-Ernly, John 104

Lake District 1, 6, 224, 253
lake-moats 22–27, 31, 204
lakes: basic construction 9–10, 100–114, 225, 258–259; construction of the different types 113–121; definitions of 3–9, 253; etymology 2–3; geology 113, 125–129, 131; geometric lakes 3, 52, 62–66, 113, 114, 253–255, 257; hybrid lakes 5, 52, 115–118, 254, 258; informal lakes 39, 57, 119; irregular lakes 5–9, 17, 57, 68–86, 93–94, 97, 114, 118–119, 129, 131, 179, 182, 192, 202, 205, 211, 232, 254–258; landscaping of 225–228; man-made water 68; marginal banks 227; natural lakes 1, 12, 119, 183,

186–187, 191, 227, 253; numbers in eighteenth century 93–94; numbers in nineteenth century 210–211, 249; puddling 104–113; river-lakes 5–6, 119–121, 128, 149–150, 162, 223, 254; semi-geometric lakes 3, 67–68, 114; separation of house and 238–241; serpentine lakes 6–9, 87–92; shapes 212–216, 249; sites for 125–131, 253–255; split-level lakes 150–152, 162; as status symbols 202–205, 257–259; topography 113, 129–131, 258; weirs 121–125
Landscape Database 12, 15, 62
landscapes: as Arcadian 196; classical buildings in 196; gender and 192–202, 257; Grand Tour 76, 179, 182–191, 205, 255–256; Italian influence 29–36, 45, 255; medieval period 18–28, 29, 31, 257; naturalistic landscape style 97; paintings 183–184, 192–202, 205, 248–249, 256; separation of house and lake 238–241; Stuart period 36–46; styles 173–176, 258; Tudor period 29–36; unbalancing of 179–182, 258; water features c. 1700 49–57
Laneham, Robert 32
Langley, Batty 69, 72, 92, 190
Langold Park 223
Lansdowne, 9th Marquis, Charles Petty-Fitzmaurice 100
Lansdowne Monument 245
Latimer House 149
lawns 144, 146, 167–168
Lawrence, John 52, 103
leats 28
Leeds Castle 22, 23, 24, 26
Leicester, 1st Earl, Robert Dudley 32
Le Nôtre, André 47, 181, 258
Lenton Sandstone Formation 115
Le Pautre, Jean 76
Lince dam 162
Lion Bridge 203
Loch Thom 259
Locko Park 171
Londesborough Park 70–72, 84, 184, 186, 190, 255
Longleat House 47–49, 56, 88–89, 128, 142, 143, 146, 181, 182, 213
Long Water 87
long water 119
Lorrain, Claude 183, 184, 205, 256
Loudon, Jane 247
Loudon, John Claudius 11, 91, 211, 225–228, 257

Louis XII, King of France 31
Louis XIV, King of France 47, 205
Ludlow, Lawrence de 28
Luton Hoo Estate 8, 202
Lyveden New Bield 32

Madeira Walk 237
Manchester, 1st Duke, Charles Montagu 54, 188
manipulated landscapes 18, 32
Margam House 20
Markham, Gervaise 43
Marnock, Robert 236
Marshall, C. 21
Matw. Willcox & Co. 223
Melton Constable Hall 148
Menagerie Lake 171
Mergogliano, Pacello de 31
Miller, Sanderson 149
mill ponds 10, 16, 27–28, 253
mills 27–28, 35, 39, 43, 120
moats 16, 22–27, 34–35, 39, 43, 49, 52, 253, 257
Montagu, Mary Wortley 11, 192–195
Moor Park, Surrey 79, 181
Moor Park, Hertfordshire 168
More, The 29
Moreton, Matthew Ducie 56
Morton Hall 8
Mowl, Tim 11, 139, 181

Nappa Hall 28
naturalistic landscape style 43, 46, 97, 189, 250
natural lakes 1, 12, 119, 183, 186–187, 191, 227, 253
naumachia 32, 198, 201–202, 257
Nesfield, William Andrews 231, 241–244, 246, 247
Newcastle-under-Lyne, 3rd Duke, Thomas Pelham-Clinton 165
Newcastle-upon-Tyne, 1st Duke, John Holles 76
Newcastle-upon-Tyne, 1st Duke, Thomas Pelham-Holles
Newark Park 128
Newdigate, Roger 112
Newnham Paddox House 134
New Park 181
New River 244
Newstead Abbey 84, 198
Northenden Weir 125
North, Roger 11, 21, 47, 101–103, 104, 106, 113, 125–126, 192, 196, 253, 258–259

Nugent, Thomas 182, 191, 198
Nunney Castle 23

Oatlands 65
Ordnance Survey (OS) drawings 11, 12, 211, 215
Ordnance Survey (OS) maps 11, 12, 121, 128, 171, 196, 209–210, 211, 223, 247
ornamental canals 52
ornamental lakes: definition of 3; dual purpose 35, 50–52, 196, 257–259; emergence 62–97; landscape paintings 192–202, 205; lining of 113; role of fishponds in development 39; as status symbols 202–205, 257–259; use of water for theatricals 33
Oulton Park 171

Packington Hall 136, 196
Paine, James 150, 203
Painshill 69, 182
Palladio, Andrea 31, 34, 76, 183
Pallett, Thomas 221
Panshanger Estate 218–223
Pantheon 204
Parker, Robert le 21
parterres 30, 47, 65, 181, 238, 247, 248
Pautre, Le 76
Payne Knight, Richard 11, 176, 216, 218, 223–225, 235
Paxton, Joseph 231
Pepys, Samuel 47
Petworth Park 39, 108, 146, 148, 162, 182
Picturesque: movement 209, 216, 223–225, 234–235, 239, 249, 254, 259; role of Vanbrugh in concept 84
piece of water 108, 110, 221
Pierrepont, Evelyn, 1st Duke of Kingston-upon-Hull 72–76, 97, 182, 184, 187, 198, 199, 255, 256
pitching 108, 110, 233
plantations 47, 49, 69, 72, 76, 86, 135, 172, 181–184, 186, 211, 234, 249, 255, 258
plesaunce 25–26
Pliny, the Younger 188, 190
Portland, 2nd Duke, William Bentinck 164
Pococke, Richard 11, 191
ponds 2, 41, 102–104, 130–131, 258–259
Pope, Alexander 84, 181, 183, 190–191, 195, 205
Poussin, Nicolas 184, 256
Price, Uvedale 11, 176, 216, 218, 223–225, 228, 235

Prideaux, Edmund 13
Prior Park 150
proto-lakes 29, 36
puddled clay 104–105, 108, 113, 128, 232, 253, 258
puddling 104–113, 112, 258–259
Pulhamite 235, 237
Pulham, James, I 235
Pulham, James, II 231, 235–238
Puritans 46

Quarrendon Manor 27, 35

Rackham, Oliver 22
Raglan Castle 22, 23, 26–27, 32, 39–43, 241
Ravensworth Castle 23
Raynham Park 116–118
Ray Wood 77–79
Redlodge Plantation 247
Rennie, John 112
Repton, Humphry 3, 11, 75, 176, 209, 216–223, 218, 250
reservoirs 112–113, 259
Rhuddlan Castle 19
Richardson, Francis 164
Richmond, Nathaniel 11, 139, 164, 168–170, 255
Richmond Park 29, 31, 195
Ridgway, Christopher 244
Rigaud, Jacques 257
Ripa, Matteo 191
River Cam 150
River Derwent 119, 121
River Glyme 68, 80, 162
river-lakes 5–6, 119–121, 128, 149–150, 162, 223, 254
River Meden 72, 218
River Mimram 218
River Nadder 47
River Onny 28
Roadford Lake 259
Robinson, Elizabeth 90
Robinson, William 245
Robins, Thomas 195
Rocque, John 86, 88, 91, 189
Roman gardens 31
Rood Ashton House 213
roof walk 62
Roque, John 2
Rosa, Salvator 184, 256
Rousham House 146
royal gardens 31
Rutland Water 55
Rycote House 49
Rysbrack, Pieter Andreas 195

Salisbury, 1st Earl, Robert Cecil, 11, 27, 38–39, 41, 45
Salisbury Plain 128–129
Saltwell Park 232
Salvin, Anthony 231, 241–246, 246, 247
Sandby, Thomas 165
Sandringham 236–237
Saye's Court 47
Scotney Castle 23
Seeswood Pool 112
semi-geometric lakes 3, 67–68, 114
serpentine canals 7–8, 92, 143, 148
serpentine lakes 6–9, 87–92, 119
serpentine paths 181
Serpentine River 7, 90–91
Serpentine, The 7–8, 68, 87–92, 198, 202
sharawadgi 188
Shardeloes Lake 168
Sheffield Park 237, 249
Shelburne, 2nd Earl, William Petty 101, 144
Sherborne Castle 190
Shortgrove Park 142, 150
Shrubland Park 242
sinuous lakes 8
sluices 9–10, 21, 28, 102–103, 121
Smith, Joshua 171, 173
Smith, Norman 11, 112
Somersham 19, 24, 203, 241
spillways 10
split-level lakes 150–152, 162
split-level ponds 162
Stanley Park 232
Staunton Harold Estate 55–56, 63, 66, 69, 72, 241
Stoke Park 139
Stokesay Castle 28
Stoneleigh Abbey 217
store ponds (*servatoria*) 18
Stourhead Estate 39, 94, 105, 184, 196, 204, 213, 256
Stowe Estate 65, 84, 110, 146, 151, 155, 184, 196, 198, 216, 255, 257
Strong, Roy 29
Stroud, Dorothy 11
Studley Royal Park 184, 204
Stukeley, William 3, 68, 84
Sudbury Hall 228, 243
Sulby Abbey 20
Switzer, Stephen: Dyrham Park 50–51; fish in ornamental lakes and pools 196; as historical source 254; *Ichnographia Rustica* 11, 92, 104, 105–106; *Introduction to a General system of Hydrostaticks & Hydraulics, An* 11, 105; irregular elements 69, 190; on making ponds

104–106, 258; siting of fishponds 125; *Universal System of Water & Water-works, A* 3, 11; use of the term 'lake' 84; water-carriage design 104
Syon House 8
Syson, Leslie 28

Tabley Old Hall 22
Tackley Manor 43, 45
Taverner, John 21, 103
Taylor, Christopher 11, 18, 22, 39
Telford, Thomas 112
Temple Newsam House 55
Temple, Sir William, 1st Baronet 76, 79
terraces 247
Tewin 217, 265
Theobalds Palace 27, 32, 33, 35, 45
Thom, Robert 112, 259
Thoresby Hall: roof walk 54; Salvin's work at 241
Thoresby Park: boating 192, 198–199, 257; formal and informal water features 189, 258; informal lakes 57; irregular lake 69, 72, 79, 84, 86, 97, 165, 184, 189; Repton's work at 218, 223–224, 235; sinuous walks 92; unbalancing in landscapes 181–182, 258; use of park by family 192–195; vistas 186
Thornbury Castle 29
Thornton Park 232
Tillemans, Peter 75, 192
topographical view 13
topography 113, 129–131, 258
Tortworth Court 56
Tottenham Park 128
tree planting 47, 49, 136, 175
Trentham Park 146, 158–162, 213–214, 242, 245
trompe l'oeil effect 244

Vanbrugh, John: bridge at Blenheim Palace 147, 162; broad water 2; clients 139; comments on the house of Blenheim 54; contributions to the development of lakes 62, 175, 255; garden-making 188; great water 2; influence on unbalanced landscapes 182; irregular lakes 2–3, 76–77, 80–84, 82, 256; lake at Blenheim 2–3, 80–84; lake at Castle Howard 76–77, 80; plan of Blenheim 82–84; sacking of 68
Vaux-le-Vicomte 258
Versailles 181, 258
viewpoint 13–15, 54, 142, 176, 196
Vignola, Jacopo Barozzi da 31

Villa d'Este 32
Villa Rotonda 34
Virginia Water 165

Waddesdon Manor 237
Wakefield Lodge 134
Waldstein, Baron 11, 35
Wallington Park and Garden 119
Walpole, Horatio 68, 120
Walpole, Robert 184, 188
Wanstead House 65, 88
Wardour Castle 148, 166–167
Warwick, 1st Earl, Francis Greville 134
Warwick Castle 134, 168
water-carriage 104
water features: *c.* 1700 49–56; *c.* 1720 57; design 18; fishponds 18–22; Italian influence 29–36, 45; lake-moats 22–27; location 52–56; medieval period 18–28, 29, 31; mill ponds 27–28; moats 22–27; Stuart period 36–46; Tudor period 29–36; water gardens 36–46
watermills 28, 35
water plays (*giochi d'acqua*) 45
water sources 9, 21, 28, 66, 103, 106, 110–112, 125–126, 130–131, 148, 162, 182, 253, 258
Watts, W. 204
Waverley Abbey 20
Webb, John 171, 223
weirs 107, 114, 119, 120, 121–125, 254
Welbeck Estate 164, 216, 217
Welford Park 63–64, 69, 258
Westbury Court 49
West Wycombe Park 257
Weymouth, 1st Viscount, Thomas Thynne 88–89, 143, 147, 198
Whateley, Thomas 3, 11, 153–155, 230
Whetham Estate 104
Whitehall Palace 29, 31
White, Thomas 164
Whitney, John 51, 196
Whittlesay, Robert 89
wildernesses 49, 69–70, 92, 179, 181
William III, King of England, Ireland, and Scotland 49
Williamson, Tom 11, 211, 216
Willis, P. 188
Wilton House 47–49, 128
Wimpole Estate 2, 65, 66, 148
Windsor Forest 195
Wingfield Manor 198–199
Wivenhoe Park 202
Woburn Abbey 216–217, 218

Wollaton Park 115
Wolsey, Thomas, Cardinal of the Catholic
    Church 29, 35
Wolterton Park 67–68, 115, 228
women 192–202
Woodbridge, Kenneth 184
Woods, Richard 11, 106–107, 139, 148,
    164, 165–168, 255
Woodstock Manor 80, 84
Worthy River 111

Worcester, 1st Earl, Edward Somerset 41
Worcester, 2nd Earl, Henry Somerset 41
Worcester, 3rd Earl, William Somerset 41
Wotton Underwood House 152–155, 162
Wrest Park 198, 199–201, 257
Wright, Stephen 164
Wright, Thomas 199, 257
Wyngaerde, Anton van den 29

Young, Arthur 3, 6, 11